CAMBRIDGE STUDIES IN
ADVANCED MATHEMATICS

EDITORIAL BOARD
D.J.H. GARLING, W. FULTON, T. TOM DIECK,
P. WALTERS

CALCULUS OF VARIATIONS

Already published

1 W.M.L. Holcombe *Algebraic automata theory*
2 K. Petersen *Ergodic theory*
3 P.T. Johnstone *Stone spaces*
4 W.H. Schikhof *Ultrametric calculus*
5 J.-P. Kahane *Some random series of functions*, 2nd edition
6 H. Cohn *Introduction to the construction of class fields*
7 J. Lambek & P.J. Scott *Introduction to higher-order categorical logic*
8 H. Matsumura *Commutative ring theory*
9 C.B. Thomas *Characteristic classes and the cohomology of finite groups*
10 M. Aschbacher *Finite group theory*
11 J.L. Alperin *Local representation theory*
12 P. Koosis *The logarithmic integral I*
13 A. Pietsch *Eigenvalues and s-numbers*
14 S.J. Patterson *An introduction to the theory of the Riemann zeta-function*
15 H.J. Baues *Algebraic homotopy*
16 V.S. Varadarajan *Introduction to harmonic analysis on semisimple Lie groups*
17 W. Dicks & M. Dunwoody *Groups acting on graphs*
18 L.J. Corwin & F.P. Greenleaf *Representations of nilpotent Lie groups and their applications*
19 R. Fritsch & R. Piccinini *Cellular structures in topology*
20 H Klingen *Introductory lectures on Siegel modular forms*
21 P. Koosis *The logarithmic integral II*
22 M.J. Collins *Representations and characters of finite groups*
24 H. Kunita *Stochastic flows and stochastic differential equations*
25 P. Wojtaszczyk *Banach spaces for analysts*
26 J.E. Gilbert & M.A.M. Murray *Clifford algebras and Dirac operators in harmonic analysis*
27 A. Frohlich & M.J. Taylor *Algebraic number theory*
28 K. Goebel & W.A. Kirk *Topics in metric fixed point theory*
29 J.F. Humphreys *Reflection groups and Coxeter groups*
30 D.J. Benson *Representations and cohomology I*
31 D.J. Benson *Representations and cohomology II*
32 C. Allday & V. Puppe *Cohomological methods in transformation groups*
33 C. Soulé et al *Lectures on Arakelov geometry*
34 A. Ambrosetti & G. Prodi *A primer of nonlinear analysis*
35 J. Palis & F. Takens *Hyperbolicity and sensitive chaotic dynamics at homoclinic bifurcations*
36 M. Auslander, I. Reiten & S. Smalo *Representation theory of Artin algebras*
37 Y. Meyer *Wavelets and operators*
38 C. Weibel *An introduction to homological algebra*
39 W. Bruns & J. Herzog *Cohen-Macaulay rings*
40 V. Snaith *Explicit Brauer induction*
41 G. Laumon *Cohomology of Drinfeld modular varieties I*
42 E.B. Davies *Spectral theory and differential operators*
43 J. Diestel, H. Jarchow & A. Tonge *Absolutely summing operators*
44 P. Mattila *Geometry of sets and measures in Euclidean spaces*
45 R. Pinsky *Positive harmonic functions and diffusion*
46 G. Tenenbaum *Introduction to analytic and probabilistic number theory*
47 C. Peskine *An algebraic introduction to complex projective geometry I*
48 Y. Meyer & R. Coifman *Wavelets and operators II*
49 R. Stanley *Enumerative combinatories*
50 I. Porteous *Clifford algebras and the classical groups*
51 M. Audin *Spinning tops*
52 V. Jurdjevic *Geometric control theory*
53 H. Voelklein *Groups as Galois groups*
54 J. Le Potier *Lectures on vector bundles*
55 D. Bump *Automorphic forms*
56 G. Laumon *Cohomology of Drinfeld modular varieties II*
59 P. Taylor *Practical foundations of mathematics*
60 M. Brodmann & R. Sharp *Local cohomology*
64 J. Jost & X. Li-Jost *Calculus of variations*

Calculus of Variations

Jürgen Jost and Xianqing Li-Jost
*Max-Planck-Institute for Mathematics in the Sciences,
Leipzig*

CAMBRIDGE UNIVERSITY PRESS
Cambridge, New York, Melbourne, Madrid, Cape Town, Singapore, São Paulo

Cambridge University Press
The Edinburgh Building, Cambridge CB2 8RU, UK

Published in the United States of America by Cambridge University Press, New York

www.cambridge.org
Information on this title: www.cambridge.org/9780521642033

© Cambridge University Press 1998

This publication is in copyright. Subject to statutory exception
and to the provisions of relevant collective licensing agreements,
no reproduction of any part may take place without the written
permission of Cambridge University Press.

First published 1998
This digitally printed version 2008

A catalogue record for this publication is available from the British Library

Library of Congress Cataloguing in Publication data

Jost, Jürgen, 1956–
Calculus of variations / Jürgen Jost and Xianqing Li-Jost.
p. cm.
Includes index.
ISBN 0 521 64203 5 (hc.)
1. Calculus of variations. I. Li-Jost, Xianqing, 1956–
II. Title.
QA315.J67   1999
515'.64–dc21   98-38618   CIP

ISBN 978-0-521-64203-3 hardback
ISBN 978-0-521-05712-7 paperback

Dedicated to Stefan Hildebrandt

Contents

Preface and summary		*page* x
Remarks on notation		xv
	Part one: One-dimensional variational problems	**1**
1	**The classical theory**	**3**
1.1	The Euler–Lagrange equations. Examples	3
1.2	The idea of the direct methods and some regularity results	10
1.3	The second variation. Jacobi fields	18
1.4	Free boundary conditions	24
1.5	Symmetries and the theorem of E. Noether	26
2	**A geometric example: geodesic curves**	**32**
2.1	The length and energy of curves	32
2.2	Fields of geodesic curves	43
2.3	The existence of geodesics	51
3	**Saddle point constructions**	**62**
3.1	A finite dimensional example	62
3.2	The construction of Lyusternik–Schnirelman	67
4	**The theory of Hamilton and Jacobi**	**79**
4.1	The canonical equations	79
4.2	The Hamilton–Jacobi equation	81
4.3	Geodesics	87
4.4	Fields of extremals	89
4.5	Hilbert's invariant integral and Jacobi's theorem	92
4.6	Canonical transformations	95

5	**Dynamic optimization**	**104**
5.1	Discrete control problems	104
5.2	Continuous control problems	106
5.3	The Pontryagin maximum principle	109

	Part two: Multiple integrals in the calculus of variations	**115**

1	**Lebesgue measure and integration theory**	**117**
1.1	The Lebesgue measure and the Lebesgue integral	117
1.2	Convergence theorems	122

2	**Banach spaces**	**125**
2.1	Definition and basic properties of Banach and Hilbert spaces	125
2.2	Dual spaces and weak convergence	132
2.3	Linear operators between Banach spaces	144
2.4	Calculus in Banach spaces	150

3	L^p **and Sobolev spaces**	**159**
3.1	L^p spaces	159
3.2	Approximation of L^p functions by smooth functions (mollification)	166
3.3	Sobolev spaces	171
3.4	Rellich's theorem and the Poincaré and Sobolev inequalities	175

4	**The direct methods in the calculus of variations**	**183**
4.1	Description of the problem and its solution	183
4.2	Lower semicontinuity	184
4.3	The existence of minimizers for convex variational problems	187
4.4	Convex functionals on Hilbert spaces and Moreau–Yosida approximation	190
4.5	The Euler–Lagrange equations and regularity questions	195

5	**Nonconvex functionals. Relaxation**	**205**
5.1	Nonlower semicontinuous functionals and relaxation	205
5.2	Representation of relaxed functionals via convex envelopes	213

6	**Γ-convergence**	**225**
6.1	The definition of Γ-convergence	225

6.2	Homogenization	231
6.3	Thin insulating layers	235
7	**BV-functionals and Γ-convergence: the example of Modica and Mortola**	**241**
7.1	The space $BV(\Omega)$	241
7.2	The example of Modica–Mortola	248
Appendix A	The coarea formula	257
Appendix B	The distance function from smooth hypersurfaces	262
8	**Bifurcation theory**	**266**
8.1	Bifurcation problems in the calculus of variations	266
8.2	The functional analytic approach to bifurcation theory	270
8.3	The existence of catenoids as an example of a bifurcation process	282
9	**The Palais–Smale condition and unstable critical points of variational problems**	**291**
9.1	The Palais–Smale condition	291
9.2	The mountain pass theorem	301
9.3	Topological indices and critical points	306
Index		319

Preface and summary

The calculus of variations is concerned with the construction of optimal shapes, states, or processes where the optimality criterion is given in the form of an integral involving an unknown function. The task of the calculus of variations then is to demonstrate the existence and to deduce the properties of some function that realizes the optimal value for this integral. Such variational problems occur in many-fold applications, in particular in physics, engineering, and economics, and the variational integral may represent some action, energy, or cost functional. The calculus of variations also has deep and important connections with other fields of mathematics. For instance, in geometrically defined classes of objects, a variational principle often permits the selection of a unique optimal representative, and the properties of this representative can frequently be used to much advantage to deduce additional information about its class. For these reasons, the calculus of variations is a rich and ample mathematical subject, and a good impression of this diversity can be obtained by reading the beautiful book by S. Hildebrandt and A. Tromba, *The Parsimonious Universe*, Springer, 1996.

In this textbook, we have attempted to present some of the many faces of the calculus of variations, and a brief summary may be useful before putting the contents into a broader perspective. At the same time, we shall also describe the logical connections between the various chapters, in order to facilitate reading for readers with a specific aim. The book is divided into two parts. The first part treats variational problems for functions of one independent variable; the second, problems for functions of several variables. The distinction between these two parts, however, is also that the first treats the more elementary and more classical aspects of the subject, while the second is concerned with some more difficult topics and uses somewhat more abstract reasoning. In this second part,

Preface and summary

also some examples are presented in detail that occurred in recent applications of the calculus of variations. This second part leads the reader to some topics and questions of current research in the calculus of variations.

The first chapter of Part I is of a somewhat introductory nature and attempts to develop some intuition for the properties of solutions of variational problems. In the basic Section 1.1, we derive the Euler–Lagrange equations that any smooth solution of a variational problem has to satisfy. The topics of the other sections of that chapter contain some regularity questions and an outline of the so-called direct methods of the calculus of variations (a subject that will be taken up in much more detail in Chapter 4 of Part II), Jacobi's theory of the second variation and stability of solutions, and Noether's theorem that deduces conservation laws from invariance properties of variational integrals. All those results will not be directly applied in subsequent chapters, but should rather serve as a motivation. In any case, basically all the chapters of Part I can be read independently, after the reader has gone through Section 1.1.

In Chapter 2, we treat one of the most important variational problems, namely that of geodesics, i.e. of finding (locally) shortest curves under smooth geometric constraints. Geodesics are of fundamental importance in Riemannian geometry and several physical applications. We shall make use of the geometric nature of this problem and develop some elementary geometric constructions, to deduce the existence not only of length-minimizing curves, but also of curves that furnish unstable critical points of the length functional. In Chapter 3, we present some more abstract aspects of such so-called saddle point constructions. At this point, however, we can only treat problems that allow the reduction to a finite dimensional situation. A deeper treatment needs additional tools and therefore has to wait until Chapter 9 of Part II. Geodesics will only occur once more in the remainder, namely as an example in Section 4.3.

Chapter 4 is concerned with one of the classical highlights of the calculus of variations, the theory of Hamilton and Jacobi. This theory is of particular importance in mechanics. Presently, its global aspects are resurging in connection with symplectic geometry, one of the most active fields of present mathematical research.

Chapter 5 is a brief introduction to dynamic optimization and control theory. The canonical equations of Hamilton and Jacobi of Section 4.1 briefly reoccur as an example of the Pontryagin maximum principle at the end of Section 5.3.

As mentioned, Part II is of a less elementary nature. We therefore need to develop some general theory first. In Chapter 1 of that part, Lebesgue integration theory is summarized (without proofs) for the convenience of the reader. While in Part I, the Riemann integral entirely suffices (with the exception of some places in Section 1.2), the function spaces that are basic for Part II, namely the L^p and Sobolev spaces, are essentially based on Lebesgue's notion of the integral. In Chapter 2, we develop some results from functional analysis about Banach and Hilbert spaces that will be applied in Chapter 3 for deriving the fundamental properties of the L^p and Sobolev spaces. (In fact, as the tools from functional analysis needed in subsequent chapters are of a quite varied nature, Chapter 2 can also serve as a brief introduction into the field of functional analysis itself.) These chapters serve the purpose of making the book self-contained, and for most readers the best strategy might be to start with Chapter 4, or at most with Chapter 3, and look up the results of the previous chapters only when they are applied. Chapter 4 is fundamental. It is concerned with the existence of minimizers of variational integrals under appropriate convexity and lower semicontinuity assumptions. We treat both the standard method based on weak compactness and a more abstract method for minimizing convex functionals that does not need the concept of weak convergence. Chapters 5–7 essentially discuss situations where those assumptions are no longer satisfied. Chapter 5 deals with the method of relaxation, while Chapters 6 and 7 present the important concept of Γ-convergence for minimizing functionals that can be represented only in an indirect manner as limits of other functionals. Such problems occur in many applications, including homogenization and phase transitions, and several such examples are treated in detail. Chapter 8 discusses bifurcation theory. We first discuss the variational aspects (Jacobi fields), taking up the constructions of Sections 1.1 and 1.3 of Part I, then develop a general functional analytic framework for analyzing bifurcation phenomena and then treat the example of minimal surfaces of revolution (catenoids) in the light of that framework. Chapter 8 is independent of Chapters 4–7, and of a more elementary nature than those. The key tool is the implicit function theorem in Banach spaces, proved in Section 2.4. The last Chapter 9 returns to the topic of the existence of non-miminizing, unstable critical points of variational integrals. While such solutions usually cannot be observed in physical applications because of their unstable nature, they are of considerable mathematical interest, for example in the context of Riemannian geometry. Chapter 9 is independent of Chapters 4–8.

The present book is self-contained, with very few exceptions. Prerequisites are only the calculus of one and several variables.

Although, as indicated, there are important connections between the calculus of variations and geometry, the present book is of an analytic nature and does not explore those connections. One such connection concerns the global aspects of the space of solutions of one-dimensional variational problems and their trajectories that started with the qualitative investigations of Poincaré and is for example represented in V.I. Arnold, *Mathematical Methods of Classical Mechanics*, GTM 60, Springer, New York, 2nd edition, 1987. Here, geometric methods are used to study variational problems. In the opposite direction, variational methods can often be used to solve geometric problems. This is the topic of geometric analysis; we refer the interested reader to J. Jost, *Riemannian Geometry and Geometric Analysis*, Springer, Berlin, 2nd edition, 1998, and the references contained therein.

There is one important omission in this textbook. Namely, the regularity theory for solutions of variational problems is not treated, with the exception of the one-dimensional case in Section 1.2 of Part I, and the simplest example of the multi-dimensional theory, namely harmonic functions (plus an easy generalization) in Section 4.5 of Part II. Therefore, the solutions of the variational problems that are discussed usually only are obtained in some Sobolev space. We think that a detailed treatment of regularity theory more properly belongs to the realm of partial differential equations, and therefore we have to refer the reader to textbooks and monographs on partial differential equations, for example D. Gilbarg and N. Trudinger, *Elliptic Partial Differential Equations of Second Order*, Springer, Berlin, 2nd edition, 1983, or J. Jost, *Partielle Differentialgleichungen*, Springer, Berlin, 1998.

In any case, the present textbook cannot cover all the many diverse aspects of the calculus of variations. For readers who are interested in a more extensive treatment, we strongly recommend M. Giaquinta and St. Hildebrandt, *Calculus of Variations*, several volumes, Springer, Berlin, 1996 ff., as well as E. Zeidler, *Nonlinear Functional Analysis and its Applications*, Vols. III and IV, Springer, New York, 1984 ff. (a second edition of Vol. IV appeared in 1995). Additional references are given in the course of the text. Since the present book, however, is neither a research monograph nor an account of the historical development of the calculus of variations, references to individual contributions are usually not given. We just list our sources, and refer the interested readers as well as the contributing mathematicans to those for references to the original contributions.

The authors thank Felicia Bernatzki, Ralf Muno, Xiao-Wei Peng, Marianna Rolf, and Wilderich Tuschmann for their help in proofreading and checking the contents and various corrections, and Michael Knebel and Micaela Krieger for their competent typing.

The present authors owe much of their education in the calculus of variations to their teacher, Stefan Hildebrandt. In particular, the presentation of the material of Chapters 1 and 4 in Part I is influenced by his lectures that the authors attended as students. For example, the regularity arguments in Section 1.2 are taken directly from his lectures. For these reasons, and for his generous support of the authors over many years, and for his profound contributions to the subject, in particular to geometric variational problems, the authors dedicate this book to him.

Remarks on notation

A dot '·' always denotes the Euclidean scalar product in $\mathbb{R}^d$, i.e. if
$$x = (x^1, \ldots, x^d) \, , \, y = (y^1, \ldots, y^d) \in \mathbb{R}^d,$$
then
$$x \cdot y = \sum_{i=1}^{d} x^i y^i = x^i y^i \quad \text{(Einstein summation convention)},$$
and
$$|x|^2 = x \cdot x.$$
For a function $u(t)$, we write
$$\dot{u}(t) = \frac{d}{dt} u(t).$$
In Part I, the independent variable is usually called t, because in many physical applications, it is interpreted as the time parameter. Here, the dependent variables are mostly called $u(t)$ or $x(t)$. In Part II, the independent variables are denoted by $x = (x^1, \ldots, x^d)$, conforming to established conventions.

We use the standard notation
$$C^k(\Omega)$$
for the space of k-times continuously differentiable functions on some open set $\Omega \subset \mathbb{R}^d$, for $k = 0$ (continuous functions), $1, 2, \ldots, \infty$ (infinitely often differentiable functions). For vector valued functions, with values in $\mathbb{R}^d$, we write
$$C^k(\Omega, \mathbb{R}^d)$$

for the corresponding spaces.

$$C_0^\infty(\Omega)$$

denotes the space of functions of class C^∞ on Ω that vanish identically outside some compact subset $K \subset \Omega$ (where K may depend on the function, of course). Occassionally, we also use the notation

$$C_0^k(\Omega)$$

for C^k functions on Ω that again vanish outside some compact subset $K \subset \Omega$.

Finally, we use the notation

$$:=$$

to indicate that the expression on the left of this symbol is defined by the expression on the right of it.

Part one
One-dimensional variational problems

1
The classical theory

1.1 The Euler–Lagrange equations. Examples

The classical calculus of variations consists in minimizing expressions of the form

$$I(u) = \int_a^b F(t, u(t), \dot{u}(t))\, dt,$$

where $F : [a, b] \times \mathbb{R}^d \times \mathbb{R}^d \to \mathbb{R}$ is given. One seeks a function $u : [a, b] \to \mathbb{R}^d$ minimizing I. More generally, one is also interested in other critical points of I. Usually, u has to satisfy some constraints, the most common one being a Dirichlet boundary condition

$$u(a) = u_1$$
$$u(b) = u_2.$$

Also, one needs to specify a class of admissible functions among which one seeks a minimizing u. For example, one might want to take the class of continuously differentiable or piecewise continuously differentiable functions. Let us consider some examples of such variational problems:

(1) We want to minimize the arc-length of the graph of a function $u : [a, b] \to \mathbb{R}$, i.e. the length of the curve $(t, u(t)) \subset \mathbb{R}^2$ among all graphs with prescribed boundary values $u(a), u(b)$. This leads to the variational problem

$$\int_a^b \sqrt{1 + \dot{u}(t)^2}\, dt \to \min.$$

Of course, one knows and easily proves that the solution is the straight line between $u(a)$ and $u(b)$, i.e. satisfies $\ddot{u}(t) \equiv 0$.

(2) Historically, the calculus of variations started with the so-called brachystochrone problem that was posed by Johann Bernoulli. Here, one wants to connect two points (t_0, y_0) and (t_1, y_1) in $\mathbb{R}^2$ by such a curve that a particle obeying Newton's law of gravitation and moving without friction travels the distance between those points in the fastest possible way. After falling the height y, the particle has speed $(2gy)^{\frac{1}{2}}$ where g is the gravitational acceleration. The time the particle needs to traverse the path $y = u(t)$ then is

$$I(u) = \int_{t_0}^{t_1} \sqrt{\frac{1 + \dot{u}(t)^2}{2gu(t)}}\, dt.$$

(3) A generalization of (1) and (2) is

$$I(u) = \int_a^b \frac{\sqrt{1 + \dot{u}(t)^2}}{\gamma(t, u(t))}\, dt,$$

where $\gamma : [a, b] \times \mathbb{R} \to \mathbb{R}$ is a given positive function. This variational problem also arises from Fermat's principle. That principle says that a light ray chooses the path that needs the shortest time to be traversed among all possible paths. If the speed of light in a given medium is $\gamma(t, u(t))$, we obtain the preceding variational problem.

If one seeks a minimum of a smooth function

$$f : \Omega \to \mathbb{R} \quad (\Omega \text{ open in } \mathbb{R}^d),$$

one knows that at a minimizing point $z_0 \in \Omega$, one necessarily has

$$Df(z_0) = 0,$$

where Df is the derivative of f. The first variation of f actually has to vanish at any stationary point, not only at minimizers. In order to distinguish a minimizer from other critical points, one has the additional necessary condition that the Hessian $D^2 f(z_0)$ is positive semidefinite and (at least for a local minimizer) the sufficient condition that it is positive definite.

In the present case, however, we do not have a function f of finitely many independent real variables, but a functional I on a class of functions. Nevertheless, we expect that a first derivative of I — something still to be defined — needs to vanish at a minimizer, and moreover that a suitably defined second derivative is positive (semi)definite.

1.1 The Euler–Lagrange equations. Examples

In order to investigate this more closely, we assume that F is of class C^1 and that we have a minimizer or, more generally, a critical point of I that also is C^1. We also assume prescribed Dirichlet boundary conditions $u(a) = u_1$, $u(b) = u_2$. In other words, we assume that u minimizes I in the class of all functions of class C^1 satisfying the prescribed boundary condition. We then have for any $\eta \in C_0^1\left([a,b], \mathbb{R}^d\right)$† and any $s \in \mathbb{R}$

$$I(u + s\eta) \geq I(u).$$

Now

$$I(u + s\eta) = \int_a^b F\left(t, u(t) + s\eta(t), \dot{u}(t) + s\dot{\eta}(t)\right) dt.$$

Since F, u, and η are assumed to be of class C^1, we may differentiate the preceding expression w.r.t. s and obtain at $s = 0$

$$\frac{d}{ds} I(u + s\eta)\big|_{s=0} \qquad (1.1.1)$$
$$= \int_a^b \left\{ F_u\left(t, u(t), \dot{u}(t)\right) \cdot \eta(t) + F_p\left(t, u(t), \dot{u}(t)\right) \cdot \dot{\eta}(t) \right\} dt,$$

where F_u is the vector of partial derivatives of F w.r.t. the components of u, and F_p the one w.r.t. the components of $\dot{u}(t)$.

We now keep η fixed and let s vary. We are thus just in the situation of a real valued $f(s), s \in \mathbb{R}, (f(s) = I(u + s\eta))$, and the condition $f'(0) = 0$ translates into

$$0 = \int_a^b \left\{ F_u\left(t, u(t), \dot{u}(t)\right) \cdot \eta(t) + F_p\left(t, u(t), \dot{u}(t)\right) \cdot \dot{\eta}(t) \right\} dt, \qquad (1.1.2)$$

and this actually then has to hold for all $\eta \in C_0^1$. We now assume that F and u are even of class C^2. Equation (1.1.2) may then be integrated by parts. Noting that we do not get a boundary term since $\eta(a) = 0 = \eta(b)$, we thus obtain

$$0 = \int_a^b \left\{ \left(F_u\left(t, u(t), \dot{u}(t)\right) - \frac{d}{dt}\left(F_p\left(t, u(t), \dot{u}(t)\right)\right) \right) \cdot \eta(t) \right\} dt \qquad (1.1.3)$$

for all $\eta \in C_0^1([a,b], \mathbb{R}^d)$. In order to proceed, we need the so-called *fundamental lemma of the calculus of variations*:

---

† This means that η is continuously differentiable as a function on $[a,b]$ with values in $\mathbb{R}^d$ and that there exist $a < a_1 \leq b_1 < b$ with $\eta(x) = 0$ if x is not contained in $[a_1, b_1]$.

Lemma 1.1.1. *If $h \in C^0\left((a,b), \mathbb{R}^d\right)$ satisfies*

$$\int_a^b h(t)\varphi(t)dt = 0 \quad \text{for all } \varphi \in C_0^\infty\left((a,b), \mathbb{R}^d\right),$$

then $h \equiv 0$ on (a,b).

Proof. Otherwise, there exists some $t_0 \in (a,b)$ with

$$h(t_0) \neq 0.$$

Thus, $h^{i_0}(t_0) \neq 0$ for some index $i_0 \in \{1, \ldots, d\}$. Since h is continuous, there exists some $\delta > 0$ with

$$a < t_0 - \delta < t_0 + \delta < b$$

and

$$\left|h^{i_0}(t)\right| > \frac{1}{2}\left|h^{i_0}(t_0)\right| \quad \text{whenever} \quad |t_0 - t| < \delta.$$

We then choose $\varphi \in C_0^\infty\left((a,b), \mathbb{R}^d\right)$ with

$$\varphi(t) = 0 \quad \text{if} \quad |t_0 - t| \geq \delta$$
$$\varphi^{i_0}(t) > 0 \quad \text{if} \quad |t_0 - t| < \delta$$
$$\varphi^{i_0}(t) = 0 \quad \text{for } i \neq i_0,\ i \in \{1, \ldots, d\}.$$

For this choice of φ, however

$$\int_a^b h(t)\varphi(t)dt = \int_{t_0 - \delta}^{t_0 + \delta} h(t)\varphi(t)dt \neq 0,$$

contradicting our assumption. Thus, necessarily $h(t_0) = 0$ for all $t_0 \in (a,b)$.

q.e.d.

Lemma 1.1.1 and (1.1.3) imply that a minimizer of I of class C^2 has to satisfy the so-called *Euler–Lagrange equations*, namely:

Theorem 1.1.1. *Let $F \in C^2([a,b] \times \mathbb{R}^d \times \mathbb{R}^d, \mathbb{R})$, and let $u \in C^2([a,b], \mathbb{R}^d)$ be a minimizer of*

$$I(u) = \int_a^b F(t, u(t), \dot{u}(t))\, dt$$

among all functions with prescribed boundary values $u(a)$ and $u(b)$. Then u is a solution of the following system of second order ordinary differential equations, the Euler–Lagrange equations

$$\frac{d}{dt}\left(F_p(t, u(t), \dot{u}(t))\right) - F_u(t, u(t), \dot{u}(t)) = 0. \tag{1.1.4}$$

1.1 The Euler–Lagrange equations. Examples

Written out, the Euler–Lagrange equations are

$$F_{pp}(t, u(t), \dot{u}(t))\ddot{u}(t) + F_{pu}(t, u(t), \dot{u}(t))\dot{u}(t)$$
$$+ F_{pt}(t, u(t), \dot{u}(t)) - F_u(t, u(t), \dot{u}(t)) = 0, \quad (1.1.5)$$

i.e. a system of d ordinary differential equations of second order that are linear in the second derivatives of the unknown function u.

Let us compute the Euler–Lagrange equations for our preceding three examples:

(1) Here $F_u = 0$, $F_p = \frac{\dot{u}(t)}{\sqrt{1+\dot{u}(t)^2}}$, and we get

$$0 = \frac{d}{dt}\frac{\dot{u}(t)}{\sqrt{1+\dot{u}(t)^2}} = \frac{\ddot{u}(t)}{\sqrt{1+\dot{u}(t)^2}} - \frac{\dot{u}(t)^2 \ddot{u}(t)}{\left(\sqrt{1+\dot{u}(t)^2}\right)^3}$$

$$= \frac{\ddot{u}(t)}{\left(\sqrt{1+\dot{u}(t)^2}\right)^3},$$

i.e.

$$\ddot{u}(t) = 0$$

meaning that u has to be a straight line, a fact that we know of course.

(3) For the general example (3), we obtain as Euler–Lagrange equations

$$0 = \frac{d}{dt}\frac{\dot{u}(t)}{\gamma(t, u(t))\sqrt{1+\dot{u}(t)^2}} + \frac{\gamma_u}{\gamma^2}\sqrt{1+\dot{u}(t)^2}$$

$$= \frac{\ddot{u}(t)}{\gamma\sqrt{1+\dot{u}(t)^2}} - \frac{\dot{u}(t)^2 \ddot{u}(t)}{\gamma\left(\sqrt{1+\dot{u}(t)^2}\right)^3} - \frac{\gamma_t}{\gamma^2}\frac{\dot{u}(t)}{\sqrt{1+\dot{u}(t)^2}}$$

$$- \frac{\gamma_u}{\gamma^2}\frac{\dot{u}(t)^2}{\sqrt{1+\dot{u}(t)^2}} + \frac{\gamma_u}{\gamma^2}\sqrt{1+\dot{u}(t)^2},$$

hence

$$0 = \ddot{u}(t) - \frac{\gamma_t}{\gamma}\dot{u}(t)\left(1 + \dot{u}(t)^2\right) + \frac{\gamma_u}{\gamma}\left(1 + \dot{u}(t)^2\right). \quad (1.1.6)$$

(2) We just need to insert $\gamma = \sqrt{2gu(t)}$ into (1.1.6) to obtain

$$0 = \ddot{u}(t) + \left(1 + \dot{u}(t)^2\right)\frac{1}{2u(t)}.$$

Actually, (2) is an example of an integrand $F(t, u, \dot{u})$ that does not depend explicitly on t, i.e. $F_t \equiv 0$. In this case

$$\frac{d}{dt}(F - \dot{u}F_p) = \dot{u}(F_u - \frac{d}{dt}F_p) = 0 \text{ by } (1.1.4),$$

and hence every solution of the Euler–Lagrange equation (1.1.4) satisfies

$$F(t, u(t), \dot{u}(t)) - \dot{u}(t)F_p(t, u(t), \dot{u}(t)) = \text{constant}. \quad (1.1.7)$$

Conversely, every solution of (1.1.7), with the exception of $\dot{u} = 0$, i.e. $u \equiv$ constant, also satisfies (1.1.4).

In the case of example (2), we have $F = \sqrt{\frac{1+\dot{u}^2}{2gu}}$, and (1.1.7) becomes $\frac{1}{2g\lambda^2} = u(1 + \dot{u}^2)$, if we denote the constant in (1.1.7) by λ.

In all examples (1)–(3), we actually had $d = 1$. If one modifies e.g. (1) and seeks a curve $g(t) = (g_1(t), \ldots, g_d(t)) \subset \mathbb{R}^d$ connecting two given points $g(a)$ and $g(b)$, our variational problem becomes

$$I(g) = \int_a^b |\dot{g}(t)| \, dt = \int_a^b \left(\sum_{i=1}^d \left(\frac{d}{dt} g_i(t) \right)^2 \right)^{\frac{1}{2}} dt.$$

The Euler–Lagrange equations in this case are

$$0 = \frac{d}{dt} \frac{\dot{g}_i(t)}{\left(\sum_{j=1}^d \dot{g}_j(t)^2 \right)^{\frac{1}{2}}} = \frac{\ddot{g}_i \sum_{j=1}^d (\dot{g}_j)^2 - \dot{g}_i \sum_{j=1}^d \dot{g}_j \ddot{g}_j}{\left(\sum_{j=1}^d (\dot{g}_j)^2 \right)^{\frac{3}{2}}} \quad (1.1.8)$$

for $i = 1, \ldots, d$.

We now recall that any smooth curve $g(t) \subset \mathbb{R}^d$ may be parameterized by arc-length, i.e.

$$\left| \frac{d}{dt} g(t) \right| \equiv 1. \quad (1.1.9)$$

We also know that a reparameterization of a curve $g(t)$ does not change its arc-length $I(g)$. Consequently, we may assume (1.1.9) in (1.1.8). The latter then becomes

$$\frac{d}{dt} \left(\frac{d}{dt} g_i(t) \right) = 0 \quad \text{for} \quad i = 1, \ldots, d \ ,$$

so that we see again that a length minimizing curve in $\mathbb{R}^d$ is a straight line.

1.1 The Euler–Lagrange equations. Examples

Often, one also meets the task of minimizing

$$I(u) = \int_a^b F(t, u(t), \dot{u}(t)) \, dt$$

subject to some constraint, for example

$$S(u) = \int_a^b G(t, u(t), \dot{u}(t)) \, dt = c_0 \quad \text{(a given constant)}. \quad (1.1.10)$$

As in the case of finite dimensional minimization problems, one then finds a Lagrange multiplier λ with

$$0 = \frac{d}{ds}\left(I(u + s\eta) + \lambda S(u + s\eta)\right)|_{s=0} \quad (1.1.11)$$

for all $\eta \in C_0^1([a, b], \mathbb{R}^d)$. This leads to the Euler–Lagrange equations

$$\frac{d}{dt}(F_p(t, u(t), \dot{u}(t)) + \lambda G_p(t, u(t), \dot{u}(t)))$$
$$- (F_u(t, u(t), \dot{u}(t)) + \lambda G_u(t, u(t), \dot{u}(t))) = 0. \quad (1.1.12)$$

Example. We wish to miminize

$$I(u) = \int_a^b \dot{u}(t)^2 \, dt$$

under the constraint

$$S(u) = \int_a^b u(t)^2 \, dt = 1, \quad (1.1.13)$$

with $u(a) = 0 = u(b)$. (1.1.12) becomes

$$\ddot{u}(t) - \lambda u(t) = 0. \quad (1.1.14)$$

Thus, λ is an eigenvalue for the differential operator d^2/dt^2 under the Dirichlet boundary conditions $u(a) = 0 = u(b)$. Of course, this example can easily be generalized.

Summary. We seek solutions of the variational problem

$$I(u) \to \min,$$

with

$$I(u) = \int_a^b F(t, u(t), \dot{u}(t)) \, dt$$

for given F and unknown $u : [a, b] \to \mathbb{R}^d$. If F and u are differentiable, one may consider some kind of partial derivative, namely

$$\delta I(u, \eta) := \frac{d}{ds} I(u + s\eta)|_{s=0}$$

for $\eta \in C_0^1([a, b], \mathbb{R}^d)$. For a minimizer u then

$$\delta I(u, \eta) = 0 \quad \text{for all such } \eta.$$

If F and u are of class C^2, this leads to the Euler–Lagrange equations

$$\frac{d}{dt} F_p(t, u(t), \dot{u}(t)) - F_u(t, u(t), \dot{u}(t)) = 0.$$

The classical strategy for solving the problem

$$I(u) \to \min$$

consists in solving the Euler–Lagrange equations and then investigating whether a solution of the equations is a minimum of I or not.

1.2 The idea of the direct methods and some regularity results

So far, our formulation of the variational problem

$$I(u) \to \min$$

has been rather vague, because we did not specify in which class of functions u we are trying to minimize I. The only things we did prescribe were boundary conditions of Dirichlet type, i.e. we prescribed the values $u(a)$ and $u(b)$ for our functions $u : [a, b] \to \mathbb{R}^d$.

Because of our derivation of the Euler–Lagrange equations in the preceding section, it would be desirable to have a solution u of class C^2. So one might want to specify in advance that one minimizes I only among functions of class C^2. This, however, directly leads to the question whether I achieves its infimum among functions of class C^2 (with prescribed Dirichlet boundary conditions, as always) or not, and if it does, whether the infimum of I in some larger class of functions, say C^1, could be strictly smaller than the one in C^2. In the light of this question, it might be preferable to minimize I in the class of all functions u for which

$$I(u) = \int_a^b F(t, u(t), \dot{u}(t)) dt$$

is meaningful. Here, we assume that $F(t, u, p)$ is continuous in u and p and measurable in t. For this purpose, one needs the class of functions for which the derivative $\dot{u}(t)$ exists almost everywhere and is finite. This is the class

$$AC([a, b])$$

of *absolutely continuous functions*. A function $u \in AC([a, b])$ satisfies for $t_1, t_2 \in [a, b]$

$$u(t_2) - u(t_1) = \int_{t_1}^{t_2} \dot{u}(t) dt.$$

Note that $F(t, u(t), \dot{u}(t))$ is a measurable function of t for $u \in AC$ by our assumptions on F and the fact that the composition of a measurable and a continuous function is measurable†. The idea of the direct methods in the calculus of variations, as opposed to the classical methods described in the preceding section then consists in minimizing I in a class of functions like $AC([a, b])$ and then trying to show that a solution u because of its minimizing character actually enjoys better regularity properties, for example to be of class C^2, provided F satisfies suitable assumptions.

This minimizing procedure will be treated later‡, since we want to return to the classical theory for a while. Nevertheless, even for the classical theory, one occasionally needs certain regularity results, and therefore we now briefly address the regularity theory. To simplify our notation, we put $I := [a, b]$§. A class of functions intermediate between C^1 and AC is

$$D^1(I, \mathbb{R}^d) := \{u : I \to \mathbb{R}^d,\ u \text{ continuous and piecewise}$$
$$\text{continuously differentiable, i.e. there exist}$$
$$a = t_0 < t_1 < \ldots < t_m = b \text{ with } u \in$$
$$C^1([t_j, t_{j+1}], \mathbb{R}^d) \text{ for } j = 0, \ldots, m - 1\}.$$

$u \in D^1$ then has left and right derivatives $\dot{u}_-(t_j)$ and $\dot{u}_+(t_j)$ even at the points where the derivative is discontinuous, and

$$\dot{u}_-(t_j) = \lim_{t \to t_j - 0} \dot{u}(t) \quad , \quad \dot{u}_+(t_j) = \lim_{t \to t_j + 0} \dot{u}(t).$$

---

† Lebesgue integration theory is summarized in Chapter 1 of Part II. The required composition property is stated there as Theorem 1.1.2. Here, this point will not be pursued or used any further.
‡ See Chapter 4 of Part II.
§ We shall use the same letter I to denote the functional to be minimized and the domain of definition of the functions, inserted into this functional. This conforms to standard notations. The reader should be aware of this and not be confused.

Examples

Example 1.2.1. $[a,b] = [-1,1], d = 1$

$$I(u) = \int_{-1}^{1} \left(1 - (\dot{u}(t))^2\right)^2 dt,$$

$$u(-1) = 1 = u(1).$$

A minimizer is

$$u(t) = |t| \in D^1(I, \mathbb{R})$$

which is not of class C^1. The minimizer of I is not unique (exercise: determine all minimizers), but none of them is of class C^1.

Example 1.2.2. $[a,b] = [-1,1], d = 1$

$$I(u) = \int_{-1}^{1} (1 - \dot{u}(t))^2 u(t)^2 dt$$

$$u(-1) = 0 \quad , \quad u(1) = 1.$$

Here, the unique minimizer is

$$u(t) = \begin{cases} 0 & \text{for } -1 \leq t \leq 0 \\ t & \text{for } 0 \leq t \leq 1 \end{cases}$$

which again is of class D^1, but not C^1.

Example 1.2.3. $[a,b] = [-1,1], d = 1$

$$I(u) = \int_{-1}^{1} (2t - \dot{u}(t))^2 u(t)^2 dt,$$

$$u(-1) = 0 \quad , \quad u(1) = 1.$$

The unique minimizer is

$$u(t) = \begin{cases} 0 & \text{for } -1 \leq t \leq 0 \\ t^2 & \text{for } 0 \leq t \leq 1 \end{cases}$$

which is of class C^1, but not of class C^2.

Theorem 1.2.1. *Let $F(t, u, p)$ be of class C^1 w.r.t. u and p and continuous w.r.t. t ($F : I \times \mathbb{R}^d \times \mathbb{R}^d \to \mathbb{R}$), and let $u \in AC(I, \mathbb{R}^d)$ be a solution of*

$$\delta I(u, \eta) = \int_a^b \{F_u(t, u, \dot{u}) \cdot \eta + F_p(t, u, \dot{u}) \cdot \dot{\eta}\} dt = 0 \quad (1.2.1)$$

1.2 Direct methods, regularity results

for all $\eta \in AC_0(I, \mathbb{R}^d)$ (i.e. $\eta \in AC(I, \mathbb{R}^d)$) and we require that if $I = [a, b]$, there exist $a < a_1 \leq b_1 < b$ with $\eta(x) = 0$ if x is not contained in $[a_1, b_1]$, as in the definition of $C_0^1([a, b], \mathbb{R}^d)$). Then for almost all points in I

$$\frac{d}{dt} F_p(t, u(t), \dot{u}(t)) = F_u(t, u(t), \dot{u}(t)) \tag{1.2.2}$$

(note, however, that the derivative on the left hand side cannot be computed by the chain rule). If $u \in C^1(I, \mathbb{R}^d)$, (1.2.2) holds for all $t \in I$, and if $u \in D^1(I, \mathbb{R}^d)$, at those points t_j where $\dot{u}(t_j)$ is discontinuous

$$\frac{d}{dt}\left(F_p(t_j, u(t_j), \dot{u}_-(t_j))\right)_- = F_u(t_j, u(t_j), \dot{u}_-(t_j)),$$

and analogously for the right derivative.

Remark. It actually suffices to assume (1.2.1) for all $\eta \in C_0^\infty(I, \mathbb{R}^d)$, because functions in AC_0 may be approximated by C_0^∞ functions. If $u \in C^1$ or D^1, the proof anyway only requires (1.2.1) for $\eta \in C_0^1$ or D_0^1, respectively (where D_0^1 is defined analogously to C_0^1).

Proof. We have, omitting the obvious arguments of F, F_u, etc.,

$$\int_a^b F_u \eta \, dt = \int_a^b \frac{d}{dt}\left(\int_a^t F_u \, dy\right) \eta \, dt = -\int_a^b \left(\int_a^t F_u \, dy\right) \dot{\eta} \, dt$$

(1.2.1) then implies

$$0 = \int_a^b \left(F_p - \int_a^t F_u \, dy\right) \dot{\eta} \, dt.$$

We now make use of:

Lemma 1.2.1. *Let $h \in L^1(I, \mathbb{R})$ satisfy*

$$\int_a^b h(t) \dot{\varphi}(t) \, dt = 0 \quad \text{for all } \varphi \in AC_0(I, \mathbb{R}). \tag{1.2.3}$$

Then there exists a constant $c \in \mathbb{R}$ with

$$h(t) = c \quad \text{for almost all } t \in I.$$

Remark. It actually suffices to assume (1.2.3) for all $\varphi \in C_0^\infty(I, \mathbb{R})$. If $h \in C^1$, one directly sees from the proof that $\varphi \in C_0^1$ suffices.

Proof. We put

$$c := \frac{1}{b-a} \int_a^b h(t) \, dt$$

and
$$\varphi(t) := \int_a^t (h(y) - c)\, dy.$$
Then
$$\varphi(a) = 0, \text{ and } \varphi(b) = \int_a^b \dot\varphi(t)dt = 0. \tag{1.2.4}$$

Equation (1.2.3) implies
$$0 = \int_a^b (h(y) - c)\, h(y)dy = \int_a^b (h(y) - c)^2\, dy$$
because of (1.2.4). This implies the claim.

q.e.d.

We now may complete the proof of Theorem 1.2.1:
By Lemma 1.2.1 there exists $c \in \mathbb{R}^d$ with
$$F_p(t, u(t), \dot u(t)) = \int_a^t F_u(y, u(y), \dot u(y))dy + c \tag{1.2.5}$$
for almost all $t \in I$. Therefore, F_p is of class AC, and differentiating (1.2.5) gives (1.2.3). The claims for $u \in C^1$ or D^1 are obvious from the proof.

q.e.d.

Theorem 1.2.2. Let $F : I \times \mathbb{R}^d \times \mathbb{R}^d$ be of class C^1, and let F_p be also of class C^1, and let $\det\left(F_{p^ip^j}(t, u(t), \dot u(t))_{i,j=1,\ldots,d}\right) \neq 0$ for all $t \in I$ and a solution $u \in C^1(I, \mathbb{R}^d)$ of
$$\delta I(u, \eta) = 0 \quad \text{for all } \eta \in C_0^1(I, \mathbb{R}^d).$$
Then u is of class C^2.

Proof. We define
$$\phi : \mathbb{R} \times \mathbb{R}^d \times \mathbb{R}^d \times \mathbb{R}^d \to \mathbb{R}$$
via
$$\phi(t, u, p, q) := F_p(t, u, p) - q.$$

Our assumption $\det F_{pp} \neq 0$ makes it possible to apply the implicit function theorem to conclude that
$$\phi(t, u, p, q) = 0$$
may be uniquely solved w.r.t. p near $u_0 = u(t_0)$, $p_0 = \dot u(t_0)$, $q_0 =$

$F(t_0, u_0, p_0)$ for any $t_0 \in I$. Thus, there exists a neighbourhood U of (t_0, u_0, q_0) such that for each $(t, u, q) \in U$, $\phi = 0$ has a unique solution $p = \varphi(t, u, q)$ and that $\varphi : U \to \mathbb{R}^d$ is of class C^1. Since we already know a solution of $\phi = 0$, namely $(t, u(t), \dot{u}(t), F_p(t, u(t), \dot{u}(t)))$, the uniqueness of the solution φ implies

$$\dot{u}(t) = \varphi(t, u(t), F_p(t, u(t), \dot{u}(t))) \quad \text{for } t \text{ near } t_0.$$

Since φ is of class C^1, so then is $\dot{u}(t)$, hence $u \in C^2$. Since $t_0 \in I$ was arbitrary, $u \in C^2(I, \mathbb{R}^d)$.

q.e.d.

Theorem 1.2.3. *Let F satisfy the assumptions of Theorem. 1.2.2, and in addition assume that F_{pp} is (positive or negative) definite on $\Omega \times \mathbb{R}^d$ where $\Omega \subset \mathbb{R}^{d+1}$ contains $\{(t, u(t)) : t \in I\}$. Let $u \in AC(I, \mathbb{R}^d)$ satisfy*

$$\delta I(u, \eta) = 0 \quad \text{for all } \eta \in AC_0(I, \mathbb{R}^d)$$

(assume that $F_u(t, u(t), \dot{u}(t))$ and $F_p(t, u(t), \dot{u}(t))$ are integrable). Then $u \in C^2(I, \mathbb{R}^d)$.

Proof. Since the uniqueness result of the implicit function theorem is only local, it cannot be applied anymore because $\dot{u}(t)$ might be discontinuous. We thus need a global argument. Thus, assume that for given $(t, u, q) \in \Omega \times \mathbb{R}^d$, there are two solutions $p_1, p_2 \in \mathbb{R}^d$ of $\phi(t, u, p, q) = 0$, i.e.

$$q = F_p(t, u, p_1) \quad \text{and} \quad q = F_p(t, u, p_2).$$

Thus

$$\int_0^1 F_{pp}(t, u, p_1 + s(p_2 - p_1)) ds \; (p_2 - p_1) = 0. \tag{1.2.6}$$

By our assumption on F_{pp}, (1.2.6) is invertible, hence $p_2 = p_1$, hence uniqueness.

Using this global uniqueness together with the existence result of the implicit function theorem, we now see that for any (t, u, q) in a sufficiently small neighbourhood of (t_0, u_0, q_0) ($t_0 \in I$, $u_0 = u(t_0)$, $q_0 = F_p(t_0, u_0, p_0)$, $p_0 = \dot{u}_0(t_0)$), there is a unique solution $\varphi(t, u, q)$ of

$$F_p(t, u, p) - q = 0$$

and φ is of class C^1. Thus, as in the proof of Theorem 1.2.2,

$$\dot{u}(t) = \varphi(t, u(t), F_p(t, u(t), \dot{u}(t)))$$

for almost all t in a neighbourhood of t_0. Since $u(t)$ and $F_p(t, u(t), \dot{u}(t))$ are absolutely continuous w.r.t. t (the latter by Theorem 1.2.1), $\dot{u}(t)$ coincides for almost all t near t_0 with an absolutely continuous function $v(t)$. We put

$$w(t) := u(t_0) + \int_{t_0}^t v(y) dy.$$

w then is of class C^1. Since u is absolutely continuous, by a theorem of Lebesgue

$$u(t) = u(t_0) + \int_{t_0}^t u'(y) dy.$$

Since $v = \dot{u}$ almost everywhere, we conclude $u = w$, hence $u \in C^1$ near t_0, which was arbitrary in I. Theorem 1.2.2 then gives $u \in C^2$.

q.e.d.

Corollary 1.2.1. *Under the assumptions of Theorem 1.2.3, any AC-solution of $\delta I(u, \eta) = 0$ for all $\eta \in AC_0(I, \mathbb{R}^d)$ is a solution of the Euler–Lagrange equations*

$$\frac{d}{dt} F_p(t, u(t), \dot{u}(t)) - F_u(t, u(t), \dot{u}(t)) = 0 \qquad (1.2.7)$$

or equivalently of

$$F_{pp}(t, u(t), \dot{u}(t)) \ddot{u}(t) + F_{pu}(t, u(t), \dot{u}(t)) \dot{u}(t)$$
$$+ F_{pt}(t, u(t), \dot{u}(t)) - F_u(t, u(t), \dot{u}(t)) = 0. \qquad (1.2.8)$$

The same holds under the assumptions of Theorem 1.2.2 for a C^1- solution of $\delta I(u, \eta) = 0$ for all $\eta \in C_0^1(I, \mathbb{R}^d)$.

q.e.d.

Theorem 1.2.4. *Let $F : I \times \mathbb{R}^d \times \mathbb{R}^d \to \mathbb{R}$ be of class C^k, and let F_p also be of class C^k, $k \in \{2, 3, \ldots, \infty\}$. Suppose u is of class C^1 and a solution of $\delta I(u, \eta) = 0$ for all $\eta \in C_0^1(I, \mathbb{R}^d)$, and suppose*

$$\det \left(F_{p^i p^j}(t, u(t), \dot{u}(t))_{i,j=1,\ldots,d} \right) \neq 0 \quad \text{for all } t \in I. \qquad (1.2.9)$$

Then $u \in C^{k+1}(I, \mathbb{R}^d)$. (The same result holds if we assume that $u \in C^1$ is a solution of the Euler–Lagrange equations (1.2.8).)

Proof. By Theorem 1.2.2, u is of class C^2, and by Corollary. 1.2.1, it

1.2 Direct methods, regularity results

solves (1.2.8). Because of (1.2.9), $F_{pp}(t, u(t), \dot{u}(t))$ is an invertible matrix, hence

$$\ddot{u}(t) = F_{pp}^{-1}(t, u(t), \dot{u}(t))$$
$$\{-F_{pu}(t, u(t), \dot{u}(t)) - F_{pt}(t, u(t), \dot{u}(t)) + F_u(t, u(t), \dot{u}(t))\}.$$
(1.2.10)

Let now $j \leq k$, and suppose inductively $u \in C^j$. The right hand side of (1.2.10) then is of class C^{j-1}. Therefore, $\ddot{u}$ is of class C^{j-1}, hence u is of class C^{j+1}.

q.e.d.

The preceding proof most clearly shows the importance of the assumption $\det(F_{p^i p^j}(t, u(t), \dot{u}(t))) \neq 0$ that already occurred in the proof of Theorem 1.2.2. Namely, it implies that the Euler–Lagrange equations (1.2.8) can be solved for $\ddot{u}$ in terms of u and $\dot{u}$.

Corollary 1.2.2. *If under the assumption of Theorem 1.2.3, F and F_p are of class C^k, then a solution u of $\delta I(u, \eta) = 0$ for all $\eta \in AC_0$ is of class C^{k+1}.*

q.e.d.

Summary. If one wants to solve

$$I(u) \to \min$$

by a direct minimization procedure, it is preferable to admit a class of comparison functions u that is as large as possible. $AC(I, \mathbb{R}^d)$ seems to be a good choice, because this is the largest class for which

$$I(u) = \int F(t, u(t), \dot{u}(t))$$

is well defined, assuming $F(t, u, p)$ to be continuous in u and p and measurable in t. However, if one then finds a minimizer u, it might not be a solution of the Euler–Lagrange equations, because it is not regular enough. If the invertibility condition $\det F_{pp} \neq 0$ is satisfied, however, one may show that a minimizer u is as regular as F allows. Namely, if F and F_p are of class C^k, $k \in \{1, 2, \ldots, \infty\}$, then u is of class C^{k+1}. Examples show that without such an invertibility condition, regularity need not hold. This invertibility condition $\det F_{pp} \neq 0$ implies that the Euler–Lagrange equations allow the expression of $\ddot{u}(t)$ in terms of $u(t)$ and $\dot{u}(t)$.

1.3 The second variation. Jacobi fields

We assume that $u \in D^1(I, \mathbb{R}^d)$ is a critical point of

$$I(u) = \int_a^b F(t, u(t), \dot{u}(t)) dt,$$

i.e.

$$\delta I(u, \eta) = 0 \quad \text{for all } \eta \in D_0^1(I, \mathbb{R}^d). \tag{1.3.1}$$

We recall that

$$\delta I(u, \eta) := \frac{d}{ds} I(u + s\eta)_{|s=0},$$

and $\delta I(u, \eta) = 0$ is equivalent to $s = 0$ being a critical point of the function

$$f(s) = I(u + s\eta).$$

If we want to decide if a given solution u minimizes I instead of just being a critical point, we immediately see that a necessary condition would be

$$f''(0) \geq 0 \tag{1.3.2}$$

for the above function f and all $\eta \in D_0^1(I, \mathbb{R}^d)$. Namely, by Taylor's theorem, since $f'(0) = 0$

$$f(s) - f(0) = \frac{1}{2} s^2 f''(0) + o(s^2) \quad \text{for } s \to 0.$$

More precisely, (1.3.2) is needed for u to minimize I when compared with $u + s\eta$ for sufficiently small s. In other words, we want u to minimize I in a D^1-neighbourhood of itself, i.e. among functions

$$v \in D^1(I, \mathbb{R}^d)$$

with

$$u(a) = v(a), u(b) = v(b) \quad \text{and} \tag{1.3.3}$$

$$\sup_{t \in I} (|u(t) - v(t)| + |\dot{u}_-(t) - \dot{v}_-(t)| + |\dot{u}_+(t) - \dot{v}_+(t)|) < \epsilon \tag{1.3.4}$$

for some $\epsilon > 0$. (Note: It is not clear that ϵ may be chosen independently of v.) We define the *second variation* of I at u in the direction $\eta \in D_0^1$ as

$$\delta^2 I(u, \eta) := \frac{d^2}{ds^2} I(u + s\eta)_{|s=0}.$$

1.3 The second variation. Jacobi fields

In order that this variation exists, we require for the rest of the section that F is of class C^2. We then compute

$$\delta^2 I(u,\eta) = \frac{d^2}{ds^2}\int_a^b F(t,u(t)+s\eta(t),\dot{u}(t)+s\dot{\eta}(t))dt_{|s=0}$$

$$= \int_a^b \{F_{p^i p^j}(t,u(t),\dot{u}(t))\dot{\eta}_i(t)\dot{\eta}_j(t)$$
$$+ 2F_{p^i u^j}(t,u(t),\dot{u}(t))\dot{\eta}_i(t)\eta_j(t)$$
$$+ F_{u^i u^j}(t,u(t),\dot{u}(t))\eta_i(t)\eta_j(t)\} \, dt. \quad (1.3.5)$$

Here, and in the sequel, we employ the standard summation conventions, e.g.

$$F_{p^i p^j}\dot{\eta}_i\dot{\eta}_j = \sum_{i,j=1}^d F_{p^i p^j}\dot{\eta}_i\dot{\eta}_j.$$

We abbreviate (1.3.5) as

$$\delta^2 I(u,\eta) = \int_a^b \{F_{pp}\dot{\eta}\dot{\eta} + 2F_{pu}\dot{\eta}\eta + F_{uu}\eta\eta\} \, dt. \quad (1.3.6)$$

Our preceding considerations imply:

Theorem 1.3.1. *Suppose $F \in C^2(I \times \mathbb{R}^d \times \mathbb{R}^d \times \mathbb{R})$ and let $u \in D^1(I,\mathbb{R}^d)$ satisfy $I(u) \leq I(v)$ for all v with (1.3.3), (1.3.4). Then*

$$\delta^2 I(u,\eta) \geq 0 \quad \text{for all } \eta \in D_0^1(I,\mathbb{R}^d). \quad (1.3.7)$$

We now put, for given u,

$$\phi(t,\eta,\pi) := F_{p^i p^j}(t,u(t),\dot{u}(t))\pi_i\pi_j + 2F_{p^i u^j}(t,u(t),\dot{u}(t))\pi_i\eta_j$$
$$+ F_{u^i u^j}(t,u(t),\dot{u}(t))\eta_i\eta_j,$$

and we define the *accessory variational problem* for $I(u) \to \min$ as

$$Q(\eta) := \int_a^b \phi(t,\eta(t),\dot{\eta}(t))dt \to \min \quad \text{among all } \eta \in D_0^1(I,\mathbb{R}^d). \quad (1.3.8)$$

If u satisfies the assumptions of Theorem 1.3.1, then

$$Q(\eta) \geq 0 \quad \text{for all } \eta \in D_0^1, \quad (1.3.9)$$

and hence $\eta = 0$ is a trivial solution of (1.3.8). We are interested in the question whether there are others. The Euler–Lagrange equations for (1.3.8) are

$$\frac{d}{dt}\phi_\pi(t,\eta(t),\dot{\eta}(t)) = \phi_\eta(t,\eta(t),\dot{\eta}(t)), \quad (1.3.10)$$

i.e.
$$\frac{d}{dt}(F_{pp}(t,u(t),\dot{u}(t))\dot{\eta}(t) + F_{pu}(t,u(t),\dot{u}(t))\eta(t))$$
$$= F_{pu}(t,u(t),\dot{u}(t))\dot{\eta}(t) + F_{uu}(t,u(t),\dot{u}(t))\eta(t). \quad (1.3.11)$$

Since u is considered as given, our first observation is that (1.3.11) is a *linear* homogeneous system of second order equations for the unknown η. These equations are called *Jacobi equations*.

Definition 1.3.1. *A solution $\eta \in C^2(I, \mathbb{R}^d)$ of the Jacobi equations (1.3.11) is called a* Jacobi field *along $u(t)$.*

Lemma 1.3.1. *Let $F \in C^3(I \times \mathbb{R}^d \times \mathbb{R}^d, \mathbb{R})$, $\det F_{pp}(t, u(t), \dot{u}(t)) \neq 0$ for all $t \in I$, $u \in C^2(I, \mathbb{R}^d)$. Then any solution of $\eta \in AC_0(I, \mathbb{R}^d)$, $\delta Q(\eta, \varphi) = 0$ for all $\varphi \in AC_0(I, \mathbb{R}^d)$ is of class C^2 and hence a Jacobi field.*

Proof. We apply Theorem 1.2.3. For that purpose, we note that
$$\phi_{\pi\pi}(t, \eta(t), \dot{\eta}(t)) = F_{pp}(t, u(t), \dot{u}(t)) \quad \text{for all } t \text{ and } \eta$$
and so the assumption $\det F_{pp}(t, u(t), \dot{u}(t)) \neq 0$, that is seemingly weaker than the one of Theorem 1.2.3, indeed suffices to apply that Theorem.
q.e.d.

We now derive the so-called *necessary Legendre condition*:

Theorem 1.3.2. *Under the assumption of Theorem 1.3.1, i.e. $u \in D^1(I, \mathbb{R}^d)$ minimizes I in the sense described there, we have that*
$$F_{pp}(t, u(t), \dot{u}(t)) \text{ is positive semidefinite for all } t \in I,$$

i.e.
$$F_{p^i p^j}(t, u(t), \dot{u}(t))\xi^i \xi^j \geq 0 \quad \text{for all } \xi = (\xi^1, \ldots, \xi^d) \in \mathbb{R}^d.$$

(At points where $\dot{u}(t)$ is discontinuous, this holds for the left and right derivatives.)

Proof. We may assume that $t_0 \in \mathring{I}$ and $\dot{u}$ is continuous at t_0. The result at the points where $\dot{u}$ jumps then follows by taking appropriate limits, and likewise at $t_0 = a, b$. We then consider $0 < \epsilon \leq \min(t_0 - a, b - t_0)$ and define $\eta \in D_0^1(I, \mathbb{R}^d)$ by

$$\eta(t) := \begin{cases} 0 & \text{for } a \leq t \leq t_0 - \epsilon \text{ and } t_0 + \epsilon \leq t \leq b \\ \epsilon \xi & \text{for } t = t_0 \\ \text{linear} & \text{for } t_0 - \epsilon \leq t \leq t_0 \text{ and for } t_0 \leq t \leq t_0 + \epsilon \end{cases}$$

1.3 The second variation. Jacobi fields

for given $\xi \in \mathbb{R}^d$. Then

$$\dot{\eta}(t) = \begin{cases} 0 & \text{for } a \leq t < t_0 \text{ or } t_0 + \epsilon < t \leq b \\ \xi & \text{for } t_0 - \epsilon < t < t_0 \\ -\xi & \text{for } t_0 < t < t_0 + \epsilon. \end{cases}$$

We apply Theorem 1.3.1 to obtain

$$0 \leq \delta^2 I(u, \eta) = \int_{t_0 - \epsilon}^{t_0 + \epsilon} F_{p^i p^j}(t, u(t), \dot{u}(t))\xi^i \xi^j \, dt + O(\epsilon^2) \quad \text{for } \epsilon \to 0,$$

since all other terms contain a factor ϵ, and we integrate over an interval of length 2ϵ. Hence

$$F_{p^i p^j}(t_0, u(t_0), \dot{u}(t_0))\xi^i \xi^j = \lim_{\epsilon \to 0} \frac{1}{2\epsilon} \int_{t_0 - \epsilon}^{t_0 + \epsilon} F_{p^i p^j}(t, u(t), \dot{u}(t))\xi^i \xi^j \, dt \geq 0.$$

q.e.d.

The Jacobi equations and the notion of Jacobi fields are meaningful for arbitrary solutions of the Euler–Lagrange equations, not only for minimizing ones. In fact, Jacobi fields are solutions of the linearized Euler–Lagrange equations. Namely:

Theorem 1.3.3. *Let $F \in C^3(I \times \mathbb{R}^d \times \mathbb{R}^d, \mathbb{R})$, and let $u_s(t)$ be a family of C^2-solutions of the Euler–Lagrange equations*

$$\frac{d}{dt} F_p(t, u_s(t), \dot{u}_s(t)) - F_u(t, u_s(t), \dot{u}_s(t)) = 0, \qquad (1.3.12)$$

with u_s depending differentiably on a parameter $s \in (-\epsilon, \epsilon)$. Then

$$\eta(t) := \frac{d}{ds} u_s(t)_{|s=0}$$

is a Jacobi field along $u = u_0$.

Proof. We differentiate (1.3.12) w.r.t. s at $s = 0$ to obtain

$$\frac{d}{dt}\left(F_{pp}(t, u(t), \dot{u}(t))\dot{\eta}(t) + F_{pu}(t, u(t), \dot{u}(t))\eta(t)\right)$$
$$- F_{pu}(t, u(t), \dot{u}(t))\dot{\eta}(t) - F_{uu}(t, u(t), \dot{u}(t))\eta(t) = 0.$$

i.e. the Jacobi equation (1.3.11). q.e.d.

Lemma 1.3.2. *Let $a \leq a_1 < a_2 \leq b$, and let F and F_p be of class C^2 in $[a_1, a_2]$, and suppose $\eta \in C^1([a_1, a_2], \mathbb{R}^d)$ is a Jacobi field on $[a_1, a_2]$ with $\eta(a_1) = 0 = \eta(a_2)$. Then*

$$\int_{a_1}^{a_2} \phi(t, \eta(t), \dot{\eta}(t)) \, dt = 0. \qquad (1.3.13)$$

Proof. Since ϕ is homogeneous of second order in (η, π), we have
$$2\phi(t, \eta, \pi) = \phi_\eta(t, \eta, \pi)\eta + \phi_\pi(t, \eta, \pi)\pi.$$
Therefore
$$2\int_{a_1}^{a_2} \phi(t, \eta, \dot\eta)dt = \int_{a_1}^{a_2} \{\phi_\eta(t, \eta, \dot\eta) \cdot \eta + \phi_\pi(t, \eta, \dot\eta) \cdot \dot\eta\}\, dt. \quad (1.3.14)$$

Comparing (1.3.10) and (1.3.11), we see that ϕ_π is of class C^1 as a function of t. We may hence integrate the last term in (1.3.14) by parts. Since $\eta(a_1) = 0 = \eta(a_2)$, we obtain
$$2\int_{a_1}^{a_2} \phi(t, \eta, \dot\eta)dt = \int_{a_1}^{a_2} \left(\phi_\eta(t, \eta, \dot\eta) - \frac{d}{dt}\phi_\pi(t, \eta, \dot\eta)\right) \cdot \eta\, dt = 0,$$
since η is a Jacobi field. q.e.d.

As before, let F be of class C^3, and let $u(t)$ be a solution of class C^2 on $[a, b]$ of the Euler–Lagrange equations
$$\frac{d}{dt}F_p(t, u(t), \dot u(t)) - F_u(t, u(t), \dot u(t)) = 0.$$

Definition 1.3.2. *Let $a \leq a_1 < a_2 \leq b$. We call the parameter value a_2 conjugate to a_1 and the point $(a_2, u(a_2))$ conjugate to $(a_1, u(a_1))$ if there exists a not identically vanishing Jacobi field η on $[a_1, a_2]$ with $\eta(a_1) = 0 = \eta(a_2)$.*

We may derive the important result of Jacobi:

Theorem 1.3.4. *Let $F \in C^3(I \times \mathbb{R}^d \times \mathbb{R}^d, \mathbb{R})$ and suppose $u \in C^2(I, \mathbb{R}^d)$. Suppose that $F_{pp}(t, u(t), \dot u(t))$ is positive definite on I. If there exists a* with $a < a^* < b$ that is conjugate to a, then u cannot be a local minimum of I. More precisely, for any $\epsilon > 0$, there exists $v \in D^1(I, \mathbb{R}^d)$ with $v(a) = u(a)$, $v(b) = u(b)$,*
$$\sup_{t \in I} (|u(t) - v(t)| + |\dot u(t) - \dot v_\pm(t)|) < \epsilon$$
and
$$I(v) < I(u).$$

Proof. Let $\eta(t)$ be a nontrivial Jacobi field on $[a, a^*]$. We put
$$\eta^*(t) := \begin{cases} \eta(t) & \text{for } a \leq t \leq a^* \\ 0 & \text{for } a^* \leq t \leq b. \end{cases}$$

1.3 The second variation. Jacobi fields

Then $\eta^* \in D_0^1(I, \mathbb{R}^d)$, and by Lemma 1.3.2

$$Q(\eta^*) = \int_a^b \phi(t, \eta^*, \dot{\eta}^*)dt = 0.$$

If u were a local minimum, then by Theorem 1.3.1

$$0 \leq \delta^2 I(u, \tilde{\eta}) = Q(\tilde{\eta}) \quad \text{for all } \tilde{\eta} \in D_0^1(I, \mathbb{R}^d).$$

Hence η^* would be a minimizer of Q, hence by Lemma 1.3.1 $\eta^* \in C^2(I, \mathbb{R}^d)$. Since $\dot{\eta}_+^*(a^*) = 0$, then

$$\dot{\eta}^*(a^*) = 0.$$

Since also $\eta^*(a^*) = 0$, and since η^* solves the Jacobi equation, a (linear) second order ordinary differential equation, the uniqueness theorem for solutions of such equations implies

$$\eta^* \equiv 0,$$

a contradiction, because by assumption η does not vanish identically. Hence u cannot be a local minimizer.

q.e.d.

In words, Theorem 1.3.4 says that a solution of the Euler–Lagrange equations cannot be minimizing beyond the first conjugate point. Turned the other way round, Theorem 1.3.4 says that if u is a local minimizer, then there cannot be any parameter value a^* with $a < a^* < b$ that is conjugate to a. It may happen, however, that b is conjugate to a. An example will be given in the next chapter.

Summary. In order to obtain necessary conditions for a solution of the Euler–Lagrange equations

$$\frac{d}{dt} F_p(t, u(t), \dot{u}(t)) = F_u(t, u(t), \dot{u}(t))$$

to minimize

$$I(u) = \int_a^b F(t, u(t), \dot{u}(t))dt,$$

one needs to study the second variation

$$Q(\eta) := \delta^2 I(u, \eta) = \frac{d}{ds^2} I(u + s\eta)\big|_{s=0} \quad \text{for } \eta \in D_0^1.$$

If, for fixed u, we consider the variational problem $Q(\eta) \to 0$, we are led to the Jacobi equations

$$\frac{d}{dt}\left(F_{pp}(t, u(t), \dot{u}(t))\dot{\eta}(t) + F_{pu}(t, u(t), \dot{u}(t))\eta(t)\right)$$
$$= F_{up}(t, u(t), \dot{u}(t))\dot{\eta}(t) + F_{uu}(t, u(t), \dot{u}(t))\eta(t)$$

for η.

Solutions η with $\eta(a) = \eta(b) = 0$ are called Jacobi fields. $a^* \in (a, b)$ for which there exists a nontrivial Jacobi field on $[a, a^*]$ is called conjugate to a, and if there exists such a^*, u cannot be locally minimizing on $[a, b]$. In other words, a solution of the Euler–Lagrange equations cannot be minimizing beyond the first conjugate point.

1.4 Free boundary conditions

We recall the definition of an n-dimensional embedded differentiable submanifold M of $\mathbb{R}^d$: For every $p \in M$, there have to exist a neighbourhood $V = V(p) \subset \mathbb{R}^d$, an open set $U \subset \mathbb{R}^n$ and an injective differentiable map $f : U \to V$ of everywhere maximal rank n (i.e. for every $z \in U$, the derivative $Df(z)$, a linear map from $\mathbb{R}^n$ to $\mathbb{R}^d$, has rank n) with

$$M \cap V = f(U).$$

An example is the sphere S^n described in detail in Section 2.1 (Example 2.1.1). The tangent space $T_p M$ of M at p then is the vector space $Df(z)(\mathbb{R}^n)$. It can be considered as a subspace of the vector space $T_p \mathbb{R}^d$, the tangent space of $\mathbb{R}^d$ at p.

As in 1.1, we now consider the variational problem

$$I(u) = \int_a^b F(t, u(t), \dot{u}(t))dt \to \min$$

with F of class C^2. This time, however, we do not impose the Dirichlet boundary condition that the values of $u(a)$ and $u(b)$ were prescribed, but the more general condition that for given submanifolds M_1, M_2 (differentiable, embedded) of $\mathbb{R}^d$, we require that

$$u(a) \in M_1, u(b) \in M_2.$$

(Dirichlet boundary conditions constitute the special case where M_1 and M_2 are points.)

In this section, we do not consider regularity questions. As an exercise,

1.4 Free boundary conditions

the reader should supply the necessary regularity assumptions on F, u, etc. at each step.

Let u be a solution. Then, as before, u has to satisfy the Euler–Lagrange equations, because if $u(a) \in M_1$, $\eta(a) = 0$, then also $u(a) + s\eta(a) \in M_1$ for any s, and likewise at b, and so we may again consider variations of the form $u + s\eta$, $\eta \in D_0^1$. This time, however, also more general variations are admissible. Namely, let $u_s(t)$ be a family of maps from I into $\mathbb{R}^d$ depending differentiably on $s \in (-\epsilon, \epsilon)$, with $u(t) = u_0(t)$ and

$$u_s(a) \in M_1 \quad , \quad u_s(b) \in M_2 \quad \text{for all } s.$$

Let

$$\eta(t) := \frac{d}{ds} u_s(t)_{|s=0},$$

Then again

$$0 = \frac{d}{ds} I(u_s)_{|s=0} = \frac{d}{ds} \int_a^b F(t, u(t), \dot{u}(t)) dt_{|s=0}$$

$$= \int_a^b \{F_p(t, u(t), \dot{u}(t)) \cdot \dot{\eta}(t) + F_u(t, u(t), \dot{u}(t)) \cdot \eta(t)\} dt$$

$$= \int_a^b \left\{-\frac{d}{dt} F_p + F_u\right\} \cdot \eta + F_p(t, u(t), \dot{u}(t)) \cdot \eta(t)|_{t=a}^{t=b}$$

$$= F_p(t, u(t), \dot{u}(t)) \cdot \eta(t)|_{t=a}^{t=b},$$

since u solves the Euler–Lagrange equations.

We now observe that $\eta(a) \in T_{u(a)} M_1$ (and likewise at b), since we may find a 'local chart' f as above with $M_1 \cap V(u(a)) = f(U)$ for a neighbourhood V of $u(a)$ and some open set $U \subset \mathbb{R}^{n_1}$ ($n_1 = \dim M_1$). By choosing ϵ smaller if necessary, we may assume $u_s(a) \in M_1 \cap V = f(U)$ for $s \in (-\epsilon, \epsilon)$. Since f is injective, there then has to exist a curve $\gamma(s) \subset U$ with $u_s(a) = f \circ \gamma(s)$ for all s. Hence $\eta(a) = \frac{d}{ds} u_s(a)_{|s=0} = Df(f^{-1}u(a))\gamma'(0)$ is indeed tangent to M_1 at $u(a)$. Moreover, any tangent vector to M_1 at $u(a)$ can be realized in this manner. Therefore, since we may choose the values of η at a and b independently of each other, we conclude

$$F_p(a, u(a), \dot{u}(a)) \cdot V = 0 \quad \text{for all } V \in T_{u(a)} M_1,$$

and likewise

$$F_p(b, u(b), \dot{u}(b)) \cdot W = 0 \quad \text{for all } W \in T_{u(b)} M_2.$$

We have thus shown:

Theorem 1.4.1. *Let u be a critical point of I among curves with $u(a) \in M_1$, $u(b) \in M_2$ (M_1, M_2 given differentiable embedded submanifolds of $\mathbb{R}^d$), i.e. $\frac{d}{ds}I(u_s)|_{s=0} = 0$ for all variations $u_s(t)$ differentiable in s with $u_s(a) \in M_1$, $u_s(b) \in M_2$ for all $s \in (-\epsilon, \epsilon)$ ($\epsilon > 0$). Then u is a solution of the Euler–Lagrange equations for I, and in addition, $F_p(a, u(a), \dot{u}(a))$ and $F_p(b, u(b), \dot{u}(b))$ are orthogonal to $T_{u(a)}M_1$ and $T_{u(b)}M_2$, respectively. In particular, if for example $M_1 = \mathbb{R}^d$, then $F_p(a, u(a), \dot{u}(a)) = 0$.*

Summary. If instead of a Dirichlet boundary condition, we more generally impose a free boundary condition that $u(a)$ and $u(b)$ are only required to be contained in given differentiable submanifolds M_1 and M_2, respectively, of $\mathbb{R}^d$, then $F_p(a, u(a), \dot{u}(a))$ and $F_p(b, u(b), \dot{u}(b))$ are orthogonal to these submanifolds for a critical point of I under those boundary conditions.

1.5 Symmetries and the theorem of E. Noether

In the variational problems of classical mechanics, one often encounters conserved quantities, like energy, momentum, or angular momentum. It was realized by E. Noether that all those conservation laws result from a general theorem stating that invariance properties of the variational integral I lead to corresponding conserved quantities. We first treat a special case.

Theorem 1.5.1. *We consider the variational integral*

$$I(u) = \int_a^b F(t, u(t), \dot{u}(t))dt,$$

with $F \in C^2([a,b] \times \mathbb{R}^d \times \mathbb{R}^d, \mathbb{R})$. We suppose that there exists a smooth one-parameter family of differentiable maps

$$h_s : \mathbb{R}^d \to \mathbb{R}^d$$

(the precise smoothness requirement is that

$$h(s, z) := h_s(z)$$

is of class $C^2((-\epsilon_0, \epsilon_0) \times \mathbb{R}^d, \mathbb{R})$ for some $\epsilon_0 > 0$), with

$$h_0(z) = z \quad \text{for all } z \in \mathbb{R}^d$$

1.5 Symmetries and the theorem of E. Noether

and satisfying

$$\int_a^b F\left(t, h_s(u(t)), \frac{d}{dt}h_s(u(t))\right) dt = \int_a^b F\left(t, u(t), \frac{d}{dt}u(t)\right) dt \quad (1.5.1)$$

for all $s \in (-\epsilon, \epsilon)$ and all $u \in C^2([a, b], \mathbb{R}^d)$.
Then, for any solution $u(t)$ of the Euler–Lagrange equations (1.1.4) for I,

$$F_p(t, u(t), \dot{u}(t)) \frac{d}{ds} h_s(u(t))|_{s=0} \quad (1.5.2)$$

is constant in $t \in [a, b]$.

Definition 1.5.1. *A quantity $C(t, u(t), \dot{u}(t))$ that is constant in t for each solution of the Euler–Lagrange equations of a variational integral $I(u)$ is called a (first) integral of motion.*

Proof of Theorem 1.5.1: Equation (1.5.1) yields for any $t_0 \in [a, b]$, using $h_0(z) = z$,

$$\begin{aligned} 0 &= \frac{d}{ds} \int_a^{t_0} F\left(t, h_s(u(t)), \frac{d}{dt}h_s(u(t))\right) dt|_{s=0} \\ &= \int_a^{t_0} \left\{ F_u(t, u(t), \dot{u}(t)) \frac{d}{ds} h_s(u(t)) \right. \\ &\quad \left. + F_p(t, u(t), \dot{u}(t)) \frac{d}{dt}\frac{d}{ds} h_s(u(t)) \right\} dt|_{s=0}. \end{aligned} \quad (1.5.3)$$

We recall the Euler–Lagrange equations (1.1.4) for u:

$$0 = \frac{d}{dt} F_p(t, u(t), \dot{u}(t)) - F_u(t, u(t), \dot{u}(t)). \quad (1.5.4)$$

Using (1.5.4) in (1.5.3) to replace F_u, we obtain

$$\begin{aligned} 0 &= \int_a^{t_0} \left\{ \frac{d}{dt} F_p(t, u(t), \dot{u}(t)) \frac{d}{ds} h_s(u(t)) \right. \\ &\quad \left. + F_p(t, u(t), \dot{u}(t)) \frac{d}{dt}\frac{d}{ds} h_s(u(t)) \right\} dt|_{s=0} \\ &= \int_a^{t_0} \frac{d}{dt} \left(F_p(t, u(t), \dot{u}(t)) \frac{d}{ds} h_s(u(t))|_{s=0} \right) dt. \end{aligned} \quad (1.5.5)$$

Therefore

$$F_p(t_0, u(t_0), \dot{u}(t_0)) \frac{d}{ds} h_s(u(t_0))|_{s=0} = F_p(a, u(a), \dot{u}(a)) \frac{d}{ds} h_s(u(a))|_{s=0} \quad (1.5.6)$$

for any $t_0 \in [a, b]$. This means that (1.5.2) is constant on $[a, b]$.

q.e.d.

Examples

Example 1.5.1. We consider for $u : \mathbb{R} \to \mathbb{R}^{3n}$, $u = (u_1, \ldots, u_n)$ with $u_i = (u_i^1, u_i^2, u_i^3)$,

$$F(t, u(t), \dot{u}(t)) = \sum_{i=1}^{n} m_i \frac{\|\dot{u}_i\|^2}{2} - V(u) \quad \left(\|\dot{u}_i\|^2 = \sum_{j=1}^{3} \dot{u}_i^j \dot{u}_i^j \right),$$

i.e. a mechanical system in $\mathbb{R}^3$ with point masses m_i, and a potential $V(u)$ that is independent of the third coordinates of the u_i. Then

$$h_s(z) = z + se_3,$$

where e_3 is the third unit vector in $\mathbb{R}^3$, leaves F invariant in the sense of Theorem 1.5.1. Since

$$\frac{d}{ds} h_s|_{s=0} = e_3,$$

we conclude that

$$\sum_{i=1}^{n} m_i \dot{u}_i^3,$$

i.e. the third component of the momentum vector of the system is conserved.

Example 1.5.2. Similarly, if a system as in Example 1.5.1 is invariant under rotations about the e_3-axis, and if h_s now denotes such rotations, then (up to a constant factor)

$$\frac{d}{ds} h_s|_{s=0} u_i = e_3 \wedge u_i.$$

Hence, the conserved quantity is the angular momentum w.r.t. the e_3-axis,

$$\sum_{i=1}^{n} F_p e_3 \wedge u_i = \sum_i (m_i \dot{u}_i) \cdot (e_3 \wedge u_i) = \sum_i (u_i \wedge m_i \dot{u}_i) \cdot e_3.$$

We now come to the general form of E. Noether's theorem

Theorem 1.5.2 (Theorem of E. Noether). *We consider the variational integral*

$$I(u) = \int_a^b F(t, u(t), \dot{u}(t)) dt$$

1.5 Symmetries and the theorem of E. Noether

with $F \in C^2([a,b] \times \mathbb{R}^d \times \mathbb{R}^d, \mathbb{R})$. We suppose that there exists a smooth one-parameter family of differentiable maps

$$\bar{h}_s = (h_s^0, h_s) : [a,b] \times \mathbb{R}^d \to \mathbb{R} \times \mathbb{R}^d$$

($s \in (-\epsilon_0, \epsilon_0)$ as before) with

$$\bar{h}_0(t,z) = (t,z) \quad \text{for all } (t,z) \in [a,b] \times \mathbb{R}^d$$

and satisfying

$$\int_{h_s^0(a)}^{h_s^0(b)} F\left(t_s, h_s(u(t_s)), \frac{d}{dt_s}h_s(u(t_s))\right) dt_s = \int_a^b F(t, u(t), \dot{u}(t)) dt \tag{1.5.7}$$

for $t_s = h_s^0(t)$, all $s \in (-\epsilon_0, \epsilon_0)$ and all $u \in C^2([a,b], \mathbb{R}^d)$. Then, for any solution $u(t)$ of the Euler–Lagrange equations (1.1.4) for I,

$$F_p(t, u(t), \dot{u}(t)) \frac{d}{ds} h_s(u(t))|_{s=0}$$

$$+ (F(t, u(t), \dot{u}(t)) - F_p(t, u(t), \dot{u}(t)) \dot{u}(t)) \frac{d}{ds} h_s^0(t)|_{s=0} \tag{1.5.8}$$

is constant in $t \in [a,b]$.

Proof. We reduce the statement to the one of Theorem 1.5.1 by artificially considering t as a dependent variable on the same footing with u. Thus, we consider the integrand

$$\bar{F}\left(t(\tau), u(t(\tau)), \frac{dt}{d\tau}, \frac{d}{d\tau} u(t(\tau))\right)$$

$$:= F\left(t, u(t), \frac{\frac{d}{d\tau} u(t(\tau))}{\frac{dt}{d\tau}}\right) \frac{dt}{d\tau} \tag{1.5.9}$$

$$= F(t, u(t), \dot{u}(t)) \frac{dt}{d\tau}.$$

Then

$$\bar{I}(t,u) := \int_{\tau_0}^{\tau_1} \bar{F}\left(t(\tau), u(t(\tau)), \frac{dt}{d\tau}, \frac{d}{d\tau} u(t(\tau))\right) d\tau$$

$$= \int_a^b F(t, u(t), \dot{u}(t)) dt, \quad \text{if } t(\tau_0) = a, \, t(\tau_1) = b \tag{1.5.10}$$

$$= I(u).$$

By our assumption, $\bar{F}$ remains invariant under replacing (t,u) by $\bar{h}_s(t,u)$. Consequently, Theorem 1.5.1 applied to $\bar{I}$ yields that

$$\bar{F}_p(t, u(t), \dot{u}(t)) \frac{d}{ds} h_s(u(t))|_{s=0} + \bar{F}_{p^0}(t, u(t), \dot{u}(t)) \frac{d}{ds} h_s^0(t)|_{s=0}$$

with p^0 standing for the place of the argument $\frac{dt}{d\tau}$ of $\bar{F}$ (while p stands as before for the arguments $\dot{u}$), is invariant. Since, by (1.5.9),

$$\bar{F}_p = F_p,$$
$$\bar{F}_{p^0} = F - F_p \dot{u}$$

at $s = 0$ (note $\frac{dt}{d\tau} = 1$ for $s = 0$ since $h_0^0(t) = t$), this implies the invariance of (1.5.8).

q.e.d.

Example 1.5.3. Suppose $F = F(u, \dot{u})$, i.e. F does not depend explicitly on t. Then

$$h_s(t, z) = (t + s, z)$$

leaves I invariant as required in Theorem 1.5.2. Therefore, the 'energy'

$$F(t, u(t), \dot{u}(t)) - F_p(t, u(t), \dot{u}(t))\dot{u}(t)$$

is conserved. We shall see another proof of this fact in Section 4.1.

Summary. The theorem of E. Noether identifies a quantity that is preserved along any solution $u(t)$ of the Euler–Lagrange equations of a variational integral, a so-called first integral of motion, with any differentiable symmetry of the integrand. For example, in classical mechanics, conservation of momentum and angular momentum correspond to translational and rotational invariance of the integral, respectively, while time invariance leads to the conservation of energy.

Exercises

1.1 For mappings $u : [a, b] \to \mathbb{R}^d$, consider

$$E(u) := \frac{1}{2} \int_a^b |\dot{u}(t)|^2 dt$$

($|\cdot|$ is the Euclidean norm of $\mathbb{R}^d$, i.e. for $z = (z^1, \ldots, z^d)$, $|z|^2 = \sum_{i=1}^d (z^i)^2$). Compute the Euler–Lagrange equations and the second variation. Also, let

$$L(u) := \int_a^b |\dot{u}(t)| dt.$$

Show that

$$L(u) \leq \sqrt{2(b-a)E(u)},$$

with equality if $|\dot{u}(t)| \equiv$ constant almost everywhere. (What is an appropriate regularity class for the mappings u that are considered here?)

1.2 Determine all minimizers of the variational integral

$$I(u) = \int_{-1}^{1} (1 - \dot{u}(t))^2 dt$$

with $u(-1) = 0 = u(1)$.

1.3 Develop a theory of Jacobi fields for variational problems with free boundary conditions. In particular, you should obtain an analogue of Jacobi's theorem.

1.4 For mappings $u : [a, b] \to \mathbb{R}^d$, consider

$$I(u) := \int_a^b \frac{1}{(1 + |u(t)|^2)} |\dot{u}(t)|^2 dt.$$

Compute the first and second variation of I and the Jacobi equation. Can you find Jacobi fields?

2
A geometric example: geodesic curves

2.1 The length and energy of curves

We let M be an n-dimensional embedded submanifold of $\mathbb{R}^d$. In this section, we assume that f is of class C^3, i.e. that all local charts are thrice differentiable. We let $c \in AC([0,T], M)$ be a curve on M. This means that c is an absolutely continuous map from the interval $[0,T]$ into $\mathbb{R}^d$ with the property that $c(t) \in M$ for every $t \in [0,T]$. The derivative of c w.r.t. t will be denoted by a dot ˙,

$$\dot{c}(t) := \frac{dc}{dt}(t).$$

The length of c is given by

$$L(c) := \int_0^T |\dot{c}(t)|\, dt = \int_0^T \left(\sum_{\alpha=1}^d (\dot{c}^\alpha)^2 \right)^{\frac{1}{2}} dt, \qquad (2.1.1)$$

where $(c^1, \ldots, c^d)$ are the coordinates of $c = c(t)$. We also define the energy of c as

$$E(c) := \frac{1}{2}\int_0^T |\dot{c}(t)|^2\, dt = \frac{1}{2}\int_0^T \sum_{\alpha=1}^d (\dot{c}^\alpha)^2\, dt. \qquad (2.1.2)$$

We let now

$$f : U \to V, \quad f(U) = M \cap V$$

be a local chart for M as defined in Section 1.4. We assume for a moment that $c([0,T])$ is contained in $f(U)$. Since f maps U bijectively onto $f(U)$, there exists a curve

$$\gamma(t) \subset U$$

2.1 The length and energy of curves

with
$$c(t) = f(\gamma(t)). \tag{2.1.3}$$

Since the derivative $Df(z)$ has maximal rank everywhere (by definition of a chart, cf. 1.4), γ is absolutely continuous, since c is, and we have the chain rule
$$\dot{c}(t) = (Df)(\gamma(t)) \circ \dot{\gamma}(t),$$
or
$$\dot{c}^\alpha(t) = \frac{\partial f^\alpha}{\partial z^i}(\gamma(t))\dot{\gamma}^i(t),$$
where the index i is summed from 1 to n. Thus
$$L(c) = \int_0^T \left(\frac{\partial f^\alpha}{\partial z^i}(\gamma(t))\dot{\gamma}^i(t) \frac{\partial f^\alpha}{\partial z^j}(\gamma(t))\dot{\gamma}^j(t) \right)^{\frac{1}{2}} dt$$
and
$$E(c) = \frac{1}{2} \int_0^T \frac{\partial f^\alpha}{\partial z^i}(\gamma(t))\dot{\gamma}^i(t) \frac{\partial f^\alpha}{\partial z^j}(\gamma(t))\dot{\gamma}^j(t) dt.$$

In these formulae, and in sequel, the index is summed from 1 to d. For $z \in U$, we put
$$g_{ij}(z) := \frac{\partial f^\alpha}{\partial z^i}(z) \frac{\partial f^\alpha}{\partial z^j}(z). \tag{2.1.4}$$

With this notation, the preceding formulae become
$$L(c) = \int_0^T \left(g_{ij}(\gamma(t))\dot{\gamma}^i(t)\dot{\gamma}^j(t) \right)^{\frac{1}{2}} dt \tag{2.1.5}$$
and
$$E(c) = \frac{1}{2} \int_0^T g_{ij}(\gamma(t))\dot{\gamma}^i(t)\dot{\gamma}^j(t) dt. \tag{2.1.6}$$

Definition 2.1.1.
$$g_{ij}(z) = \frac{\partial f^\alpha}{\partial z^i}(z) \frac{\partial f^\alpha}{\partial z^j}(z)$$
is called the metric tensor of M w.r.t. the chart $f\ U \to V$.

We note that $(g_{ij}(z))_{i,j=1,\ldots,n}$ is symmetric, i.e.
$$g_{ij}(z) = g_{ji}(z) \quad \text{for all } i,j$$
and positive definite, i.e.
$$g_{ij}(z)\eta^i\eta^j > 0 \quad \text{whenever } \eta = (\eta^1, \ldots, \eta^n) \neq 0 \in \mathbb{R}^n.$$

Remark 2.1.1. The use of local charts for M seems to have the obvious disadvantage that the expressions for length and energy of curves become more complicated. The advantage of this approach, namely not to consider curves on M as curves in $\mathbb{R}^d$ satisfying a constraint, is that this constraint now is automatically fulfilled. All curves represented in local charts lie on M. This more than compensates for the complication in the formulae for L and E.

Our aim will be to find curves of shortest length or of smallest energy on M, i.e. to minimize the functionals L and E among curves on M. For this purpose it will be useful to observe certain invariance properties of L and E. First of all, whenever $i : \mathbb{R}^d \to \mathbb{R}^d$ is a Euclidean isometry, i.e. $i(y) = Ay + b$ with $A \in O(d)$, the orthogonal group, and $b \in \mathbb{R}^d$, then

$$L(i(c)) = L(c) \tag{2.1.7}$$

$$E(i(c)) = E(c) \tag{2.1.8}$$

for any curve $c : [0, T] \to \mathbb{R}^d$.

Secondly, L is parameterization invariant in the sense that whenever

$$\tau : [0, S] \to [0, T]$$

is a diffeomorphism (i.e. τ is bijective, and both τ and its inverse τ^{-1} are everywhere differentiable), then

$$L(c) = L(c \circ \tau), \quad \text{for any curve } c : [0, T] \to \mathbb{R}^d. \tag{2.1.9}$$

Namely

$$\begin{aligned} L(c \circ \tau) &= \int_0^S \left| \frac{d}{ds}(c \circ \tau)(s) \right| ds \\ &= \int_0^S \left| \left(\frac{d}{dt}c\right)(\tau(s)) \right| \left| \frac{d\tau}{ds}(s) \right| ds \\ &= \int_0^T |\dot{c}(t)| \, dt. \end{aligned}$$

E, however, is not parameterization invariant. By the Schwarz inequality, we have instead

$$L(c) = \int_0^T 1 \cdot |\dot{c}(t)|\, dt \leq \left(\int_0^T dt\right)^{\frac{1}{2}} \cdot \left(\int_0^T |\dot{c}(t)|^2\, dt\right)^{\frac{1}{2}} = \sqrt{2T}\sqrt{E(c)}, \tag{2.1.10}$$

with equality iff

$$|\dot{c}(t)| \equiv \text{constant} \quad \text{for almost all } t. \tag{2.1.11}$$

We have shown:

Lemma 2.1.1. *For every $c \in AC([0,T], \mathbb{R}^d)$*

$$L(c) \leq \sqrt{2T}\sqrt{E(c)},$$

with strict inequality, unless

$$|\dot{c}(t)| \equiv \text{constant} \quad \text{almost everywhere.}$$

If

$$|\dot{c}(t)| \equiv \text{constant} \quad \text{almost everywhere},$$

we say that the curve c is parameterized proportionally to arc-length, and if

$$|\dot{c}(t)| \equiv 1,$$

we say that it is parameterized by arc-length. We recall that a *Jordan curve*, i.e. an injective curve $c : [0,T] \to \mathbb{R}^d$, is rectifiable if it is absolutely continuous (which we always assume), and this implies that it may be parameterized by arc-length, i.e. there exists a diffeomorphism

$$\tau : [0, L(c)] \to [0, T]$$

with

$$\left|\frac{d}{ds}(c \circ \tau)(s)\right| \equiv 1 \quad \text{for almost all } s,$$

i.e. the reparameterized curve

$$\tilde{c} = c \circ \tau$$

is parameterized by arc-length. From Lemma 2.1.1, we obtain:

Corollary 2.1.1. *Let $c : [0, L(c)] \to \mathbb{R}^d$ be a curve parameterized on $[0, L(c)]$. Among all reparameterizations*

$$\tau : [0, L(c)] \to [0, L(c)]$$

(i.e. we keep the interval of definition fixed, namely $[0, L(c)]$), the parameterization by arc-length leads to the smallest energy. Namely, if $c : [0, L(c)] \to \mathbb{R}^d$ is parameterized by arc-length

$$L(c) = 2E(c), \qquad (2.1.12)$$

whereas for any other parameterization of c on the same interval,

$$L(c) < 2E(c). \qquad (2.1.13)$$

We now return to those curves c that are confined to lie on M, in order to discover a third invariance. Namely, we compare the two expressions (2.1.1) and (2.1.5) for the length of c, and similarly (2.1.2) and (2.1.6) for its energy. (2.1.1) is obviously independent of the chart $f : U \to V$ and its metric tensor, and therefore (2.1.5) has to be independent of them, too. In order to study this more closely, let

$$\tilde{f} : \tilde{U} \to \tilde{V}$$

be another chart with

$$c([0, T]) \subset \tilde{f}(\tilde{U}).$$

Then there exists a curve $\tilde{\gamma}$ in $\tilde{U}$ with $c(t) = \tilde{f}(\tilde{\gamma}(t))$ for all t. Putting

$$\tilde{g}_{kl}(z) := \frac{\partial \tilde{f}^\alpha}{\partial z^k}(z) \frac{\partial \tilde{f}^\alpha}{\partial z^l}(z) \quad \text{for } z \in \tilde{U},$$

we then also have

$$L(c) = \int_0^T \left(\tilde{g}_{kl}(\tilde{\gamma}(t)) \dot{\tilde{\gamma}}^k \dot{\tilde{\gamma}}^l(t) \right)^{\frac{1}{2}} dt. \qquad (2.1.14)$$

In order to study this invariance property more closely, we define

$$\varphi := \tilde{f}^{-1} \circ f : f^{-1}\left(f(U) \cap \tilde{f}(\tilde{U})\right) \to \tilde{f}^{-1}\left(f(U) \cap \tilde{f}(\tilde{U})\right)$$

(see Figure 2.1).

φ is called a coordinate transformation. φ is a diffeomorphism, i.e. a bijective map between open subsets of $\mathbb{R}^n$ whose derivative $D\varphi(z)$ has maximal rank ($= n$) at every z. Then from

$$f \circ \gamma(t) = c(t) = \tilde{f} \circ \tilde{\gamma}(t),$$

$$\tilde{\gamma}(t) = \varphi(\gamma(t)), \quad \text{hence} \quad \dot{\tilde{\gamma}}^k(t) = \frac{\partial \varphi^k}{\partial z^j}(\gamma(t)) \dot{\gamma}^j(t) \qquad (2.1.15)$$

and from

$$\tilde{f}(\varphi(z)) = f(z)$$

2.1 The length and energy of curves

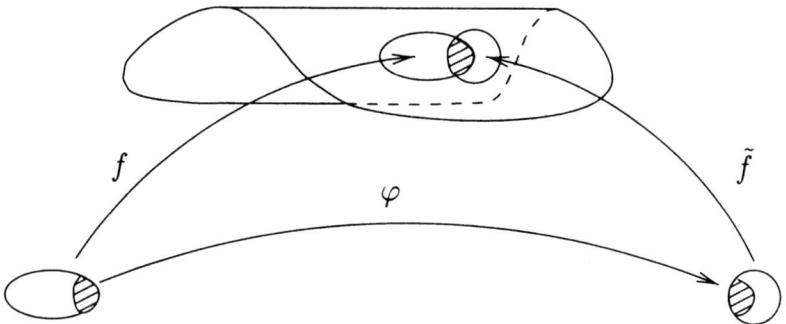

Figure 2.1.

we get

$$g_{ij}(z) = \tilde{g}_{kl}(\varphi(z))\frac{\partial \varphi^k}{\partial z^i}(z)\frac{\partial \varphi^l}{\partial z^j}(z). \tag{2.1.16}$$

From (2.1.15) and (2.1.16), we see

$$g_{ij}\left(\gamma(t)\right)\dot{\gamma}^i(t)\dot{\gamma}^j(t) = \tilde{g}_{kl}\left(\tilde{\gamma}(t)\right)\dot{\tilde{\gamma}}^k(t)\dot{\tilde{\gamma}}^l(t), \tag{2.1.17}$$

and this shows again the equivalence of (2.1.5) and (2.1.1), and likewise for the corresponding expressions of the energy. The important transformation formula (2.1.16) shows how the metric tensor transforms under coordinate transformations. This invariance property of L and E makes it possible to express the length and energy of an arbitrary curve c on M that is not necessarily contained in the image of a single chart as follows:

One finds a subdivision

$$t_0 = 0 < t_1 < \ldots < t_{m-1} < t_m = T$$

of $[0, T]$ with the property that

$$c\left([t_{\nu-1}, t_\nu]\right)$$

is contained in the image of a single chart

$$f_\nu : U_\nu \to V_\nu$$

for each $\nu = 1, \ldots, m$. Let $\left(g^\nu_{ij}(z)\right)_{i,j=1,\ldots,n}$ be the metric tensor of M

w.r.t. the chart f_ν. Then

$$L(c) = \sum_{\nu=1}^{m} L\left(c_{|[t_{\nu-1}, t_\nu]}\right)$$

$$= \sum_{\nu=1}^{m} \int_{t_{\nu-1}}^{t_\nu} \left(g_{ij}^\nu(\gamma_\nu(t))\dot{\gamma}_\nu^i(t)\dot{\gamma}_\nu^j(t)\right)^{\frac{1}{2}} dt$$

where $c(t) = f_\nu \circ \gamma_\nu(t)$ for $t \in [t_{\nu-1}, t_\nu]$. By the preceding considerations, this does not depend on the choice of charts f_ν. For this reason, one usually just says that for a curve c on M

$$L(c) = \int_0^T \left(g_{ij}(\gamma(t))\dot{\gamma}^i(t)\dot{\gamma}^j(t)\right)^{\frac{1}{2}} dt, \qquad (2.1.18)$$

where γ is the representation for c w.r.t. a local chart, and $(g_{ij})_{i,j=1,\ldots,n}$ is the metric tensor of M w.r.t. this chart. Similarly

$$E(c) = \frac{1}{2} \int_0^T g_{ij}(\gamma(t))\dot{\gamma}^i(t)\dot{\gamma}^j(t) dt. \qquad (2.1.19)$$

We now assume that the charts for M are twice differentiable and return to the question of finding shortest curves on M, for example between two given points. By Corollary. 2.1.1, it is preferable to minimize E instead of L, because a minimizer for E contains more information than one for L; namely, minimizers for E are precisely those minimizers for L that are parameterized proportionally to arc-length. Thus, minimizing E not only selects shortest curves but also convenient parameterizations of such curves.

We now compute the Euler–Lagrange equations for E as given by (2.1.19):

$$0 = \frac{d}{dt} E_{\dot{\gamma}^i} - E_{\gamma^i} \quad \text{for } i = 1, \ldots, m$$

$$\Leftrightarrow 0 = \frac{d}{dt}\left(2g_{ij}(\gamma(t))\dot{\gamma}^j(t)\right) - \left(\frac{\partial}{\partial z^i} g_{kj}\right)(\gamma(t))\dot{\gamma}^k(t)\dot{\gamma}^j(t)$$

(the factor 2 in the first term results from the symmetry $g_{ij} = g_{ji}$)

$$\Leftrightarrow 0 = 2g_{ij}\ddot{\gamma}^j + 2\frac{\partial}{\partial z^k} g_{ij} \dot{\gamma}^k \dot{\gamma}^j - \frac{\partial}{\partial z^i} g_{kj} \dot{\gamma}^k \dot{\gamma}^j. \qquad (2.1.20)$$

We now introduce some further notation:

$$\left(g^{ij}\right)_{i,j=1,\ldots,n}$$

2.1 The length and energy of curves

is the matrix inverse to $(g_{ij})_{i,j=1,\ldots,n}$, i.e.

$$g^{ij}g_{jk} = \delta_k^i := \begin{cases} 1 & \text{for } i = k \\ 0 & \text{for } i \neq k \end{cases} \quad \text{for all } i, k,$$

$$g_{ij,k} := \frac{\partial}{\partial z^k} g_{ij},$$

and finally the *Christoffel symbols*

$$\Gamma_{jk}^i := \frac{1}{2} g^{il}(g_{jl,k} + g_{kl,j} - g_{jk,l}).$$

Equation (2.1.20) then becomes

$$\begin{aligned} 0 &= \ddot{\gamma}^i + \tfrac{1}{2} g^{il}\left(2 g_{lj,k} \dot{\gamma}^k \dot{\gamma}^j - g_{kj,l} \dot{\gamma}^k \dot{\gamma}^j\right) \\ &= \ddot{\gamma}^i + \tfrac{1}{2} g^{il}\left(g_{jl,k} + g_{kl,j} - g_{jk,l}\right) \dot{\gamma}^k \dot{\gamma}^j \end{aligned}$$

by using symmetries. Thus:

Lemma 2.1.2. *The Euler–Lagrange equations for the energy E for curves on M are*

$$0 = \ddot{\gamma}^i(t) + \Gamma_{jk}^i(\gamma(t)) \dot{\gamma}^j(t) \dot{\gamma}^k(t) \quad \text{for } i = 1, \ldots, n. \tag{2.1.21}$$

The theorem of Picard–Lindelöf about solutions of ordinary differential equations implies:

Lemma 2.1.3. *For any $z \in U$, $v \in \mathbb{R}^n$, the system (2.1.21) has a unique solution $\gamma(t)$ with*

$$\gamma(0) = z, \; \dot{\gamma}(0) = v \quad \text{for } t \in [-\epsilon, \epsilon] \text{ and some } \epsilon > 0.$$

Moreover, $\gamma(t)$ depends differentiably on the initial values z, v.

Definition 2.1.2. *The solutions of (2.1.21) are called* geodesics *on M.*

Examples

Example 2.1.1. The sphere

$$S^n := \left\{ (x^1, \ldots, x^{n+1}) \in \mathbb{R}^{n+1}, \sum_{i=1}^{n+1} (x^i)^2 = 1 \right\} \subset \mathbb{R}^{n+1}$$

is a differentiable manifold of dimension n. In order to construct local charts, we put

$$\Omega_1 := S^n \setminus \{(0, 0, \ldots, 0, 1)\}, \; \Omega_2 := S^n \setminus \{(0, 0, \ldots, 0, -1)\}$$

and define
$$g_1 : \Omega_1 \to \mathbb{R}^n \ , \ g_2 : \Omega_2 \to \mathbb{R}^n$$
as
$$g_1(x^1,\ldots,x^{n+1}) = \left(\frac{x^1}{1-x^{n+1}},\ldots,\frac{x^n}{1-x^{n+1}}\right)$$
and
$$g_2(x^1,\ldots,x^{n+1}) = \left(\frac{x^1}{1+x^{n+1}},\ldots,\frac{x^n}{1+x^{n+1}}\right)$$

(g_1 and g_2 are the stereographic projections from the south and north pole, respectively). We then obtain charts
$$f_1 = g_1^{-1} : \mathbb{R}^n \to S^n \setminus \{(0,\ldots,0,1)\}$$
$$f_2 = g_2^{-1} : \mathbb{R}^n \to S^n \setminus \{(0,\ldots,0,-1)\}.$$

More explicitly, f_1 can be computed as follows:
With
$$(z^1,\ldots,z^n) = \left(\frac{x^1}{1-x^{n+1}},\ldots,\frac{x^n}{1-x^{n+1}}\right),$$
$$1 = x^\alpha x^\alpha = z^i z^i (1-x^{n+1})^2 + x^{n+1} x^{n+1},$$
hence
$$x^{n+1} = \frac{z^i z^i - 1}{z^i z^i + 1}$$
and then
$$x^j = \frac{2z^j}{1+z^i z^i} \quad (j=1,\ldots,n).$$

Thus
$$f_1(z^1,\ldots,z^n) = \left(\frac{2z^1}{1+z^i z^i},\ldots,\frac{2z^n}{1+z^i z^i}, \frac{z^i z^i - 1}{1+z^i z^i}\right).$$

For the metric tensor, we compute
$$\frac{\partial f_1^j}{\partial z^k} = \frac{2\delta_{jk}}{1+z^i z^i} - \frac{4z^j z^k}{(1+z^i z^i)^2} \quad (j,k=1,\ldots,n)$$
$$\frac{\partial f_1^{n+1}}{\partial z^k} = \frac{4z^k}{(1+z^i z^i)^2}.$$

2.1 The length and energy of curves

Hence
$$g_{ij}(z) = \frac{\partial f^\alpha}{\partial z^i}\frac{\partial f^\alpha}{\partial z^j} = \frac{4}{(1+|z|^2)^2}\delta_{ij}. \tag{2.1.22}$$

Actually, the metric tensor w.r.t. the chart f_2 is given by the same formula. In order to compute the expression for geodesics, we also need to compute the Christoffel symbols. It turns out that adding a little generality will actually facilitate the computations. We consider a metric of the form
$$g_{ij} = \frac{1}{\phi^2}\delta_{ij}, \tag{2.1.23}$$

where $\phi : \mathbb{R}^n \to \mathbb{R}^+$ is positive and differentiable. Then
$$g^{ij} = \phi^2 \delta_{ij}. \tag{2.1.24}$$

We also put
$$\varphi := \log \phi.$$

Then
$$g_{ij,k} = \frac{\partial g_{ij}}{\partial z^k} = -\delta_{ij}\frac{2}{\phi^3}\frac{\partial \phi}{\partial z^k} = -\delta_{ij}\frac{2}{\phi^2}\frac{\partial \varphi}{\partial z^k}.$$

Next
$$\begin{aligned}\Gamma^k_{ij} &= \frac{1}{2}g^{kl}\left(g_{il,j} + g_{jl,i} - g_{ij,l}\right) \tag{2.1.25}\\ &= \frac{1}{2}\phi^2\left(g_{ik,j} + g_{jk,i} - g_{ij,k}\right)\\ &= -\delta_{ik}\frac{\partial \varphi}{\partial z^j} - \delta_{jk}\frac{\partial \varphi}{\partial z^i} + \delta_{ij}\frac{\partial \varphi}{\partial z^k}.\end{aligned}$$

Thus, Γ^k_{ij} vanishes if all three indices i, j, k are distinct, and for all i, j
$$\Gamma^i_{ji} = \Gamma^i_{ij} = -\frac{\partial \varphi}{\partial z^j}, \quad \text{and} \quad \Gamma^i_{jj} = \frac{\partial \varphi}{\partial z^i} \quad \text{for } i \neq j. \tag{2.1.26}$$

In the present case,
$$\varphi = \log(1+|z|^2) - \log 2$$

hence
$$\frac{\partial \varphi}{\partial z^j} = \frac{2z^j}{1+|z|^2}.$$

Geodesic curves

Therefore, the equations for geodesics become

$$0 = \ddot{\gamma}^i + 2\sum_{j=1}^{n} \Gamma^i_{ij}(\gamma)\dot{\gamma}^i\dot{\gamma}^j - \Gamma^i_{ii}(\gamma)\dot{\gamma}^i\dot{\gamma}^i + \sum_{\substack{j=1\\j\neq i}}^{n} \Gamma^i_{jj}(\gamma)\dot{\gamma}^j\dot{\gamma}^j$$

(using the symmetry $\Gamma^i_{ij} = \Gamma^i_{ji}$)

$$= \ddot{\gamma}^i - 2\sum_{j=1}^{n} \frac{2\gamma^j}{1+|\gamma|^2}\dot{\gamma}^i\dot{\gamma}^j + \sum_{j=1}^{n} \frac{2\gamma^i}{1+|\gamma|^2}\dot{\gamma}^j\dot{\gamma}^j. \qquad (2.1.27)$$

We now claim that the geodesic $\gamma(t)$ through the origin, i.e. $\gamma(0) = 0$, with $\dot{\gamma}(0) = a \in \mathbb{R}^n$ is given by

$$\gamma(t) = a\alpha(t), \qquad (2.1.28)$$

where $\alpha : \mathbb{R} \to \mathbb{R}$ then satisfies $\alpha(0) = 0, \dot{\alpha}(0) = 1$. Making the ansatz (2.1.28) in (2.1.27) leads to

$$0 = a^i\ddot{\alpha} - 2\sum_{j=1}^{n} \frac{2a^j\alpha}{1+\alpha^2|a|^2}a^ia^j\dot{\alpha}^2 + \sum_{j=1}^{n} \frac{2a^i\alpha}{1+\alpha^2|a|^2}a^ja^j\dot{\alpha}^2$$

$$= a^i\left(\ddot{\alpha} - \frac{2|a|^2\alpha}{1+|a|^2\alpha^2}\dot{\alpha}^2\right) \quad i = 1, \ldots, n.$$

Since we may assume $a \neq 0$ (otherwise the solution with $\dot{\gamma}(0) = a$ is a point curve, hence uninteresting), this equation holds, if $\alpha(t)$ satisfies the ordinary differential equation (ODE)

$$0 = \ddot{\alpha} - \frac{2|a|^2\alpha}{1+|a|^2\alpha^2}\dot{\alpha}^2. \qquad (2.1.29)$$

The theorem of Picard–Lindelöf implies that (2.1.29) has a unique solution in a neighbourhood of $t = 0$. We then have found a solution $\gamma(t)$ of (2.1.27) of the desired form (2.1.28). The image of $\gamma(t)$ is a straight line through 0. By Lemma 2.1.3, we have thus found all solutions through 0. The images of the straight lines under the chart f_1 are the great circles on S^n through the south pole. We can now use a symmetry argument to conclude that all the geodesic lines on S^n are given by the great circles on S^n. Namely, the south pole does not play any distinguished role, and we could have constructed a local chart by stereographic projection from any other point on S^n as well, and the metric tensor would have assumed the same form (2.1.22). More generally, one may also argue as follows: We want to find the geodesic arc $\gamma(t)$ on S^n with $\gamma(0) = p_0$, $\dot{\gamma}(0) = V_0$ for some $p_0 \in S^n$, $V_0 \in T_{p_0}S^n$. Let $c_0(t)$ be the great circle on

S^n parameterized such that $c_0(0) = p_0$, $\dot{c}_0(0) = V_0$. c_0 is contained in a unique two-dimensional plane through the origin in $\mathbb{R}^{n+1}$. Let i denote the reflection across this plane. This is an isometry of $\mathbb{R}^{n+1}$ mapping S^n onto itself. It therefore maps geodesics on S^n onto geodesics, because we have observed that the length and energy functionals are invariant under isometries, and so isometries have to map critical points to critical points. Now i maps p_0 and V_0 to themselves. If γ were not invariant under i, $i \circ \gamma$ would be another geodesic with initial values p_0, V_0, contradicting the uniqueness result of Lemma 2.1.3. Therefore, $i \circ \gamma = \gamma$, and therefore $\gamma = c_0$.

We draw some conclusions:

The geodesic arc through two given points need not be unique. Namely, let p, q be antipodal points on S^n, e.g. north and south pole. Then there exist infinitely many great circles that pass through both p and q.

We shall later on see that the first conjugate point of a point $p \in S^n$ along a great circle is the antipodal point q of p. One also sees by explicit comparison that a geodesic arc on S^n ceases to be minimizing beyond the first conjugate point, in accordance with Theorem 1.3.4.

2.2 Fields of geodesic curves

Let M be an embedded, differentiable submanifold of $\mathbb{R}^d$, or, more generally, a Riemannian manifold of dimension n†, again of class C^3. Let M_0 be a submanifold of M; this means that M_0 itself is a differentiable submanifold of $\mathbb{R}^d$, respectively a Riemannian manifold, and that the inclusion $i : M_0 \hookrightarrow M$ is a differentiable embedding. We assume that M_0 has dimension $n - 1$, and that it is also of class C^3.

Theorem 2.2.1. *For any x_0 in M_0, there exist a neighbourhood V of x_0 in M, and a chart $f : U \to V$ with the following properties:*

(i) *U contains the origin of $\mathbb{R}^n$, $f(0) = x_0$.*

(ii) *$M_0 \cap V = f(U \cap \{x^n = 0\})$*

(iii) *The curves $x^i = c_i$, c_i = constant, $i = 1, \ldots, n-1$, are geodesics parameterized by arc-length. The arcs $\xi_1 \leq x^n \leq \xi_2$ on any such*

---

† We do not introduce the concept of an abstract Riemannian manifold here, but some readers may know that concept already, and in fact it provides the natural setting for the theory of geodesics. On the other hand, the embedding theorem of J.Nash says that any Riemannian manifold can be isometrically embedded into some Euclidean space $\mathbb{R}^d$, hence considered as a submanifold of $\mathbb{R}^d$. Therefore, from that point of view, no generality is gained by considering Riemannian manifolds instead of submanifolds of $\mathbb{R}^d$.

curve between the hypersurfaces $x^n = \xi_1$ and $x^n = \xi_2$ are all of the same length $\xi_1 - \xi_2$.

(iv) The metric tensor on U satisfies

$$g_{nn} = 1, \quad g_{in} = 0 \quad \text{for all } i = 1, \ldots, n-1 \qquad (2.2.1)$$

(*The second relation means that the curves* $x^i = c_i$, $i = 1, \ldots, n-1$, *intersect the hypersurfaces* $x^n = $ constant *orthogonally.*)

Proof. Since M_0 is a hypersurface, for every $p \in M_0$, there exist two unit normal vectors $n_\pm(p)$ to M_0 at p, i.e.

$$n_\pm(p) \in T_p M,$$

$$\|n_\pm(p)\| = 1$$

$$\langle n_\pm(p), v \rangle = 0 \quad \text{for all } v \in T_p M_0 \subset T_p M.$$

In a sufficiently small neighbourhood V of x_0, we may assume that such a normal vector $n(p)$ may be chosen so that it depends smoothly on $p \in M_0 \cap V =: V_0$. We assume that there is a local chart $\varphi_0 : U_0 \to V_0$ for M_0 ($U_0 \subset \mathbb{R}^{n-1}$), possibly choosing V smaller, if necessary. For every $p \in M_0 \cap V$, we then consider the geodesic arc $\gamma_p(t)$ with

$$\gamma_p(0) = p,$$
$$\dot{\gamma}_p(0) = n(p). \qquad (2.2.2)$$

This geodesic exists for $|t| \leq \epsilon = \epsilon(p)$ by Lemma 2.1.3. By choosing V smaller if necessary, we may assume that $\epsilon > 0$ is independent of p. Instead of $\gamma_p(t)$, we write $\gamma(p, t)$. Since the solution of (2.2.2) depends differentiably on its initial values (see Lemma 2.1.3), hence on p, the map

$$f : U_0 \times (-\epsilon, \epsilon) \to M$$

$$(x, t) \to \gamma(\varphi(x), t)$$

is likewise differentiable, where $\varphi : U_0 \to V_0$ is a local chart for M_0. We may assume

$$x_0 = \varphi(0),$$

by composing φ with a diffeomorphism if necessary. At $(0, 0) \in U_0 \times (-\epsilon, \epsilon)$, the Jacobian of f is spanned by the linearly independent vectors $\frac{\partial \varphi}{\partial x^1}, \ldots, \frac{\partial \varphi}{\partial x^{n-1}}, n(\varphi(x))$ (note that $\gamma(\varphi(x), 0) = \varphi(x)$ and $n(\varphi(x))$

2.2 Fields of geodesic curves

are orthogonal to all the vectors $\frac{\partial \varphi}{\partial x^j} \in T_{\varphi(x)} M_0$, $j = 1, \ldots, n-1$). Therefore, by the inverse function theorem, f yields a chart in some neighbourhood U of $(0,0) \in U_0 \times (-\epsilon, \epsilon)$. f obviously satisfies (i), (ii) (after redefining V). (iii) also holds by construction (putting $x^n = t$). Next, $g_{nn} \equiv 1$, since the curves $x^i = c_i$, namely $f(c_1, \ldots, c_{n-1}, t)$, $t \in (-\epsilon, \epsilon)$, are geodesics parameterized by arc-length, hence $g_{nn} = \langle \frac{\partial f}{\partial t}, \frac{\partial f}{\partial t} \rangle \equiv 1$. Finally, the system of equations for these curves to be geodesic is

$$\frac{\partial^2 x^k}{(\partial x^n)^2} + \Gamma^k_{ij} \frac{\partial x^i}{\partial x^n} \frac{\partial x^j}{\partial x^n} \qquad (x^n = t) \quad \text{for } k = 1, \ldots, n.$$

Hence in particular

$$\Gamma^k_{nn} = 0 \quad \text{for } k = 1, \ldots, n.$$

Now

$$\Gamma^k_{nn} = \frac{1}{2} g^{kl} (2 g_{nl,n} - g_{nn,l}) = g^{kl} g_{nl,n},$$

since $g_{nn} \equiv 1$. Therefore

$$g_{nk,n} \equiv 0 \quad \text{for all } k = 1, \ldots, n.$$

Since furthermore $g_{nk}(x^1, \ldots, x^{n-1}, 0) = 0$, because the geodesic arc $x^n = t$, $x^i = c_i = $ constant, is orthogonal to the surface $\varphi(x^1, \ldots, x^{n-1}) = f(x^1, \ldots, x^{n-1}, 0)$, we obtain

$$g_{nk} = 0.$$

q.e.d.

Definition 2.2.1. *The coordinates whose existence is affirmed by Theorem 2.2.1 are called* geodesic parallel coordinates *based on the hypersurface M_0.*

Theorem 2.2.2. *Let $f : U \to V$ be a chart with the properties described in Theorem 2.2.1. In particular, the curves $x^i = c_i$, $c_i = $ constant, for $i = 1, \ldots, n-1$ are geodesic arcs. Then any such curve is the shortest connection of its endpoints when compared with all curves contained entirely in $\bar{U}$ and having the same endpoints.*

Proof. We consider the geodesic

$$\gamma(t) = \{x^i = c_i, x^n = t, -\epsilon \le t \le \epsilon\},$$

where $U = U_0 \times (-\epsilon, \epsilon)$. Let $\tilde{\gamma}(t)$, $t_1 \le t \le t_2$ be another curve in $\bar{U}$ with $\tilde{\gamma}(t_1) = \gamma(-\epsilon)$, $\tilde{\gamma}(t_2) = \gamma(\epsilon)$. We have to prove

$$L(\tilde{\gamma}) \ge L(\gamma), \tag{2.2.3}$$

with strict inequality, unless $\tilde{\gamma}$ is a reparameterization of γ. Now

$$L(\tilde{\gamma}) = \int_{t_1}^{t_2} \left(\sum_{i,j=1}^{n-1} g_{ij}\left(\tilde{\gamma}(t)\right) \dot{\tilde{\gamma}}^i(t) \dot{\tilde{\gamma}}^j(t) + \left(\dot{\tilde{\gamma}}^n(t)\right)^2 \right)^{\frac{1}{2}} dt, \qquad (2.2.4)$$

since $g_{nn} \equiv 1$, $g_{in} \equiv 0$ for $i = 1, \ldots, n-1$ by Theorem 2.2.1(iv),

$$\geq \int_{t_1}^{t_2} \left|\dot{\tilde{\gamma}}^n(t)\right| dt \geq \tilde{\gamma}^n(t_2) - \tilde{\gamma}^n(t_1) = \gamma^n(\epsilon) - \gamma^n(-\epsilon)$$
$$= L(\gamma).$$

The first inequality is strict, unless $\dot{\tilde{\gamma}}^i$ is constant for $i = 1, \ldots, n-1$, and the second one is strict, unless $\tilde{\gamma}^n(t)$ is monotonic.

q.e.d.

Following Weierstraß, we say that the geodesics

$$\gamma(t) = \{x^i = c_i, x^n = t, -\epsilon \leq t \leq \epsilon\}$$

constitute a *field of geodesics*. Theorem 2.2.2 essentially says that any geodesic arc in this field is shorter than any other curve with the same endpoints in the region covered by the field. Both properties are essential. Namely geodesic arcs on S^n that are longer than a great semicircle show that geodesics not embedded in a field need not minimize the length between their endpoints. And geodesic arcs on a cylinder, contained in meridians, but longer than a semicircle show that there may be shorter curves not contained in the field.

We observe that if $\gamma(t)$ solves (2.1.21), so does $\gamma(\lambda t)$ for $\lambda = $ constant. We fix $z_0 \in U$ and denote the geodesic arc γ of Lemma 2.1.2 with

$$\gamma(0) = z_0, \dot{\gamma}(0) = v$$

by γ_v. Then by the above observation

$$\gamma_v(t) = \gamma_{\lambda v}\left(\frac{t}{\lambda}\right) \quad \text{for } \lambda \neq 0. \qquad (2.2.5)$$

Thus $\gamma_{\lambda v}$ is defined on $\left[\frac{-\epsilon}{\lambda}, \frac{\epsilon}{\lambda}\right]$, if γ is defined on $[-\epsilon, \epsilon]$. Since γ_v depends differentiably on v, and since $v \in \mathbb{R}^n$, $|v| = 1$, is compact, there exists $\epsilon_0 > 0$ with the property that for all v with $|v| = 1$, γ_v is defined on $[-\epsilon_0, \epsilon_0]$. From (2.2.5), we then conclude that for any $w \in \mathbb{R}^n$ with $|w| \leq \epsilon_0$, γ_w is defined on $[-1, 1]$. For later purposes, we also note that by Lemma 2.1.3, ϵ_0 may be chosen to depend continuously on z_0.

2.2 Fields of geodesic curves

We now define a map

$$e = e_{z_0} : \{w \in \mathbb{R}^n : |w| \leq \epsilon_0\} \to U$$
$$w \mapsto \gamma_w(1).$$

Then $e(0) = z_0$. We compute the derivative of e at 0 as

$$\begin{aligned}
De(0)(v) &= \frac{d}{dt}\gamma_{tv}(1)|_{t=0} \\
&= \frac{d}{dt}\gamma_v(t)|_{t=0} \quad \text{by (2.2.5)} \\
&= \dot{\gamma}_v(0) \\
&= v.
\end{aligned}$$

Hence, the derivative of e at $0 \in \mathbb{R}^n$ is the identity, and the inverse mapping theorem implies:

Theorem 2.2.3. *e maps a neighbourhood of $0 \in \mathbb{R}^n$ diffeomorphically (i.e. e is bijective, and both e and e^{-1} are differentiable) onto a neighbourhood of $z_0 \in U$.* q.e.d.

We want to normalize our chart $f : U \to V$ for M. First of all, we may assume

$$z_0 = 0 \tag{2.2.6}$$

for the point $z_0 \in U$ under consideration. Secondly, the transformation formula (2.1.16) implies that we may perform a linear change of coordinates (i.e. replace f by $f \circ A$, where $A \in GL(n, \mathbb{R})$) in order to achieve

$$g_{ij}(0) = \delta_{ij}. \tag{2.2.7}$$

We assume that $f : U \to V$ satisfies these normalizations. We then replace f by $f \circ e$ defined on $\{w \in \mathbb{R}^n : |w| \leq \epsilon_0\}$.

Theorem 2.2.4. *In this new chart, the metric tensor satisfies*

$$g_{ij}(0) = \delta_{ij} \tag{2.2.8}$$

$$\Gamma^i_{jk}(0) = 0 = g_{ij,k}(0) \quad \text{for all } i, j, k. \tag{2.2.9}$$

Proof. By (2.1.16), $g_{ij} = \delta_{ij}$ holds, since the metric tensor w.r.t. the chart f satisfies this property and $De(0)$ is the identity by the proof of Theorem 2.2.3. In order to verify (2.2.9), we observe that in our new chart, the straight lines tv ($v \in \mathbb{R}^n$, $t|v| < \epsilon$) are geodesics. Namely, tv is mapped to $\gamma_{tv}(1) = \gamma_v(t)$ (see (2.2.5)), where $\gamma_v(t)$ is the geodesic with

initial direction v. We thus insert $\gamma(t) = tv$ into the geodesic equation (2.1.21). Then $\ddot{\gamma} = 0$, hence
$$\Gamma^i_{jk}(tv)v^j v^k = 0 \quad \text{for } i = 1, \ldots, n.$$
In particular, inserting $t = 0$, we get
$$\Gamma^i_{jk}(0)v^j v^k = 0 \quad \text{for all } v \in \mathbb{R}^n, i = 1, \ldots, n.$$
We use $v = e_l$, where $(e_l)_{l=1,\ldots,n}$ is an orthonormal basis of $\mathbb{R}^n$. Then
$$\Gamma^i_{ll}(0) = 0 \quad \text{for all } i \text{ and } l.$$
We next insert $v = \frac{1}{2}(e_l + e_m)$, $l \neq m$. The symmetry $\Gamma^i_{jk} = \Gamma^i_{kj}$ (which directly follows from the definition of Γ^i_{jk} and the symmetry $g_{jk} = g_{kj}$) then yields
$$\Gamma^i_{lm}(0) = 0 \quad \text{for all } i, l, m.$$
The vanishing of $g_{ij,k}$ for all i, j, k then is an easy exercise in linear algebra. q.e.d.

Definition 2.2.2. *The local coordinates $x^1, \ldots, x^n$ constructed before Theorem 2.2.4 are called* **Riemannian normal coordinates**.

We let $x^1, \ldots, x^n$ be Riemannian normal coordinates. We transform them into polar coordinates $r, \varphi^1, \ldots, \varphi^{n-1}$ in the standard manner (e.g. if $n = 2$, $x^1 = r\cos\varphi^1$, $x^2 = r\sin\varphi^1$). This coordinate transformation is of course singular at 0. We now express the metric tensor w.r.t. these polar coordinates. We write g_{rr} instead of g_{11}, and we write $g_{r\varphi}$ instead of g_{1l}, $l = 2, \ldots, n$, and $g_{\varphi\varphi}$ instead of $(g_{kl})_{k,l=2,\ldots,d}$. In particular, by Theorem 2.2.4 and the transformation rule (2.1.16)
$$g_{rr}(0) = 1, g_{r\varphi}(0) = 0. \tag{2.2.10}$$
The lines through the origin are geodesics by the construction of Riemannian normal coordinates, and in polar coordinates, they now become the curves $\varphi = (\varphi^1, \ldots, \varphi^{n-1}) \equiv$ constant; thus they can be written as
$$\gamma(t) = (t, \varphi_0) \quad \text{with fixed } \varphi_0.$$
Therefore, the geodesic equation (2.1.21) gives
$$\Gamma^i_{rr} = 0 \quad \text{for all } i$$
(where of course Γ^i_{rr} stands for Γ^i_{11}), i.e.
$$\frac{1}{2}g^{il}(2g_{rl,r} - g_{rr,l}) = 0 \quad \text{for all } i,$$

2.2 Fields of geodesic curves

hence
$$2g_{rl,r} - g_{rr,l} = 0 \quad \text{for all } l. \tag{2.2.11}$$

Putting $r = l$ gives
$$g_{rr,r} = 0,$$

and with (2.2.10) then
$$g_{rr} \equiv 1. \tag{2.2.12}$$

Using this in (2.2.11) gives
$$g_{r\varphi,r} = 0,$$

hence with (2.2.10) again
$$g_{r\varphi} \equiv 0. \tag{2.2.13}$$

We have thus shown:

Theorem 2.2.5. *In the preceding coordinates, so called Riemannian polar coordinates, that are obtained by transforming Riemannian normal coordinates into polar coordinates, the metric tensor has the form*

$$\begin{pmatrix} 1 & 0 & \cdots & 0 \\ 0 & & & \\ \vdots & & g_{\varphi\varphi}(r,\varphi) & \\ 0 & & & \end{pmatrix},$$

where $g_{\varphi\varphi}$ stands for the $(n-1) \times (n-1)$-matrix of the components of the metric tensor w.r.t. the angular variables $\varphi^1, \ldots, \varphi^{n-1}$.

Note that this generalizes the situation for Euclidean polar coordinates. The Euclidean metric on $\mathbb{R}^2$, written in polar coordinates, e.g. takes the form
$$\begin{pmatrix} 1 & 0 \\ 0 & r^2 \end{pmatrix}.$$

Note that Theorem 2.2.5, in contrast to Theorem 2.2.4, is valid on the whole chart, not only at the origin.

Corollary 2.2.1. *Riemannian polar coordinates are geodesic parallel coordinates based on the hypersurfaces $r = $ constant $(r \neq 0$, since $r = 0$ corresponds to a single point, and not a hypersurface).*

Proof. By Theorem 2.2.5, all properties stated in Theorem 2.2.1 hold.
q.e.d.

50   Geodesic curves

By Corollary 2.2.1 and Theorem 2.2.1, the curves $\varphi \equiv$ constant, $r_1 \leq r \leq r_2$, are shortest connections between their end points among all curves lying in the chart. We are now going to observe that this holds even globally, i.e. also in comparison with curves that may leave the chart:

Theorem 2.2.6. *For each $p \in M$, there exists $\epsilon_0 > 0$ with the property that Riemannian polar coordinates centered at p may be introduced with domain*

$$\{(r,\varphi) : 0 \leq r \leq \epsilon_0\},$$

ϵ_0 may be chosen to depend continuously on p. We denote the subset of M corresponding to this coordinate domain by $B(p,\epsilon_0)$. For any ϵ with $0 \leq \epsilon \leq \epsilon_0$ and any $q \in \partial B(p,\epsilon)$, there exists precisely one geodesic of shortest length $(= \epsilon)$ from p to q. Namely, if q has coordinates (ϵ, φ_0), this geodesic arc is given by $\gamma(t) = (t, \varphi_0)$, $0 \leq t \leq \epsilon$.

Proof. The first claim follows from Theorem 2.2.3, since Riemannian polar coordinates are based on the diffeomorphism e (see the constructions before Theorems 2.2.4 and 2.2.5). As already noted before Theorem 2.2.3, Lemma 2.1.3 implies that we may choose ϵ_0 as a continuous function of p. In order to verify the second claim, let $c(t)$ be a curve from p to q, with $c(0) = p$. Let

$$t_0 := \sup\{t \geq 0 : c(\tau) \in B(p,\epsilon) \quad \text{for } 0 \leq \tau \leq t\}.$$

Since w.l.o.g. $\epsilon > 0$ and c is continuous, t_0 is positive. We are going to show that

$$L\left(c_{|[0,t_0]}\right) \geq \epsilon. \tag{2.2.14}$$

Since the curve (t, φ_0), $0 \leq t \leq \epsilon$, has length ϵ as easily follows from Theorem 2.2.5, this will imply the claim. In order to verify (2.2.14), we proceed as follows:

$$L\left(c_{|[0,t_0]}\right) = \int_0^{t_0} \left(g_{ij}\left(c(t)\right) \dot{c}^i(t) \dot{c}^j(t)\right)^{\frac{1}{2}} dt$$

(identifying $c_{|[0,t_0]}$ with its coordinate representation)

$$\geq \int_0^{t_0} \left(g_{rr} \dot{r}\dot{r}\right)^{\frac{1}{2}} dt$$

2.3 The existence of geodesics

by Theorem 2.2.5 and since $g_{\varphi\varphi}$ is positive definite (writing $c(t) = (r(t), \varphi(t))$)

$$= \int_0^{t_0} |\dot{r}| \, dt, \quad \text{again by Theorem 2.2.5}$$

$$\geq \int_0^{t_0} \dot{r} \, dt = r(t_0) = \epsilon.$$

Here, equality only holds if $g_{\varphi\varphi}\dot\varphi\dot\varphi \equiv 0$, i.e. $\varphi(t) \equiv$ constant, $\dot{r} \geq 0$, i.e. if $c|_{[0,t_0]}$ is a straight line through the origin. The second claim now easily follows. q.e.d.

Corollary 2.2.2. *If M is compact, there exists $\epsilon_0 > 0$ with the property that for every $p \in M$, there exist Riemannian polar coordinates with domain*

$$\{(r, \varphi) : 0 \leq r \leq \epsilon_0\}.$$

Proof. This follows from Theorem 2.2.6, since the constructions employed for polar coordinates depend continuously on p (see essentially the construction of the diffeomorphism e). q.e.d.

2.3 The existence of geodesics

Definition 2.3.1. *Let M be a connected differentiable submanifold of Euclidean space $\mathbb{R}^d$, or, more generally†, a connected Riemannian manifold. The distance between $p, q \in M$ is*

$$d(p, q) := \inf\{L(c)|\ c : [a, b] \to M$$
$$\text{rectifiable curve with} \quad c(a) = p, c(b) = q\}.$$

Theorem 2.3.1. *Let M (as in Definition 2.3.1) be compact. There exists $\epsilon_0 > 0$ with the property that any two points $p, q \in M$ with*

$$d(p, q) \leq \epsilon_0$$

can be connected by a unique shortest geodesic arc (i.e. of length $d(p,q)$). This geodesic arc depends continuously on p and q.

Proof. We take ϵ_0 as described in Corollary 2.2.2. This gives a unique shortest geodesic arc from p to q which furthermore depends continuously on q. Exchanging the rôles of p and q then yields continuous dependence on p, too. q.e.d.

† See footnote on p. 43.

We now proceed to establish a global result:

Theorem 2.3.2. *Let M be a compact connected differentiable submanifold of $\mathbb{R}^d$, or, more generally, a compact connected Riemannian manifold. Then any two points $p, q \in M$ can be connected by a shortest geodesic arc (i.e. of length $d(p,q)$).*

Proof. Let $(c_n)_{n \in \mathbb{N}}$ be a minimizing sequence. We may assume w.l.o.g. that all c_n are parameterized on the interval $[0,1]$ and proportionally to arc-length. Thus
$$c_n(0) = p, c_n(1) = q,$$
$$L(c_n) \to d(p,q) \quad \text{for } n \to \infty.$$

For each n, we may find
$$t_{0,n} = 0 < t_{1,n} < \ldots < t_{m,n} = 1$$
with
$$L\left(c_n|_{[t_{j-1,n}, t_{j,n}]}\right) \leq \epsilon_0,$$
with ϵ_0 given by Theorem 2.3.1. By Theorem 2.3.1, there exists a unique shortest geodesic arc between $c_n(t_{j-1,n}) =: p_{j-1,n}$ and $c_n(t_{j,n}) =: p_{j,n}$. We replace $c_n|_{[t_{j-1,n}, t_{j,n}]}$ by this shortest geodesic arc and obtain a new minimizing sequence, again denoted by c_n, that now is piecewise geodesic. Since the length of the c_n are bounded because of the minimizing property, we may actually assume that m is independent of n. Since M is compact, after selecting a subsequence of c_n, the points $p_{j,n}$ converge to limit points p_j, $(j = 0, \ldots, m)$ as $n \to \infty$. $c_n|_{[t_{j-1,n}, t_{j,n}]}$, the unique shortest geodesic arc between $p_{j-1,n}$ and $p_{j,n}$, then converges to the unique shortest geodesic arc between p_{j-1} and p_j (for this point, one verifies that limits of geodesic arcs are again geodesic arcs, that limits of shortest arcs are again shortest arcs, that $d(p_{j-1}, p_j) \leq \epsilon_0$, and one uses Theorem 2.3.1). We thus obtain a piecewise geodesic limit curve c, with $c(0) = p$, $c(1) = q$, and
$$L(c) = \lim_{n \to \infty} L(c_n),$$
since we have for the geodesic pieces
$$L\left(c|_{[t_{j-1}, t_j]}\right) = \lim_{n \to \infty} L\left(c_n|_{[t_{j-1,n}, t_{j,n}]}\right)$$
for all j ($t_j = \lim_{n \to \infty} t_{j,n}$). Since the c_n constitute a minimizing sequence,
$$L(c) = d(p,q),$$

and c thus is of shortest possible length. This implies that c is geodesic. Namely, otherwise we could find $0 \leq s_1 < s_2 \leq 1$ with $L\left(c_{|[s_1,s_2]}\right) \leq \epsilon_0$, but with $c_{|[s_1,s_2]}$ not being geodesic. Replacing $c_{|[s_1,s_2]}$ by the shortest geodesic arc between $c(s_1)$ and $c(s_2)$ would yield a shorter curve (cf. Theorem 2.2.6.), contradicting the minimizing property of c.

q.e.d.

Thus, any two points on a compact M may be connected by a shortest geodesic. We now pose the question whether they can be connected by more than one geodesic, not necessarily the shortest. On S^n, for example, this is clearly the case. Actually, the answer is that it is the case on any compact M. That result needs a topological result that is not available to us here, however. Therefore, we will restrict ourselves to a special case which, however, already displays the crucial geometric idea of the construction for the general case, too.

Theorem 2.3.3. *Let M be a differentiable submanifold of Euclidean space $\mathbb{R}^d$, (or more generally†, a Riemannian manifold), diffeomorphic to the sphere S^2. The latter condition means that there exists a bijective map*

$$h : S^2 \to M$$

that is differentiable in both directions. Then any two points p, $q \in M$ can be connected by at least two geodesics.

Proof. M is compact and connected since diffeomorphic to S^2 which is compact and connected. Let us assume $p \neq q$. We leave it to the reader to modify our constructions in order that they also apply to the case $p = q$. (In that case, Thm 2.3.3 asserts the existence of a nonconstant geodesic $c : [0,1] \to M$ with $c(0) = p = c(1)$.) One may then construct a diffeomorphism

$$h_0 : S^2 \to M$$

with the following properties:
Let $S^2 = \{(x^1, x^2, x^3) \in \mathbb{R}^3 : |x| = 1\}$. Then

$$p = h_0(0,0,1), \quad q = h_0(0,0,-1)$$

and a shortest geodesic arc $c : [0,1] \to M$ with $c(0) = p$, $c(1) = q$ is given by

$$c(t) = h_0(0, \sin \pi t, \cos \pi t).$$

† See footnote on p. 43.

Let us point out that these normalizations are not at all essential, but only convenient for our constructions. We look at the family of curves

$$\gamma(t, s) = h_0(\sin 2\pi s \sin \pi t, \cos 2\pi s \sin \pi t, \cos \pi t), \quad 0 \leq s, t \leq 1. \quad (2.3.1)$$

Then
$$\gamma(t, 0) = \gamma(t, 1) = c(t) \quad \text{for all } t$$
and
$$\gamma(0, s) = c(0), \quad \gamma(1, s) = c(1) \quad \text{for all } s.$$

We find some number K with

$$L(\gamma(\cdot, s)) \leq K \quad \text{for all } s. \quad (2.3.2)$$

Redefining the parameter t, we may also assume that all curves $\gamma(\cdot, s)$ are parameterized proportionally to arc-length. By Theorem 2.3.1, there exists $\epsilon_0 > 0$ such that the shortest geodesic between any $p, q \in M$, with $d(p, q) \leq \epsilon_0$ is unique. Let

$$0 = t_0 < t_1 < \ldots < t_m = 1$$

be a partition of $[0, 1]$ with

$$t_j - t_{j-1} < \frac{\epsilon_0}{K} \quad \text{for } j = 1, \ldots, m. \quad (2.3.3)$$

Let another partition $(\tau_1, \ldots, \tau_m)$ satisfy

$$\tau_0 = t_0 < \tau_1 < t_1 < \tau_2 < \ldots < \tau_m < t_m = \tau_{m+1}$$

and
$$\tau_j - \tau_{j-1} < \frac{\epsilon_0}{K} \quad \text{for } j = 1, \ldots, m+1. \quad (2.3.4)$$

If $\gamma : [0, 1] \to M$ is any curve parameterized proportionally to arc-length with
$$L(\gamma) \leq K,$$
we then have for $j = 1, \ldots, m$

$$d(\gamma(t_{j-1}), \gamma(t_j)) \leq L\left(\gamma|_{[t_{j-1}, t_j]}\right) < K \frac{\epsilon_0}{K} = \epsilon_0.$$

Therefore, by Theorem 2.3.1, the shortest geodesic from $\gamma(t_{j-1})$ to $\gamma(t_j)$ is unique. We then define $r_1(\gamma)$ to be that piecewise geodesic curve for which $r_1(\gamma)|_{[t_{j-1}, t_j]}$ coincides with the shortest geodesic from $\gamma(t_{j-1})$ to $\gamma(t_j)$, $j = 1, \ldots, m$. Likewise, we let $r_2(\gamma)$ by that piecewise geodesic curve for which $r_2(\gamma)|_{[\tau_{j-1}, \tau_j]}$ coincides with the — again unique — shortest geodesic from $\gamma(\tau_{j-1})$ to $\gamma(\tau_j)$, $j = 1, \ldots, m+1$. We now observe:

2.3 The existence of geodesics

Lemma 2.3.1. *Suppose $d(\gamma(t_j), \gamma(t_{j-1})) \leq \epsilon_0$ and $d(\gamma(\tau_j), \gamma(\tau_{j-1})) \leq \epsilon_0$ for all j.*

$$r(\gamma) := r_2 \circ r_1(\gamma)$$

satisfies

$$L(r(\gamma)) \leq L(\gamma) \qquad (2.3.5)$$

with equality iff γ is geodesic.

Proof. By uniqueness of the shortest geodesic between $\gamma(t_{j-1})$ and $\gamma(t_j)$, we have

$$L(r_1(\gamma)) \leq L(\gamma)$$

with equality only in case

$$r_1(\gamma) = \gamma.$$

Likewise, for every curve γ', $L\left(\gamma'_{|[\tau_{j-1},\tau_j]}\right) \leq \epsilon_0$ for all j,

$$L(r_2(\gamma')) \leq L(\gamma')$$

with equality only in case

$$r_2(\gamma') = \gamma'.$$

Therefore

$$L(r(\gamma)) \leq L(\gamma)$$

with equality only if

$$r(\gamma) = \gamma.$$

If $r(\gamma) = \gamma$, however, $\gamma_{|[t_{j-1},t_j]}$ and $\gamma_{|[\tau_{j-1},\tau_j]}$ are geodesic for every j, and hence γ is geodesic itself. (If $r_1(\gamma) = \gamma$, then γ is piecewise geodesic with corners at most at the t_j, and if $r_2(r_1(\gamma)) = r_1(\gamma)$, then $r_1(\gamma)$ is geodesic with corners at most at the τ_j. Thus, if $r(\gamma) = \gamma$, γ cannot have any corners at all.)

q.e.d.

Lemma 2.3.2. *Let $\gamma : [0,1] \to M$ be a curve parameterized proportionally to arc-length and with $L(\gamma) \leq K$. Then a subsequence of $r^n(\gamma)$ ($= r \circ \ldots \circ r(\gamma)$) converges uniformly to a geodesic with the same endpoints as γ.*

Proof. Each curve $r^n(\gamma)$, $n \in \mathbb{N}$, is a piecewise geodesic with corners $r^n\gamma(\tau_1), \ldots, r^n\gamma(\tau_m)$ and endpoints $r^n\gamma(\tau_0) = \gamma(0)$, $r^n\gamma(\tau_{m+1}) = \gamma(1)$. The individual segments are the unique shortest connections between these points. Therefore, each such curve is uniquely determined by the m-tupel

$$\lambda^n := (r^n\gamma(\tau_1), \ldots, r^n\gamma(\tau_m)) \in \underbrace{M \times \ldots \times M}_{m \text{ times}}.$$

Since M is compact, a subsequence of λ^n converges to some limit

$$(p_1, \ldots, p_m) \in M \times \ldots \times M.$$

$r^n(\gamma)$ then converges uniformly towards the piecewise geodesic γ_0 with endpoints $\gamma_0(0) = \gamma(0), \gamma_0(1) = \gamma(1)$ and nodes $\gamma_0(\tau_i) = p_i$ ($i = 1, \ldots, m$) with segments $\gamma_0|_{[\tau_{j-1}, \tau_j]}$ being the shortest geodesic arcs between their endpoints. This follows from the continuous dependence of the occurring geodesic arcs on their endpoints (Theorem 2.3.1). We denote the convergent subsequence of $(r^n(\gamma))_{n \in \mathbb{N}}$ by $(\gamma_\nu)_{\nu \in \mathbb{N}}$. For all $\nu \in \mathbb{N}$ then

$$\gamma_{\nu+1} = r^{n(\nu)}\gamma_\nu \quad \text{with } n(\nu) \in \mathbb{N}.$$

By the minimizing property of the subsegments of the γ_ν,

$$L\left(\gamma_\nu|_{[\tau_{j-1}, \tau_j]}\right) = d\left(\gamma_\nu(\tau_{j-1}), \gamma_\nu(\tau_j)\right),$$

hence

$$L(\gamma_\nu) = \sum_{j=1}^{m+1} d\left(\gamma_\nu(\tau_{j-1}), \gamma_\nu(\tau_j)\right).$$

Since $\gamma_\nu(\tau_j)$ converges to $p_j = \gamma_0(\tau_j)$, $L(\gamma_\nu)$ converges to

$$L(\gamma_0) = \sum_{j=1}^{m+1} d\left(\gamma_0(\tau_{j-1}), \gamma_0(\tau_j)\right)$$

for $\nu \to \infty$. Then also

$$\begin{aligned} L(\gamma_0) = \lim_{\nu \to \infty} L(\gamma_{\nu+1}) &= \lim_{\nu \to \infty} L(r^{n(\nu)}\gamma_\nu) \\ &\leq \lim_{\nu \to \infty} L(\gamma_\nu) \quad \text{by Lemma 2.3.1} \\ &= L(\gamma_0), \end{aligned}$$

and equality has to hold throughout. Moreover, $r(\gamma_\nu)$ converges to $r(\gamma_0)$,

2.3 The existence of geodesics

and

$$L\left(r\left(\gamma_{0}\right)\right) = \lim_{\nu \to \infty} L\left(r(\gamma_{\nu})\right)$$
$$\geq \lim_{\nu \to \infty} L\left(r^{n(\nu)}\gamma_{\nu}\right) \quad \text{by Lemma 2.3.1 again}$$
$$= L(\gamma_{0}).$$

Lemma 2.3.1 then implies that γ_0 is geodesic.

q.e.d.

We now return to the proof of Theorem 2.3.3:

We apply the preceding curve shortening process to all curves $\gamma(\cdot, s)$, $s \in [0,1]$, simultaneously. For each s, a subsequence of $r^n \gamma(\cdot, s)$ then converges to a geodesic from p to q. We want to exclude the situation that all those limit geodesics coincide with c. Let

$$\kappa_0 := L(c),$$

and

$$\kappa_1 := \sup_{0 \leq s \leq 1} \lim_{n \to \infty} L(r^n \gamma(\cdot, s)).$$

Since $\gamma(\cdot, 0) = c(\cdot)$ is geodesic, $r^n \gamma(\cdot, 0) = \gamma(\cdot, 0)$ for all n, hence $\kappa_1 \geq \kappa_0$. We distinguish two cases:

(1) $\kappa_1 > \kappa_0$

Since $\gamma(\cdot, s)$ is continuous in s, so is $r^n \gamma(\cdot, s)$ for every $n \in \mathbb{N}$. We now claim:

Whenever

$$\sup_s L(r^n \gamma(\cdot, s)) \leq \kappa_1 + \epsilon \tag{2.3.6}$$

there exists $s_n \in [0,1]$ with

$$L(r^n \gamma(\cdot, s_n)) - L(r^{n+1} \gamma(\cdot, s_n)) \leq 2\epsilon \tag{2.3.7}$$

and

$$L(r^n \gamma(\cdot, s_n)) \geq \kappa_1 - \epsilon. \tag{2.3.8}$$

Indeed, otherwise

$$\sup_s L(r^{n+1} \gamma(\cdot, s)) \leq \kappa_1 - \epsilon,$$

contradicting the definition of κ_1 (note that $\sup_s L(r^{n+1} \gamma(\cdot, s))$

is monotonically decreasing in n by Lemma 2.3.1). By definition of κ_1, there exists a subsequence $(\epsilon_n)_{n \in \mathbb{N}} \to 0$ with

$$\sup_s L(r^n \gamma(\cdot, s)) \leq \kappa_1 + \epsilon_n.$$

A subsequence of $(r^n \gamma(\cdot, s_n))_{n \in \mathbb{N}}$ has to converge to some limit curve $\bar{c}$ as above, and because of (2.3.7) with $\epsilon = \epsilon_n$, we conclude as in the proof of Lemma 2.3.2 that

$$L(r(\bar{c})) = L(\bar{c}),$$

and $\bar{c}$ is hence geodesic by Lemma 2.3.1. Because of (2.3.8) and continuity of L in the limit as in the proof of Lemma 2.3.2, we get

$$L(\bar{c}) = \kappa_1.$$

Since c and $\bar{c}$ are both defined on $[0, 1]$ and have different lengths, they have to be different curves. Thus, $\bar{c}$ is the desired second geodesic.

(2) $\kappa_1 = \kappa_0$

We are going to show that in this case, there even exist infinitely many geodesics from p to q. For that purpose, we consider the curve

$$\hat{\gamma}(s) := \gamma(\frac{1}{2}, s).$$

This is a closed curve with $\hat{\gamma}(0) = \hat{\gamma}(1) = c(\frac{1}{2})$ (see Figure 2.2).

Since h_0 is a diffeomorphism and $r^n \gamma(t, s)$ is obtained through a process that can easily be made continuous from

$$\gamma(t, s) = h_0(\sin 2\pi s \sin \pi t, \cos 2\pi s \sin \pi t, \cos \pi t),$$

$r^n \gamma(t, s)$ has to map $[0, 1] \times [0, 1]$ surjectively onto M. Therefore, for every $n \in \mathbb{N}$ and every $s \in [0, 1]$, there exists $\sigma_n(s)$ with

$$\hat{\gamma}(s) \in r^n \gamma(\cdot, \sigma_n(s)) =: \gamma_{n,s}(\cdot)$$

(in other words, $r^n \gamma(\cdot, \sigma_n(s))$ is a curve passing through $\hat{\gamma}(s)$). $\gamma_{n,s}(\cdot)$ then is a curve with

$$\gamma_{n,s}(0) = c(0) = p, \gamma_{n,s}(1) = c(1) = q,$$

and because of $\kappa_1 = \kappa_0$, we obtain

$$\lim_{n \to \infty} L(\gamma_{n,s}(\cdot)) \leq \sup_{0 \leq s \leq 1} \lim_{n \to \infty} L(r^n \gamma(\cdot, s)) = \kappa_0. \tag{2.3.9}$$

2.3 The existence of geodesics

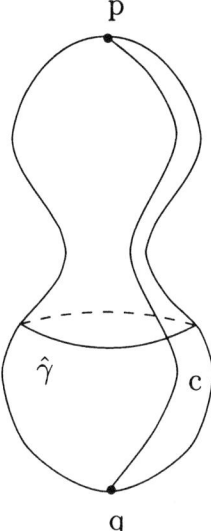

Figure 2.2.

After selection of a subsequence, $(\gamma_{n,s}(\cdot))_{n \in \mathbb{N}}$ again converges to some limit curve $c_s(\cdot)$ with

$$c_s(0) = p, c_s(1) = q$$

and

$$\hat{\gamma}(s) \in c_s(\cdot).$$

By (2.3.5),

$$L(c_s(\cdot)) \leq \kappa_0,$$

and since κ_0 is the infimum of the energies of all curves from p to q ($\kappa_0 = L(c)$, and c is minimizing), $c_s(\cdot)$ is a minimizing curve itself, hence geodesic.

Therefore, we have shown that for every s, there exists a geodesic from p to q that passes through $\hat{\gamma}(s)$. Hence there exist infinitely many geodesics from p to q, as claimed.

q.e.d.

Remarks:

(1) Lemmas 2.3.1 and 2.3.2 do not need that M is diffeomorphic to S^2. Compactness suffices.

(2) We may construct the curves $\gamma_{n,s}(\cdot)$ at the end of the proof also in case $\kappa_1 > \kappa_0$. In that case, however, limits of such curves need not be geodesic anymore.

(3) See Section 3.1 for an abstract version of the argument at the end of the preceding proof.

Exercises

2.1 For curves $\gamma(t) = (\gamma^1, \gamma^2) : \mathbb{R} \to \{(x^1, x^2) \in \mathbb{R}^2 | x^2 > 0\}$, consider
$$E(\gamma) := \frac{1}{2} \int \frac{1}{(\gamma^2(t))^2} |\dot\gamma(t)|^2 dt.$$
Compute the Euler–Lagrange equations and determine all solutions.

2.2 For curves
$$\gamma(t) = (\gamma^1, \ldots, \gamma^d) : \mathbb{R} \to \{(x^1, \ldots, x^d) \in \mathbb{R}^d | \sum_{i=1}^{d}(x^i)^2 < 1\},$$
consider
$$E(\gamma) := \frac{1}{2} \int \frac{1}{(1 - |\gamma(t)|^2)^2} |\dot\gamma(t)|^2 dt.$$
Compute the Euler–Lagrange equations and determine all solutions.

2.3 Determine all geodesics between two given points on a cylinder
$$\{(x, y, z) \in \mathbb{R}^3 : x^2 + y^2 = 1\}.$$

2.4 Let Σ be a surface of revolution in $\mathbb{R}^3$, i.e.
$$\Sigma = \{(x, y, z) \in \mathbb{R}^3 : x^2 + y^2 = f(z)\}$$
for a smooth, positive $f : \mathbb{R} \to \mathbb{R}$. What can you say about geodesics on Σ? For example, are the curves $(x, y) = $ constant geodesics? When are the curves $z = $ constant geodesics?

2.5 Determine Riemannian polar coordinates on the sphere S^n with a domain of definition that is as large as possible.

2.6 Let p be the center of Riemannian polar coordinates on M, with domain of definition $\{v \in \mathbb{R}^d : ||v|| \le \varrho\}$. Let $c : [0, \epsilon] \to M$ be a geodesic with $c(0) = p$ that is parameterized by arc-length, $0 < \epsilon < \varrho$. Show that $c([0, \epsilon])$ does not contain a point that is conjugate to p.

Exercises

2.7 Let M be a differentiable submanifold of $\mathbb{R}^d$ that is diffeomorphic to S^2. Show that for any $p \in M$, there exists a nonconstant geodesic $c : [0,1] \to M$ with $c(0) = c(1) = p$.

2.8 Try to find other topological classes of manifolds with the property that there always exists more than one geodesic connection between any two points.

3
Saddle point constructions

3.1 A finite dimensional example

Let $F : \mathbb{R}^d \to \mathbb{R}$ be a function of class C^1 which is bounded from below and which is 'proper' in the following sense:

$$F(x) \to \infty \quad \text{for } |x| \to \infty. \tag{3.1.1}$$

Since F is bounded from below, (3.1.1) is equivalent to: For every $s \in \mathbb{R}$,

$$\{x \in \mathbb{R}^d : F(x) \leq s\} \quad \text{is compact.} \tag{3.1.2}$$

Therefore, F assumes its infimum. Namely, we take any

$$s_0 > \inf_{x \in \mathbb{R}^d} F(x).$$

Then

$$\{x \in \mathbb{R}^d : F(x) \leq s_0\}$$

is compact and nonempty, and since F is continuous, it has to assume its infimum on that set. We now assume that F even has two relative minima, x_1, x_2 in $\mathbb{R}^d$, and that they are strict in the following sense: For $x = x_1, x_2$, we have

$$\exists \delta_0 \, \forall y \quad \text{with} \quad 0 < |y - x| \leq \delta_0 : F(y) > F(x). \tag{3.1.3}$$

Theorem 3.1.1. *Under the above assumptions, F has a third critical point x_3 (i.e. $\nabla F(x_3) = 0$) with*

$$F(x_3) > \max(F(x_1), F(x_2)) =: \kappa_0$$

Proof. We consider curves $\gamma : [0, 1] \to \mathbb{R}^d$ with

$$\gamma(0) = x_1 \, , \, \gamma(1) = x_2. \tag{3.1.4}$$

3.1 A finite dimensional example

We first observe that there exists $\alpha > 0$ with the property that for any such curve, there exists $t_0 \in (0,1)$ with

$$F(\gamma(t_0)) \geq \kappa_0 + \alpha. \tag{3.1.5}$$

In order to verify this, we may assume w.l.o.g.

$$F(x_1) \leq F(x_2).$$

We then choose δ with

$$0 < \delta \leq \min(\delta_0, \frac{1}{2}|x_1 - x_2|). \tag{3.1.6}$$

For every y with $|y - x_2| = \delta$ then by (3.1.3)

$$F(y) > F(x_2),$$

and since $\{|y - x_2| = \delta\}$ is compact, F assumes its minimum on this set, hence for some $\alpha > 0$

$$\min_{|y-x_2|=\delta} F(y) \geq F(x_2) + \alpha = \kappa_0 + \alpha. \tag{3.1.7}$$

Since for every curve γ with (3.1.4) we have

$$|\gamma(1) - x_2| = 0, |\gamma(0) - x_2| = |x_1 - x_2|,$$

there has to exist some $t_0 \in [0,1]$ with

$$|\gamma(t_0) - x_2| = \delta \quad (\text{recall } (3.1.6)).$$

By (3.1.7) then

$$F(\gamma(t_0)) \geq \kappa_0 + \alpha,$$

and (3.1.5) follows indeed.

We now define

$$\kappa_1 := \inf_\gamma \sup_{t \in [0,1]} F(\gamma(t)),$$

where γ again is a curve in $\mathbb{R}^d$ with $\gamma(0) = x_1, \gamma(1) = x_2$. By (3.1.5)

$$\kappa_1 > \kappa_0. \tag{3.1.8}$$

Our intention now is to find a critical point x_3 of F with

$$F(x_3) = \kappa_1.$$

Since

$$F(x_1), F(x_2) \leq \kappa_0,$$

x_3 will then be necessarily be different from x_1 and x_2. As a step towards the existence of such a point x_3, we claim

$$\forall \epsilon > 0 \quad \exists \delta > 0 \quad \forall \text{ curves } \gamma \text{ with } \quad \gamma(0) = x_1, \gamma(1) = x_2$$

with

$$\sup_{t \in [0,1]} F(\gamma(t)) \leq \kappa_1 + \delta \tag{3.1.9}$$

$\exists t_0 \in [0,1]$ with:

$$F(\gamma(t_0)) \geq \kappa_1 - \epsilon \tag{3.1.10}$$

$$|(\nabla F)(\gamma(t_0))| < \epsilon. \tag{3.1.11}$$

Suppose this is not the case. Then

$$\exists \epsilon_0 > 0 \quad \forall n \in \mathbb{N} \quad \exists \text{ curve } \gamma_n \text{ between } x_1 \text{ and } x_2 \text{ with}$$

$$\sup_t F(\gamma_n(t)) \leq \kappa_1 + \frac{1}{n} \tag{3.1.12}$$

$$\forall t_0 \quad \text{with} \quad F(\gamma_n(t_0)) \geq \kappa_1 - \epsilon_0 \tag{3.1.13}$$

$$|(\nabla F)(\gamma_n(t_0))| \geq \epsilon_0. \tag{3.1.14}$$

For $s > 0$, we define a new curve $\gamma_{n,s}$ by

$$\gamma_{n,s}(t) := \gamma_n(t) - s(\nabla F)(\gamma_n(t)).$$

Since x_1 and x_2 are minima, $\nabla F(x_1) = 0 = \nabla F(x_2)$, and so

$$\gamma_{n,s}(0) = x_1, \gamma_{n,s}(1) = x_2,$$

so that the curves $\gamma_{n,s}$ are valid comparison curves. By our properness assumption (3.1.2) and (3.1.12), $\gamma_n(t)$ stays in a bounded subset of $\mathbb{R}^d$, and ∇F will then be bounded on that bounded set, and hence for any $s_0 > 0$ and all $0 \leq s \leq s_0$, the curves $\gamma_{n,s}(t)$ stay in some bounded set, too. This set is independent of n (as long as $0 \leq s \leq s_0$, for fixed $s_0 > 0$). By Taylor's formula

$$F(\gamma_{n,s}(t)) = F(\gamma_n(t)) - s \nabla F(\gamma_n(t)) \cdot \nabla F(\gamma_n(t)) + o(s).$$

Since F is continuously differentiable and $\gamma_{n,s}(t)$ is contained in a bounded set, $o(s)$ can be estimated independently of n and t (as long as $0 \leq s \leq s_0$). In particular, after possibly choosing $s_0 > 0$ smaller,

$$F(\gamma_{n,s}(t)) \leq F(\gamma_n(t)) - \frac{s}{2} |\nabla F(\gamma_n(t))|^2 \tag{3.1.15}$$

3.1 A finite dimensional example

for all n, s with $0 \leq s \leq s_0$, and t with

$$|\nabla F(\gamma_n(t))| \geq \epsilon_0. \tag{3.1.16}$$

Thus, in particular,

$$F(\gamma_{n,s_0}(t)) \leq F(\gamma_n(t)) - \frac{s_0}{2}\epsilon_0^2 \tag{3.1.17}$$

for all such t and all n. We now simply choose n so large that

$$\frac{1}{n} < \frac{s_0}{2}\epsilon_0^2. \tag{3.1.18}$$

Then by our assumption, all t_0 with $F(\gamma_n(t_0)) \geq \kappa_1 - \epsilon_0$ satisfy (3.1.14), and hence for all such t_0

$$\begin{aligned} F(\gamma_{n,s_0}(t_0)) &\leq F(\gamma_n(t_0)) - \frac{s_0}{2}\epsilon_0^2 \\ &\leq \kappa_1 + \frac{1}{n} - \frac{s_0}{2}\epsilon_0^2 \quad \text{by (3.1.12)} \\ &< \kappa_1 \quad \text{by (3.1.18)}. \end{aligned} \tag{3.1.19}$$

Having proved (3.1.19), there are now various ways to construct a path $\tilde{\gamma}$ from x_1 to x_2 with

$$F(\tilde{\gamma}(t)) < \kappa_1 \quad \text{for all } t \in [0,1]. \tag{3.1.20}$$

One way is to refine the above construction by letting s depend on t as follows: we choose a smooth function

$$\sigma(t) : [0,1] \to [0, s_0]$$

with

$$\sigma(t) = 0 \quad \text{whenever} \quad F(\gamma_n(t)) \leq \kappa_1 - \epsilon_0$$

and

$$\sigma(t) = s_0 \quad \text{whenever} \quad F(\gamma_n(t)) \geq \kappa_1 - \frac{\epsilon_0}{2}.$$

We then look at the path $\tilde{\gamma}(t) = \gamma_{n,\sigma(t)}(t)$. Then for t with $F(\gamma_n(t)) \leq \kappa_1 - \epsilon_0$

$$F(\tilde{\gamma}(t)) = F(\gamma_n(t)) \leq \kappa_1 - \epsilon_0,$$

for t with $\kappa_1 - \epsilon_0 \leq F(\gamma_n(t)) \leq \kappa_1 - \frac{\epsilon_0}{2}$

$$F(\tilde{\gamma}(t)) \leq F(\gamma_n(t)) - \frac{\sigma(t)}{2}\epsilon_0^2 \leq \kappa_1 - \frac{\epsilon_0}{2} - \frac{\sigma(t)}{2}\epsilon_0^2$$

(cf. (3.1.15), (3.1.16), (3.1.14)), and finally for all t with $F(\gamma_n(t)) \geq \kappa_1 - \frac{\epsilon_0}{2}$

$$F(\tilde{\gamma}(t)) = F(\gamma_{n,s_0}(t)) < \kappa_1 \quad \text{(cf. (3.1.19))}.$$

Thus, (3.1.20) holds indeed. This, however, contradicts the definition of κ_1. Therefore, the assumption that our claim was not correct led to a contradiction, and the claim holds. It is now simple to prove the theorem. Namely, we let $\epsilon_n \to 0$ for $n \to \infty$, and for $\epsilon = \epsilon_n$, we find $\delta = \delta_n$ as in the claim. We than choose a curve γ_n from x_1 to x_2 with

$$\sup_{t \in [0,1]} F(\gamma_n(t)) \leq \kappa_1 + \min(\epsilon_n, \delta_n). \tag{3.1.21}$$

According to the claim, there exists $t_n \in [0,1]$ with

$$F(\gamma_n(t_n)) \geq \kappa_1 - \epsilon_n \tag{3.1.22}$$

$$|(\nabla F)(\gamma_n(t_n))| \leq \epsilon_n. \tag{3.1.23}$$

After selection of a subsequence, $(\gamma_n(t_n))_{n \in \mathbb{N}}$ then converges to some point x_3, because of (3.1.2) and (3.1.21). x_3 then satisfies by continuity of F and ∇F

$$F(x_3) = \kappa_1 \tag{3.1.24}$$

$$\nabla F(x_3) = 0. \tag{3.1.25}$$

Thus, x_3 is the desired critical point.

q.e.d.

Theorem 3.1.1 may be refined as follows:

Theorem 3.1.2. *Let F as above again have two relative minima, not necessarily strict anymore. Then either F has a critical point x_3 with*

$$F(x_3) > \max(F(x_1), F(x_2)) = \kappa_0,$$

or it has infinitely many critical points.

Proof. For the argument of the proof of Theorem 3.1.1, we only need

$$\inf_\gamma \sup_{t \in [0,1]} F(\gamma(t)) > \kappa_0, \tag{3.1.26}$$

where the infimum again is taken over curves $\gamma : [0,1] \to \mathbb{R}^d$ with $\gamma(0) = x_1$, $\gamma(1) = x_2$. So, suppose that (3.1.26) does not hold. We then want to

show the existence of infinitely many critical points. As in the proof of Theorem 3.1.1, we may assume
$$F(x_1) \leq F(x_2).$$
The argument at the beginning of the proof of Theorem 3.1.1 then shows that (3.1.26) holds if x_2 is a strict relative minimum. If x_2 is a relative minimum, which is not strict, for all sufficiently small $\delta > 0$, say $\delta \leq \delta_0$, we have
$$F(x_2) \leq F(x) \quad \text{for all } x \quad \text{with} \quad |x - x_2| \leq \delta_0 \qquad (3.1.27)$$
and there always exists some x_δ with $0 < |x_\delta - x_2| \leq \delta$ and
$$F(x_\delta) = F(x_2). \qquad (3.1.28)$$
We then put $\delta_1 = \delta_0/2$. Then x_{δ_1} is a relative minimum of F by (3.1.27), (3.1.28), hence a critical point. Having found a critical point x_{δ_n} with $0 < |x_{\delta_n} - x_2| < |x_{\delta_{n-1}} - x_2|$, we put
$$\delta_{n+1} = \frac{1}{2}|x_{\delta_n} - x_2|$$
and find a critical point $x_{\delta_{n+1}}$ with
$$0 < |x_{\delta_{n+1}} - x_2| \leq \delta_{n+1}.$$
Thus, $x_{\delta_{n+1}}$ is a critical point of F different from all preceding ones.

q.e.d.

Remark. It is not very hard to sharpen the statement of Theorem 3.1.2 from 'infinitely many' to 'uncountably many'.

3.2 The construction of Lyusternik–Schnirelman

In this section, we want to prove the following theorem, in order to exhibit some important global construction in the calculus of variations, introduced by Lyusternik–Schnirelman. The result presented is much more elementary than the theorem of Lyusternik–Schnirelman, which says that on any surface with a Riemannian metric, e.g. a surface embedded in some Euclidean space, diffeomorphic to the two-dimensional sphere, there exist at least three closed geodesics without self-intersections. The more elementary character of our setting allows us to bypass essential geometric difficulties encountered in a detailed proof of the Lyusternik–Schnirelman Theorem.

68   *Saddle point constructions*

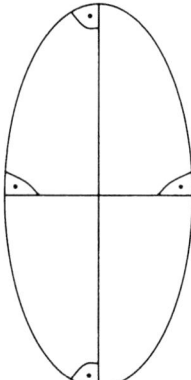

Figure 3.1.

Theorem 3.2.1. *Let γ be a closed convex Jordan curve† of class C^1 in the plane $\mathbb{R}^2$. (γ then divides the plane into a bounded region $\mathring{A}$, and an unbounded one, by the Jordan curve Theorem. That γ is convex means that the straight line between any two points of γ is contained in the closure A of $\mathring{A}$.) Then there exist at least two such straight lines between points on γ meeting γ orthogonally at both end points (see Figure 3.1).*

Proof. We start by finding one such line. Let $\mathcal{L}$ be the set of all straight lines l in A with $\partial l \subset \gamma$. We say that a sequence $(l_n)_{n \in \mathbb{N}} \subset \mathcal{L}$ converges to $l \in \mathcal{L}$, if the end points of the l_n converge to those of l. In order to have a closed space, we allow lines to be trivial i.e. to consist of a single point on γ only. We denote the space of these point curves on γ by $\mathcal{L}_0$. We let $I := [0, 1]$ be the unit interval. We consider continuous maps

$$v : I \to \mathcal{L}$$

with the following two properties:

(i) $v(0) = v(1)$.
(ii) To any such family, we may assign two subregions $A_1(t)$ and $A_2(t)$ of A in a certain manner. Namely, we let $A_1(t)$ and $A_2(t)$ be the two regions into which $v(t)$ divides A. Having chosen $A_1(0)$ and $A_2(0)$, $A_1(t)$ and $A_2(t)$ then are determined by the continuity

† A closed Jordan curve is a curve $\gamma : [0, T] \to \mathbb{R}^d$ with $\gamma(0) = \gamma(T)$ that is injective on $[0, T)$. Cf. the definition of a Jordan curve on p. 35.

3.2 The construction of Lyusternik–Schnirelman   69

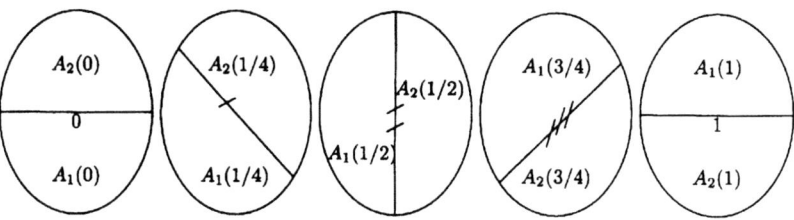

Figure 3.2.

requirement. We then require

$$A_1(1) = A_2(0).$$

We let V_1 be the class of all such families v.

The construction is visualized in Figure 3.2. (0 corresponds to $0 \in I$, / to $\frac{1}{4}$, // to $\frac{1}{2}$, /// to $\frac{3}{4}$, 1 to 1)

Actually, in order to simplify the visualization, if $v(0)$ is a point curve (on γ), i) may be relaxed to just requiring that $v(1)$ also is a point curve (on γ), not necessarily coinciding with $v(0)$ (see Figure 3.3). Namely, any point curves can be connected through point curves, i.e. with vanishing length.

We denote by $L(l)$ the length of $l \in \mathcal{L}$ and define

$$\kappa_1 := \inf_{v \in V_1} \sup_{t \in I} L(v(t)).$$

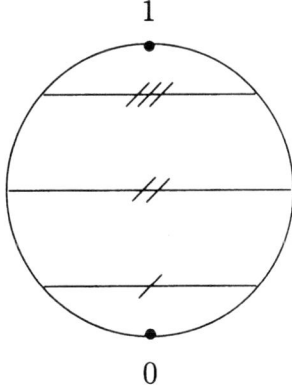

Figure 3.3.

We want to show that

$$\kappa_1 > 0.$$

For this purpose, let $\rho > 0$ be the inner radius of γ, i.e. the largest ρ for which there exists a disc

$$B(x_0, \rho) \subset A$$

for some $x_0 \in A$ ($B(x_0, \rho) := \{x \in \mathbb{R}^2 : |x - x_0| \leq \rho\}$). Then

$$\kappa_1 \geq \kappa_1' := \inf_{v \in V_1} \sup_{t \in I} L(v(t) \cap B(x_0, \rho)).$$

We let $A_i'(t) := A_i(t) \cap B(x_0, \rho)$, $i = 1, 2$. Because of (ii) and the continuous dependence of $A_i(t)$ and hence also of $A_i'(t)$ on t, there exists some $t_0 \in I$ with

$$\text{Area}\,(A_1'(t_0)) = \text{Area}\,(A_2'(t_0)).$$

Thus $v(t_0)$ divides $B(x_0, \rho)$ into two subregions of equal area. $v(t_0)$ then has to be a diameter of $B(x_0, \rho)$, i.e.

$$L\,(v(t_0) \cap B(x_0, \rho)) = 2\rho.$$

Therefore

$$\kappa_1 \geq \kappa_1' = 2\rho > 0$$

and κ_1 is positive indeed. We are now going to show by a line of reasoning already familiar from Sections 2.3 and 3.1 that κ_1 is realized by a critical point l of L among all lines with end points in γ, i.e. by l meeting γ orthogonally (see Theorem 1.4.1). For that purpose we shall assume for the moment that γ is of class C^3. Later on, we shall reduce the case where γ is only C^1 to the present one by an approximation argument. We now claim

$$\forall \epsilon > 0 \ \exists \delta > 0 : \quad \forall v \in V_1 \text{ with}$$

$$\sup_{t \in I} L\,(v(t)) \leq \kappa_1 + \delta$$

$$\exists t_0 \in I \quad \text{with} \quad L\,(v(t_0)) \geq \kappa_1 - \epsilon$$

$$\text{and} \quad |\cos(\alpha_1(v(t_0)))|\,,\,|\cos(\alpha_2(v(t_0)))| < \epsilon,$$

where $\alpha_1(l)$ and $\alpha_2(l)$ are the angles of l at its endpoints with γ.

3.2 The construction of Lyusternik–Schnirelman 71

Otherwise

$$\exists \epsilon_0 > 0 : \forall n \in \mathbb{N} \quad \exists v_n \in V_1 \text{ with}$$
$$\sup_t L(v_n(t)) \leq \kappa_1 + \tfrac{1}{n}$$
$$\forall t_0 \text{ with } L(v_n(t_0)) \geq \kappa_1 - \epsilon_0$$
$$|\cos \alpha_1 (v_n(t_0))| \geq \epsilon_0$$
$$\text{or} \quad |\cos \alpha_2 (v_n(t_0))| \geq \epsilon_0.$$

The idea to reach a contradiction from that assumption is simple, once the following Lemma is proved:

Lemma 3.2.1. *For every planar closed Jordan curve γ of class C^3, there exists $\beta > 0$ with the following property: Whenever $x \in \mathbb{R}^2$ satisfies*

$$\mathrm{dist}(x, \gamma) := \inf_{y \in \gamma} |x - y| < \beta$$

there exists a unique $y \in \gamma$ with $\mathrm{dist}(x, \gamma) = |x - y|$.

Proof. We consider γ as an embedded submanifold of the Euclidean plane $\mathbb{R}^2$. γ is then covered by the images of charts $f : U \to V$ of the type constructed in Theorem 2.2.1. Here, U and V are open in $\mathbb{R}^2$, and

$$\gamma \cap V = f\left(U \cap \{x^2 = 0\}\right).$$

Furthermore, the curves $x^1 = $ constant in U correspond to geodesics, i.e. straight lines in V perpendicular to γ, and they form shortest connections to $\gamma \cap V$. By shrinking U, if necessary, we may assume that it is of the form $(-\xi, \xi) \times (-\eta, \eta)$, with $\xi > 0$, $\eta > 0$. Since γ is compact, it can be covered by finitely many such charts

$$f_i : (-\xi_i, \xi_i) \times (-\eta_i, \eta_i) \to V_i \quad , i = 1, \ldots, m.$$

If we then restrict f_i to $(-\xi_i, \xi_i) \times \left(\frac{-\eta_i}{2}, \frac{\eta_i}{2}\right)$, the lines $x^1 = $ constant, $\frac{-\eta_i}{2} < x^2 < \frac{\eta_i}{2}$, then correspond to shortest geodesics to γ, since the part of γ not contained in V_i is not contained in the image of f_i, and hence has distance at least $\frac{\eta_i}{2}$ from the image of the smaller set $(-\xi_i, \xi_i) \times \left(\frac{-\eta_i}{2}, \frac{\eta_i}{2}\right)$. This is indicated in Figure 3.4 where the broken lines correspond to $x^2 = \frac{\eta_i}{2}$, and this is depicted for two different indices i.

Therefore, $\beta := \min_{i=1,\ldots,n} \left(\frac{\eta_i}{2}\right)$ satisfies the claim.

q.e.d.

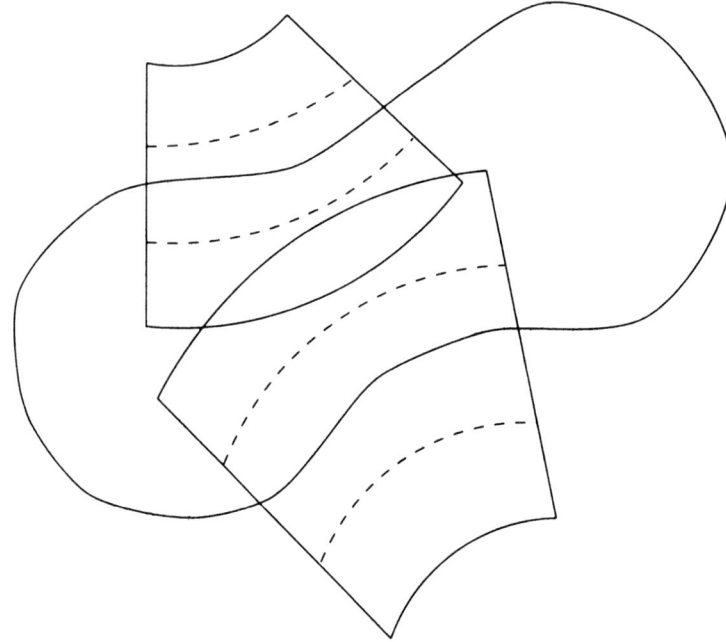

Figure 3.4.

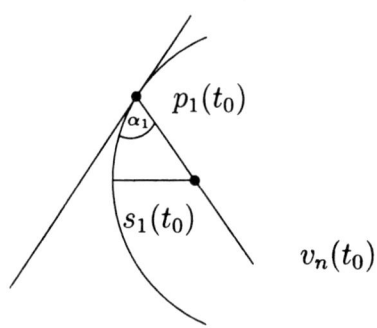

Figure 3.5.

We now return to the proof of Theorem 3.2.1:

Without loss of generality $\epsilon_0 \leq \beta \leq \frac{\kappa_1}{4}$. Assume e.g.

$$\cos \alpha_1 \left(v_n \left(t_0 \right) \right) \geq \epsilon_0.$$

The following construction is depicted in Figure 3.5. Choose $s_1(t_0) \in$

3.2 The construction of Lyusternik-Schnirelman

$v_n(t_0)$ with

$$|s_1(t_0) - p_1(t_0)| = \beta,$$

where $p_1(t_0)$ is the endpoint of $v_n(t_0)$ where it forms the angle $\alpha_1(t_0)$ with γ. We replace the subarc $v_n^1(t_0)$ of $v_n(t_0)$ between $p_1(t_0)$ and $s_1(t_0)$ by the shortest line segment $v_n'(t_0)$ from $s_1(t_0)$ to γ. By the theorem of Pythagoras and the convexity of γ

$$L\left(v_n'(t_o)\right) \leq L\left(v_n^1(t_0)\right) \sin \alpha_1(v_n(t_0))$$
$$\leq L\left(v_n^1(t_0)\right) \sqrt{1 - \epsilon_0^2}.$$

We then let

$$v_n^s(t_0)$$

be the straight line from the second endpoint $p_2(t_0)$ of $v_n(t_0)$ to the point where $v_n'(t_0)$ meets γ. Then, letting $v_n^2(t_0)$ denote the segment of $v_n(t_0)$ between $s_1(t_0)$ and $p_2(t_0)$, by the triangle inequality

$$L\left(v_n^s(t_0)\right) \leq L\left(v_n'(t_0)\right) + L\left(v_n^2(t_0)\right)$$
$$\leq L\left(v_n^1(t_0)\right) \sqrt{1 - \epsilon_0^2} + L\left(v_n^2(t_0)\right)$$
$$= \beta\sqrt{1 - \epsilon_0^2} + L\left(v_n^2(t_0)\right).$$

Since $L\left(v_n^1(t_0)\right) = \beta$, we have

$$L\left(v_n^2(t_0)\right) \leq \kappa_1 - \beta + \frac{1}{n}.$$

We then choose n so large that

$$\beta\sqrt{1 - \epsilon_0^2} + \kappa_1 - \beta + \frac{1}{n} \leq \kappa_1 - \eta$$

for some $\eta > 0$. Hence

$$L\left(v_n^s(t_0)\right) \leq \kappa_1 - \eta.$$

We now continuously select points $s_1(t), s_2(t)$ on $v_n(t)$ for every $t \in I$ with

$$|p_i(t) - s_i(t)| = \beta, \quad \text{whenever} \quad L\left(v_n(t)\right) \geq \kappa_1 - \beta$$

and

$$p_i(t) = s_i(t), \quad \text{whenever} \quad L\left(v_n(t)\right) \leq \kappa_1 - 2\beta \quad i = 1, 2$$

and

$$|p_i(t) - s_i(t)| \leq \beta \quad \text{for all } t.$$

We then choose again the shortest lines from $s_i(t)$ to γ and replace $v_n(t)$ by the straight line $v_n^s(t)$ between those points, where these two shortest lines meet γ. By our geometric argument above

$$L(v_n^s(t)) \leq \kappa_1 - \eta \quad \text{for some } \eta > 0,$$

whenever

$$L(v_n(t)) \geq \kappa_1 - \epsilon_0.$$

Since also always

$$L(v_n^s(t)) \leq L(v_n(t)),$$

we may then construct a family $v_n^s \in V_1$ with

$$\sup_{t \in I} L(v_n^s(t)) \leq \kappa_1 - \min(\eta, \beta)$$

contradicting the definition of κ_1. Consequently, our claim is correct. We then find a sequence $(t_n)_{n \in \mathbb{N}} \subset I$ and $(v_n)_{n \in \mathbb{N}} \subset V_1$ with

$$\sup_{t \in I} L(v_n(t)) \leq \kappa_1 + \frac{1}{n}$$

$$L(v_n(t_n)) \geq \kappa_1 - \frac{1}{n}$$

$$|\cos(\alpha_1(v_n(t_n)))|, |\cos(\alpha_2(v_n(t_n)))| \leq \frac{1}{n}.$$

A subsequence of $(v_n(t_n))_{n \in \mathbb{N}}$ then converges to a straight line l_1 in A of length κ_1 meeting γ orthogonally at its endpoints.

In order to construct a second line l_2 meeting γ orthogonally at its endpoints, we proceed as follows:

We denote by V_2 the class of all continuous maps

$$v : I \times I \to \mathcal{L}$$

with

$$v(\{0\} \times I) \text{ and } v(\{1\} \times I) \subset \mathcal{L}_0 \tag{3.2.1}$$

and with the following property:

For all continuous maps

$$\tau : I \to I \times I$$

$$\tau(s) = (t_1(s), t_2(s))$$

3.2 The construction of Lyusternik-Schnirelman

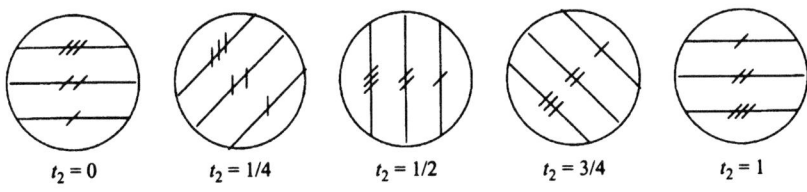

| $t_2 = 0$ | $t_2 = 1/4$ | $t_2 = 1/2$ | $t_2 = 3/4$ | $t_2 = 1$ |

Figure 3.6.

with
$$t_1(1) = 1 - t_1(0), t_2(0) = 0, t_2(1) = 1, \qquad (3.2.2)$$

we have
$$v \circ \tau \in V_1.$$

Let us exhibit an example of such a $v \in V_2$ (see Figure 3.6). We consider the $v_1 \in V_1$ of Figure 3.5 where $v_1(0)$ and $v_1(1)$ were point curves on γ, and we rotate v_1 via the parameter t_2 so that at $t_2 = 1$ we have the same picture as at $t_2 = 0$, but with t_1 interchanged with $1 - t_1$. Equation (3.2.2) then holds.

We note that $I \times I$ becomes a Möbius strip, when we identify the parameter t_1 on the line $t_2 = 1$ with the parameter $1 - t_1$ on the line $t_2 = 0$.

We define
$$\kappa_2 := \inf_{v \in V_2} \sup_{t \in I^2} L(v(t)).$$

Then
$$\kappa_2 \geq \kappa_1,$$

and κ_2 again is realized by some straight line l_2 in A meeting γ orthogonally at its endpoints. We consider two cases:

(1) $\kappa_2 > \kappa_1$.
    Then $L(l_2) = \kappa_2 > \kappa_1 = L(l_1)$, and l_2 hence is different from l_1
(2) $\kappa_2 = \kappa_1$.
    We claim that in this case, we even get infinitely many solutions of our problem, i.e. lines in A meeting γ orthogonally. Namely, we let $v_0 \in V_2$ be any critical family, i.e. satisfying
$$\sup_{t \in I^2} L(v_0(t)) = \kappa_2.$$

(It is not hard to see that in the present case such a $v_0 \in V_2$ indeed exists.)

We then have for any $\tau : I \to I^2$ with (3.2.2)

$$\sup_{s \in I} L(v_0(\tau(s))) \leq \kappa_2. \tag{3.2.3}$$

On the other hand, since $v_0 \circ \tau \in V_1$,

$$\kappa_1 \leq \sup_{s \in I} L(v_0(\tau(s))), \tag{3.2.4}$$

and since $\kappa_1 = \kappa_2$, we have equality in (3.2.3) and (3.2.4). This means that $v_0 \circ \tau$ is a critical family for κ_1, and it then has to contain a solution l_τ of our problem.

Let $S \subset \{(s,t) \in I \times I|\ L(v_0(s,t)) = \kappa_2)\}$ denote the set in $I \times I$ corresponding to all solutions induced by v_0. After carrying out the identification prescribed by (3.2.2), which makes $I \times I$ into a Möbius strip, we see that the complement of S in this Möbius strip is not path connected. Namely, otherwise we could find τ satisfying (3.2.2) for which $\tau(I)$ avoids S, and for such a τ, $v_0 \circ \tau$ would then not contain a solution, as S is the set of all solutions in the family v_0. This, however, contradicts what has just been said (see Figure 3.7). In fact, S has to carry a one dimensional cycle† on the Möbius strip. Otherwise, S would be contractible (in the Möbius strip) and one could reparameterize v_0 on I^2 so that the set of solutions corresponds to a finite number of points. But this is incompatible with $\kappa_2 = \kappa_1$ as we have just seen. Since for each path τ as in (3.2.2) with $\tau(I) \subset S$, $v_0 \circ \tau \in V_1$ is nonconstant by (3.2.1) and (3.2.2), we obtain an uncountable number of solutions.

We thus have shown our result if γ is of class C^3. If γ is only of class C^1, we choose a sequence of curves γ_n of class C^3 approximating γ. This means that there are parameterizations $\gamma_n(\tau)$, $\gamma(\tau)$ by arc-length with

$$\lim_{n \to \infty} \sup_\tau \left(|\gamma_n(\tau) - \gamma(\tau)| + \left| \frac{d}{d\tau}\gamma_n(\tau) - \frac{d}{d\tau}\gamma(\tau) \right| \right) = 0.$$

We then let $l_{1,n}$ and $l_{2,n}$ be the corresponding solutions for γ_n. After selection of subsequences, $l_{1,n}$ and $l_{2,n}$ then converge to solutions l_1, l_2 for γ, and those l_1 and l_2 realize the critical values κ_1 and κ_2, respectively. Since the argument to produce infinitely many solutions in case $\kappa_1 = \kappa_2$

---

† We have to employ here some constructions from algebraic topology. A reference is any good book on that subject, e.g. M.Greenberg, *Lectures on Algebraic Topology*, Benjamin, Reading, Mass., 1967, pp. 33–45, 186. While this is somewhat technical we strongly urge the reader to try to understand the essential geometric idea of the preceding construction.

3.2 The construction of Lyusternik–Schnirelman

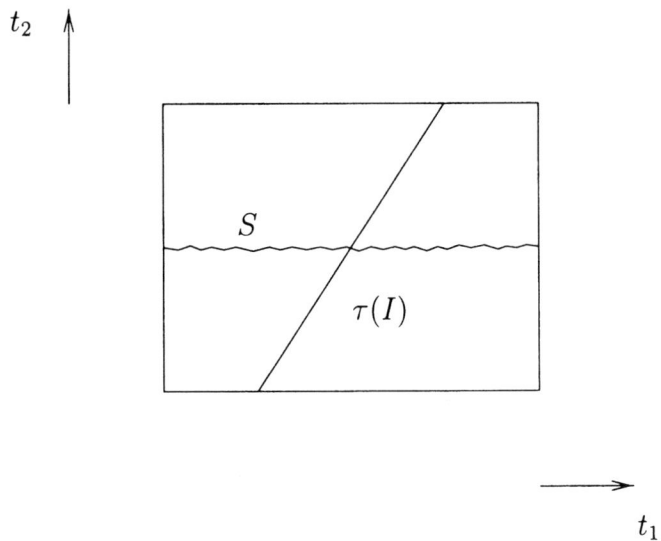

Figure 3.7.

did not depend on a higher differentiability assumption on γ, it is still applicable here, and we thus can complete the proof as before.

q.e.d.

The variational content of Theorem 3.2.1 is that we produce two geodesics in $\mathbb{R}^2$ that meet a given convex Jordan curve orthogonally. In fact, this statement generalizes to any closed convex Jordan curve on some surface, enclosing a domain homeomorphic to the unit disk.

In Sections 2.3, 3.2, we could only treat variational problems that could be reduced to finite dimensional problems, because we did not yet develop tools to show the existence of critical points of functionals defined on infinite dimensional spaces. We shall develop such tools in Part II, and consequently in Chaper 9 of Part II, we shall be able to present general results about the existence of unstable critical points in the spirit of the preceding results. The crucial notion will be the Palais–Smale condition that guarantees that the type of reasoning presented in Section 3.1 extends to certain functionals defined on infinite dimensional spaces. Also, the reasoning employed in Section 3.2 that infinitely many critical points can be found if two suitable critical values coincide will be given an axiomatic treatment in Section 9.3 of Part II.

Exercises

3.1 Let $F \in C^1(M, \mathbb{R})$ (M an embedded, connected, differentiable submanifold of $\mathbb{R}^d$) be bounded from below and proper (i.e. for all $s \in \mathbb{R}$, $\{x \in M : F(x) \leq s\}$ is compact), and suppose F has two relative minima x_1, x_2. Let

$$\kappa_0 := \max(F(x_1), F(x_2)).$$

Show that F either possesses a critical point x_3 with $F(x_3) > \kappa_0$, or that it has uncountably many critical points.

3.2 Let $F \in C^1(\mathbb{R}^d, \mathbb{R})$ be bounded from below and proper, and suppose it has three strict relative minima x_1, x_2, x_3. Try to identify conditions under which F then has to possess more than two additional critical points, e.g. three or four.

3.3 Let A be a compact convex subset of the unit sphere $S^2 \subset \mathbb{R}^3$, and suppose ∂A is a smooth curve γ; the convexity condition here means that for any two points in A, one can find precisely one geodesic arc inside A that connects them. Show the existence of at least two geodesic arcs in A that meet γ orthogonally at both endpoints.

4
The theory of Hamilton and Jacobi

4.1 The canonical equations

We let t be a real parameter varying between t_1 and t_2. We consider the variational integral

$$I = \int_{t_1}^{t_2} L\left(t, x^1(t), \ldots, x^n(t), \dot{x}^1(t), \ldots, \dot{x}^n(t)\right) dt \qquad (4.1.1)$$

for the unknown functions $x(t) = \left(x^1(t), \ldots, x^n(t)\right)$ with fixed endpoints $x(t_1)$ and $x(t_2)$. Here,

$$\dot{x}^i := \frac{dx^i}{dt}.$$

We assume that L is of class C^2. The Euler–Lagrange equations for I are

$$\frac{d}{dt} L_{\dot{x}^i} - L_{x^i} = 0 \qquad (i = 1, \ldots, n). \qquad (4.1.2)$$

We assume the invertibility condition

$$\det L_{\dot{x}^i \dot{x}^j} \neq 0. \qquad (4.1.3)$$

As shown in 1.2, this implies that solutions of (4.1.2) are of class C^2. (4.1.3) also implies that we may perform a Legendre transformation. Namely, by the implicit function theorem, we may then locally solve

$$p_i = L_{\dot{x}^i} \qquad (4.1.4)$$

w.r.t. $\dot{x}^i$, i.e.

$$\dot{x}^i = \dot{x}^i(t, x, p) \qquad (p = (p_1, \ldots, p_n)). \qquad (4.1.5)$$

The expressions p_i are called momenta. The Hamiltonian H is defined as
$$H(t,x,p) := \dot{x}^i p_i - L(t,x,\dot{x}). \tag{4.1.6}$$

We obtain
$$H_{x^i} = p_j \frac{\partial \dot{x}^j}{\partial x^i} - L_{\dot{x}^j} \frac{\partial \dot{x}^j}{\partial x^i} - L_{x^i},$$

and with (4.1.4) then
$$H_{x^i} = -L_{x^i}.$$

and with (4.1.2) and (4.1.4) then
$$H_{x^i} = -\dot{p}_i. \tag{4.1.7}$$

Also
$$H_{p_i} = p_j \frac{\partial \dot{x}^j}{\partial p_i} + \dot{x}^i - L_{\dot{x}^j} \frac{\partial \dot{x}^j}{\partial p_i},$$

and thus again with (4.1.4)
$$H_{p_i} = \dot{x}^i. \tag{4.1.8}$$

(4.1.7) and (4.1.8) constitute a so-called *canonical system*. We are going to see that (4.1.7) and (4.1.8) also arise as Euler–Lagrange equations of the variational problem obtained by expressing L in (4.1.1) through H via (4.1.6). Namely,
$$I = \int_{t_1}^{t_2} \left(\dot{x}^j p_j - H(t,x,p) \right) dt, \tag{4.1.9}$$

where the unknown functions are $x(t)$ and $p(t)$, has Euler–Lagrange equations (4.1.7) and (4.1.8), and so does
$$I = -\int_{t_1}^{t_2} \left(x^j \dot{p}_j + H(t,x,p) \right) dt. \tag{4.1.10}$$

Before proceeding, we observe that if H does not depend explicitely on t, i.e. $H = H(x,p)$, then H is a constant of motion, i.e. constant along any solution $x(t)$ of the equations, Namely,
$$\frac{d}{dt} H(x(t), p(t)) = H_{x^i} \dot{x}^i + H_{p_i} \dot{p}_i = 0 \tag{4.1.11}$$

by (4.1.7) and (4.1.8).

Example. For $L = \frac{1}{2}|\dot{x}| - V(x)$, we have

$$H = \frac{1}{2}|p|^2 + V(x),$$

and the canonical equations become

$$\dot{x} = p$$
$$\dot{p} = -V_x.$$

This example, which describes the Newtonian motion of a particle of unit mass subject to a potential V, is helpful for remembering the signs in the canonical equations.

4.2 The Hamilton–Jacobi equation

Assumption. There is given a set $\Omega \subset \mathbb{R}^{n+1} = \{(t, x^1, \ldots, x^n)\}$ with the property that for any points $A, B \in \Omega$, $A = (\sigma, \kappa^1, \ldots, \kappa^n)$, $B = (s, q^1, \ldots, q^n)$, there is a unique solution $x(t) = (x^1(t), \ldots, x^n(t))$ of (4.1.2) contained in Ω with $(\sigma, x(\sigma)) = A$, $(s, x(s)) = B$.

Thus, Ω is covered by solutions of (4.1.2), and those can be considered as functions of their endpoints. Thus

$$x^i = f^i(t; s, q^1, \ldots, q^n; \sigma, \kappa^1, \ldots, \kappa^n) \tag{4.2.1}$$

and also

$$p_i = g_i(t; s, q^1, \ldots, q^n; \sigma, \kappa^1, \ldots, \kappa^n) = L_{\dot{x}^i}. \tag{4.2.2}$$

In particular,

$$\kappa^i = f^i(\sigma; s, q^1, \ldots, q^n; \sigma, \kappa^1, \ldots, \kappa^n) \tag{4.2.3}$$

$$q^i = f^i(s; s, q^1, \ldots, q^n; \sigma, \kappa^1, \ldots, \kappa^n).$$

We also define

$$\varphi_i := g_i(\sigma; s, q^1, \ldots, q^n; \sigma, \kappa^1, \ldots, \kappa^n) = L_{\dot{\kappa}^i}(\sigma, \kappa, \dot{\kappa}) \tag{4.2.4}$$

$$v_i := g_i(s; s, q^1, \ldots, q^n; \sigma, \kappa^1, \ldots, \kappa^n) = L_{\dot{q}^i}(s, q, \dot{q}).$$

In the sequel, $\dot{f}^i$ etc. will mean a derivative w.r.t. the first independent variable, $f^{i'}$ etc. a derivative w.r.t. the second one. Inserting (4.2.1), (4.2.2) into I, we obtain

$$I = I(s, q, \sigma, \kappa) \tag{4.2.5}$$

and call this expression the *geodesic distance* betweeen A and B. In this connection, I is called *eiconal*. Recalling (4.1.9), we may write

$$I = \int_\sigma^s \left(p_i \dot{x}^i - H\left(t, x, p\right) \right) dt. \tag{4.2.6}$$

We want to compute the derivatives of $I(s, q, \sigma, \kappa)$.

$$I_s = v_i \dot{q}^i - H(s, q, v) + \int_\sigma^s \left(g_i' f^i + g_i \dot{f}^{i'} - H_{x_i} f^{i'} - H_{p_i} g_i' \right) dt.$$

Equations (4.1.7) and (4.1.8) yield $H_{x^i} = -\dot{g}_i$, $H_{p_i} = f^i$, and thus

$$I_s = v_i \dot{q}^i - H(s, q, v) + \int_\sigma^s \left(g_i f^{i'} \right)^{\cdot} dt$$
$$= v_i \dot{q}^i - H(s, q, v) + g_i f^{i'} \Big|_{t=\sigma}^{t=s}.$$

Equation (4.2.3) yields

$$\dot{f}^i + f^{i'} = 0 \quad \text{for } t = s$$
$$f^{i'} = 0 \quad \text{for } t = \sigma,$$

and thus altogether

$$I_s = v_i \dot{q}^i - H(s, q, v) - v_i \dot{q}^i$$
$$= -H(s, q, v)$$
$$= L(s, q, \dot{q}) - \dot{q}^i L_{\dot{q}^i}. \tag{4.2.7}$$

Next

$$I_{q^j} = \int_\sigma^s \left(\frac{\partial g_i}{\partial q^j} f^i + g_i \frac{\partial f^i}{\partial q^j} - H_{x^i} \frac{\partial f^i}{\partial q^j} - H_{p_i} \frac{\partial g_i}{\partial q^j} \right) dt$$
$$= g_i \frac{\partial f^i}{\partial q_j} \Big|_{t=\sigma}^{t=s} \quad \text{again by (4.1.7), (4.1.8)}$$
$$= g_j(s; s, q^1, \ldots, q^n; \sigma, \kappa^1, \ldots, \kappa^n) \quad \text{by (4.2.3)}$$
$$\text{and } \frac{\partial \kappa^i}{\partial q^j} = 0, \frac{\partial q^i}{\partial q^j} = \delta_{ij}.$$

Thus

$$I_{q^j} = v_j = L_{\dot{q}^j}(s, q, \dot{q}). \tag{4.2.8}$$

4.2 The Hamilton–Jacobi equation

Analogously,

$$I_\sigma = H(\sigma, \kappa, \varphi) = -L(\sigma, \kappa, \dot{\kappa}) + \dot{\kappa}^i L_{\dot{\kappa}^i}. \tag{4.2.9}$$

$$I_{\kappa^j} = -\varphi_j = -L_{\dot{\kappa}^j}(\sigma, \kappa, \dot{\kappa}). \tag{4.2.10}$$

Inserting (4.2.8) into (4.2.7), we obtain

$$I_s + H(s, q, I_q) = 0. \tag{4.2.11}$$

Thus, the geodesic distance as function of the endpoint satisfies (4.2.11), a *Hamilton–Jacobi equation*. In the present context that equation then is also called eiconal equation. We observed at the end of Section 4.1 that H is constant along solutions if it does not depend on t explicitly. In that case, (4.2.11) implies that I then depends linearly on s. It may be useful for understanding the preceding formulae if we derive them without the use of the Legendre transformation. Thus

$$I = \int_\sigma^s L(t, x(t), \dot{x}(t))\, dt = \int_\sigma^s L(t, f, \dot{f})\, dt$$

and

$$I_s = L(s, q, \dot{q}) + \int_\sigma^s \left(L_{x^i} f^{i'} + L_{\dot{x}^i} \dot{f}^{i'}\right) dt.$$

The Euler–Lagrange equations give

$$L_{x^i} = \frac{d}{dt} L_{\dot{x}^i},$$

and so

$$I_s = L(s, q, \dot{q}) + \int_\sigma^s \frac{d}{dt}\left(L_{\dot{x}^i} f^{i'}\right) dt$$
$$= L(s, q, \dot{q}) + L_{\dot{x}^i} f^{i'}\Big|_{t=\sigma}^{t=s}.$$

As before, we obtain from (4.2.3)

$$f^{i'} = -\dot{f}^i \quad \text{for } t = s$$
$$f^{i'} = 0 \quad \text{for } t = \sigma,$$

hence

$$I_s = L(s, q, \dot{q}) - L_{\dot{q}^i} \dot{q}^i,$$

i.e. (4.2.7). Likewise,

$$I_{q^j} = \int_\sigma^s \left(L_{x^i} \frac{\partial f^i}{\partial q^j} + L_{\dot{x}^i} \frac{\partial \dot{f}^i}{\partial q^j} \right) dt$$

$$= L_{\dot{x}^i} \frac{\partial f^i}{\partial q^j} \bigg|_{t=\sigma}^{t=s}$$

$$= L_{\dot{q}^j},$$

i.e. (4.2.8). Thus, the Hamilton–Jacobi equation (4.2.11) is

$$I_s - L(s, q, \dot{q}) + I_{q^i} \dot{q}^i = 0. \tag{4.2.12}$$

We have seen in the preceding how solutions of the canonical equations yield solutions of the Hamilton–Jacobi equations. We now want to establish a converse result.

Let $\varphi(t, x^1, \ldots, x^n)$ be a solution of the Hamilton–Jacobi equation which we now write as

$$p_0 + H(t, x^1, \ldots, x^n, p_1, \ldots, p_n) = 0 \tag{4.2.13}$$

with

$$p_0 = \varphi_t$$

$$p_i = \varphi_{x^i}.$$

Definition 4.2.1. *If*

$$\varphi = G(t, x^1, \ldots, x^n, \lambda_1, \ldots, \lambda_n) \quad \text{with } G \in C^2 \tag{4.2.14}$$

and

$$\det \left(G_{x^i \lambda_j} \right)_{i,j=1,\ldots,n} \neq 0 \tag{4.2.15}$$

is a family of solutions of (4.2.13) depending on n parameters $\lambda_1, \ldots, \lambda_n$, we call

$$\varphi = G(t, x^1, \ldots, x^n, \lambda_1, \ldots, \lambda_n) + \lambda \tag{4.2.16}$$

(where λ is a free real parameter) a complete integral of (4.2.13).

We have the following theorem of Jacobi:

Theorem 4.2.1. *Let $\varphi = G(t, x^1, \ldots, x^n, \lambda_1, \ldots, \lambda_n) + \lambda$ be a complete integral of (4.2.13). Then one may obtain a family of solutions of the*

4.2 The Hamilton–Jacobi equation

canonical equations

$$H_{p_i} = \dot{x}^i \tag{4.2.17}$$

$$H_{x^i} = -\dot{p}_i \tag{4.2.18}$$

depending on $2n$ parameters $\lambda_1, \ldots, \lambda_n, \mu^1, \ldots, \mu^n$ by solving

$$G_{\lambda_i} = \mu^i \tag{4.2.19}$$

$$G_{x^i} = p_i. \tag{4.2.20}$$

Proof. Because of (4.2.15), (4.2.19) may be solved w.r.t. x^i,

$$x^i = x^i(t, \lambda_1, \ldots, \lambda_n, \mu^1, \ldots, \mu^n).$$

Inserting this into (4.2.20) then yields

$$p_i = p_i(t, \lambda_1, \ldots, \lambda_n, \mu^1, \ldots, \mu^n).$$

We have to show that x^i and p_i satisfy the canonical equations. For this purpose, we differentiate (4.2.13) w.r.t. x^i and obtain:

$$G_{tx_i} + H_{p_k} G_{x^k x^i} + H_{x^i} = 0. \tag{4.2.21}$$

Differentiating (4.2.13) w.r.t. λ_i, we obtain

$$G_{t\lambda_i} + H_{p_k} G_{x^k \lambda_i} = 0, \tag{4.2.22}$$

since the terms containing $\frac{\partial x^k}{\partial \lambda_i}$ cancel by (4.2.21). Differentiating (4.2.19) w.r.t. t, we obtain

$$G_{\lambda_i t} + G_{\lambda_i x^k} \frac{\partial x^k}{\partial t} = 0. \tag{4.2.23}$$

Comparing (4.2.22) and (4.2.23) and recalling (4.2.15) yields (4.2.17). Differentiating (4.2.20) w.r.t. t, we obtain

$$\frac{\partial p_i}{\partial t} = G_{x^i t} + G_{x^i x^k} \frac{\partial x^k}{\partial t}. \tag{4.2.24}$$

Comparing (4.2.24) and (4.2.21) and using the relation (4.2.17) just derived, we then obtain (4.2.18).

q.e.d.

The canonical equations are a system of ODE whereas the Hamilton–Jacobi equation is a 1st order partial differential equation (PDE). The preceding considerations show the equivalence of these equations. While in general, one may consider a PDE as being more difficult than a system of ODE, in applications, one may often find a solution of the canonical equations by solving the Hamilton–Jacobi equation. Here, it is typically of great help that the Hamilton–Jacobi equation does not depend on the unknown function itself, but only on its derivatives.

Let us consider the following *example* of geometric optics:

$$I = \int_{t_1}^{t_2} \varphi(t,x)\sqrt{1+\dot{x}^2}\,dt \quad (\varphi(t,x) > 0),$$

already explained in Example (3) of Section 1.1 in a slightly different notation. The physical meaning is that $x(t)$ is considered as the graph of a light ray travelling in a medium with light velocity $\frac{c}{\varphi(t,x)}$, where c is the velocity of light in vacuum. In this example, putting

$$L(t,x,\dot{x}) = \varphi(t,x)\sqrt{1+\dot{x}^2}, \tag{4.2.25}$$

we have

$$p = L_{\dot{x}} = \frac{\varphi \dot{x}}{\sqrt{1+\dot{x}^2}}$$

$$H = p\dot{x} - L = -\sqrt{\varphi^2 - p^2}. \tag{4.2.26}$$

$I(s,q,\sigma,\kappa)$ here is the time that a light ray needs to travel from $A = (\sigma,\kappa)$ to $B = (s,q)$. The Hamilton–Jacobi equation $I_s + H(s,q,I_q) = 0$ becomes the eiconal equation

$$I_s^2 + I_q^2 = \varphi^2. \tag{4.2.27}$$

The surfaces $I(s,q) = $ constant are called wave fronts.

Another simple *example* comes from a quadratic

$$L(t,x,\dot{x}) = \frac{1}{2}(\dot{x}^2 + \alpha x^2) \quad (\alpha = \text{constant}). \tag{4.2.28}$$

Then

$$p = L_{\dot{x}} = \dot{x}, \; H = p \cdot \dot{x} - L = \frac{1}{2}(p^2 - \alpha x^2), \tag{4.2.29}$$

and the Hamilton–Jacobi equation becomes

$$I_t + \frac{1}{2}(I_x^2 - \alpha x^2) = 0. \tag{4.2.30}$$

If we substitute $I = \rho(t)x^2$, we are led to the Riccati equation

$$\dot{\rho} + 2\rho^2 - \frac{\alpha}{2} = 0. \tag{4.2.31}$$

If we substitute $I = -\lambda t + \psi(x)$ with a parameter λ, we obtain from (4.2.30)

$$-\lambda + \frac{1}{2}\left(\psi'(x)^2 - \alpha x^2\right) = 0,$$

i.e.

$$\psi'(x) = \sqrt{\alpha x^2 + 2\lambda}$$

and a solution

$$I = -\lambda t + \int_0^x \sqrt{\alpha \xi^2 + 2\lambda}\, d\xi. \tag{4.2.32}$$

The equation

$$I_\lambda(t, x, \lambda) = \mu$$

means

$$-t + \int_0^x \frac{d\xi}{\sqrt{\alpha \xi^2 + 2\lambda}} = \mu.$$

This can be solved for x; let us assume for example $\alpha < 0$; then the solution is

$$x = \sqrt{\frac{2\lambda}{-\alpha}} \sin\left(\sqrt{-\alpha}\,(t+\mu)\right).$$

x of course solves the Euler–Lagrange equation for (4.2.28)

$$\ddot{x} = \alpha x.$$

A physical realization is the *harmonic oscillator*, where $x(t)$ is the displacement of an oscillating spring, with $\alpha = -\frac{k}{m}$ (m = mass, k = spring constant). Since

$$p = I_x\,,\ I_t + H(x, I_x) = 0,$$

we obtain from (4.2.32)

$$\lambda = H(x, p),$$

i.e. λ is the energy of the spring.

4.3 Geodesics

We consider the case where L is homogeneous of degree 1, i.e.

$$L = L_{\dot{x}^i}\dot{x}^i. \tag{4.3.1}$$

Then

$$\det L_{\dot{x}^i \dot{x}^j} = 0, \tag{4.3.2}$$

and we cannot perform a Legendre transformation as in Section 4.1. We have
$$H = -L + \dot{x}^i L_{\dot{x}^i} = 0, \tag{4.3.3}$$
and the computations of Section 4.2 yield (writing $L_{\dot{x}^i}$ instead of p_i etc.)
$$I_s = L(s, q, \dot{q}) - \dot{q}^i L_{\dot{q}^i} = 0 \tag{4.3.4}$$
$$I_{q^j} = L_{\dot{q}^j}.$$

An example are the geodesic lines considered in Chapter 2. Here,
$$L = \sqrt{Q}$$
with
$$Q = g_{ij}(x^1, \ldots, x^n)\dot{x}^i \dot{x}^j. \tag{4.3.5}$$
The Euler–Lagrange equations are
$$\frac{d}{dt}\left(\frac{1}{\sqrt{Q}} Q_{\dot{x}^i}\right) - \frac{1}{\sqrt{Q}} Q_{x^i} = 0. \tag{4.3.6}$$
Since t does not occur explicitly in (4.3.5) and since I is invariant under transformations of t, we may choose t such that
$$Q \equiv 1, \tag{4.3.7}$$
i.e. that solutions are parameterized by arc-length. Equation (4.3.6) then becomes
$$\frac{d}{dt} Q_{\dot{x}^i} - Q_{x^i} = 0. \tag{4.3.8}$$
Conversely, along a solution of (4.3.8), we have $Q \equiv$ constant, justifying our choice of t. Namely, Q is homogeneous of degree 2 w.r.t. the variables $\dot{x}^i$, hence
$$Q_{\dot{x}^i} \dot{x}^i = 2Q. \tag{4.3.9}$$
Differentiating (4.3.9) w.r.t. t along a solution,
$$\left(\frac{d}{dt} Q_{\dot{x}^i}\right) \dot{x}^i + Q_{\dot{x}^i} \ddot{x}^i = 2\frac{d}{dt} Q = 2Q_{x^i} \dot{x}^i + 2Q_{\dot{x}^i} \ddot{x}^i,$$
and (4.3.8) indeed yields
$$\frac{d}{dt} Q = 0 \quad \text{along a solution.}$$

4.4 Fields of extremals

As already demonstrated in 2.1, (4.3.8) are the Euler–Lagrange equations for

$$E = \frac{1}{2}\int_\sigma^s Q(x(t), \dot{x}(t))\, dt = \frac{1}{2}\int_\sigma^s g_{ij}(x(t))\, \dot{x}^i(t)\dot{x}^j(t)\, dt. \qquad (4.3.10)$$

We recall (Lemma 2.1.1) that the Schwarz inequality implies

$$\int_\sigma^s \sqrt{Q}\, dt \le (s-\sigma)\left(\int_\sigma^s Q\, dt\right)^{\frac{1}{2}}$$

with equality precisely if $Q \equiv$ constant, and the extremals of E are precisely those extremals of I parameterized proportionally to arc-length. In contrast to I, E is no longer invariant under transformations of t. Therefore, for solutions of the Euler–Lagrange equations corresponding to E, the parameterization is determined up to a constant factor. The Hamiltonian for E is

$$H = Q_{\dot{x}^i}\dot{x}^i - Q = Q \quad \text{because of (4.3.9)}. \qquad (4.3.11)$$

Moreover,

$$p_i = Q_{\dot{x}^i} = 2g_{ij}\dot{x}^j. \qquad (4.3.12)$$

Thus

$$H = \frac{1}{4}g^{ij}p_i p_j \quad (\text{with } g^{ij} = (g_{ij})^{-1}). \qquad (4.3.13)$$

The Hamilton–Jacobi equation becomes

$$E_t + \frac{1}{4}g^{ij}E_{x^i}E_{x^j} = 0 \quad \text{cf. (4.3.13), (4.2.11), (4.3.10)} \qquad (4.3.14)$$

and the canonical equations are

$$\dot{x}^i = \frac{1}{2}g^{ij}p_j \quad \text{cf. (4.1.8), (4.3.13)} \qquad (4.3.15)$$

$$\dot{p}_i = -\frac{1}{4}\frac{\partial g^{kj}}{\partial x^i}p_k p_j \quad \text{cf. (4.1.7), (4.3.11), (4.3.5)}.$$

As observed at the end of Section 4.2, E depends linearly on t.

4.4 Fields of extremals

Let $\Omega \subset \mathbb{R}^{n+1}$ satisfy the assumptions of 4.2, $T \in C^1(\Omega, \mathbb{R})$. The equation

$$T(\sigma, \kappa^1, \ldots, \kappa^n) = 0 \qquad (4.4.1)$$

then defines a possibly degenerate hypersurface Σ (assume $\Sigma \neq \emptyset$). Given $B = (s, q^1, \ldots, q^n) \in \Omega$, we seek $A = (\sigma, \kappa^1, \ldots, \kappa^n) \in \Sigma$ that minimizes

$$I(s, q^1, \ldots, q^n, \sigma, \kappa^1, \ldots, \kappa^n)$$

as a function of $(\sigma, \kappa^1, \ldots, \kappa^n)$ satisfying (4.4.1). At such a minimizing A, we have with some Lagrange multiplier λ

$$I_\sigma + \lambda T_\sigma = 0 \qquad (4.4.2)$$
$$I_{\kappa^j} + \lambda T_{\kappa^j} = 0 \quad (j = 1, \ldots, n).$$

Unless the situation is degenerate ($\lambda = 0$ or $T_\sigma = T_{\kappa^i} = 0$ for all i), this means that the vector $(I_\sigma, I_{\kappa^1}, \ldots, I_{\kappa^n})$ is proportional to the gradient of T, hence orthogonal to Σ. From (4.2.9), (4.2.10), we then obtain

$$-H(\sigma, \kappa, \varphi) = L(\sigma, \kappa, \dot{\kappa}) - \dot{\kappa}_i L_{\dot{\kappa}_i} = \lambda T_\sigma \qquad (4.4.3)$$

$$\varphi_j = L_{\dot{\kappa}^j} = \lambda T_{\kappa^j}.$$

These are equations for the tangent vector $(\dot{\kappa}^1, \ldots, \dot{\kappa}^n)$ of the solution from A to B. A solution satisfying (4.4.3) is called orthogonal to Σ. We want to use the following:

Assumption. *Through each point of Ω, there is precisely one solution orthogonal to Σ.*

For each $B = (s, q^1, \ldots, q^n)$, we thus find a unique $A = (\sigma(s, q), \kappa(s, q)) \in \Sigma$ minimizing $I(s, q, \sigma, \kappa)$. We call

$$J(s, q) := I(s, q, \sigma(s, q), \kappa(s, q))$$

the *geodesic distance* from the hypersurface Σ.

Theorem 4.4.1. *Given such a field of solutions orthogonal to Σ, the geodesic distance satisfies*

$$J_s = -H(s, q, L_{\dot{q}}) \qquad (4.4.4)$$

and

$$J_{q^j} = L_{\dot{q}^j}, \qquad (4.4.5)$$

hence also the eiconal equation

$$J_s + H(s, q, J_q) = 0. \qquad (4.4.6)$$

Proof.

$$J_s = I_s + I_\sigma \sigma_s + I_{\kappa^i} \kappa_s^i \qquad (4.4.7)$$

$$J_{q^j} = I_{q^j} + I_\sigma \sigma_{q^j} + I_{\kappa^i} \frac{\partial \kappa^i}{\partial q^j}$$

$T(\sigma(s,q), \kappa(s,q)) = 0$ implies

$$\sigma_s T_\sigma + \kappa_s^i T_{\kappa^i} = 0$$

and likewise

$$\sigma_{q^j} T_\sigma + \frac{\partial \kappa^i}{\partial q^j} T_{\kappa^i} = 0.$$

If we then use (4.4.2), we obtain in (4.4.7)

$$J_s = I_s$$

$$J_{q^j} = I_{q^j},$$

and the result follows from (4.2.7), (4.2.8), (4.2.11).

q.e.d.

Conversely

Theorem 4.4.2. *If $J(s,q)$ is a solution of (4.4.6) of class C^2, there exists a field of solutions orthogonal to the hypersurfaces $J(s,q) = $ constant, and J is the geodesic distance from the hypersurface $J = 0$.*

Proof. Let J satisfy (4.4.6). We put

$$p_i := J_{q^i}(s,q). \qquad (4.4.8)$$

The following system of ODE

$$\dot{q}^i = H_{p_i}(s, q^j, J_{q^j}) \qquad (4.4.9)$$

then defines an n-parameter family of curves. By (4.4.8), we have along any such curve

$$\dot{p}_i = J_{q^i s} + J_{q^i q^j} \dot{q}^j,$$

and (4.4.6) gives

$$J_{sq^i} + H_{q^i} + H_{p_j} J_{q^j q^i} = 0.$$

Recalling (4.4.9), we obtain

$$\dot{p}_i = -H_{q^i}. \qquad (4.4.10)$$

Equations (4.4.9) and (4.4.10) state that the curves $q(s)$ constitute a field of solutions. (4.4.6) and (4.4.8) yield

$$-H = J_s$$
$$p_j = J_{q^j}.$$

This means that (4.4.3) is satisfied for $T = J$ with $\lambda = 1$, and the solutions are orthogonal to the hypersurfaces $J = $ constant.

q.e.d.

Theorem 4.4.1 gives solutions of the Hamilton–Jacobi equation (4.4.6) depending on an arbitrarily given function $T \in C^1(\mathbb{R}^{n+1})$ (namely, we obtain those solutions that start on $T = 0$), whereas Theorem 4.4.2 implies that all solutions are obtained in that way. The surfaces $J = $ constant are called parallel surfaces of the field. In the special case where the hypersurface $T = 0$ degenerates into a point, we recover the considerations of Section 4.2.

4.5 Hilbert's invariant integral and Jacobi's theorem

For a solution $J(t, x^1, \ldots, x^n)$ of the Hamilton–Jacobi equation, we put again

$$p_i := J_{x^i}(t, x^1, \ldots, x^n).$$

If $A = (\sigma, \kappa^1, \ldots, \kappa^n)$ and $B = (s, q^1, \ldots, q^n)$ are connected by an arbitrary differentiable path $x^i(\tau)$, the integral

$$J(B) - J(A) = \int_\sigma^s \frac{d}{dt} J(\tau, x(\tau)) \, d\tau$$
$$= \int_\sigma^s \left(J_{x^i} \frac{dx^i}{d\tau} + J_\tau \right) d\tau$$

does not depend on this particular path, but only on the end points A and B. We rewrite this integral as

$$\int_\sigma^s \left(p_i \frac{dx^i}{d\tau} - H(\tau, x(\tau), p(\tau)) \right) d\tau \qquad (4.5.1)$$

and call it *Hilbert's invariant integral*. Conversely now let functions $p_i(\tau, x^1, \ldots, x^n)$ be given in a region $\Omega \subset \mathbb{R}^{n+1}$ for which the integral (4.5.1) does not depend on the path $x(\tau)$ connecting $A = (\sigma, x(\sigma))$ and

4.5 Hilbert's invariant integral and Jacobi's theorem

$B = (s, x(s))$. Thus, we may define $J : \Omega \to \mathbb{R}$ by

$$J(B) - J(A) = \int_\sigma^s \left(p_i \frac{dx^i}{d\tau} - H(\tau, x(\tau), p(\tau)) \right) d\tau. \tag{4.5.2}$$

Since this integral does not depend on the path connecting A and B, we must have

$$J_{x^i} = p_i \tag{4.5.3}$$

$$J_t = -H(t, x, p).$$

J then solves the Hamilton–Jacobi equation. By Theorem 4.4.2, any solution of the Hamilton–Jacobi equation is the geodesic distance function for a field of solutions of the canonical equations. Thus, any invariant integral of the form (4.5.1) yields a field of solutions.

Let us now reconsider Jacobi's Theorem 4.2.1. Let

$$\varphi = G(t, x^1, \ldots, x^n, \lambda_1, \ldots, \lambda_n) + \lambda \tag{4.5.4}$$

be a complete integral of

$$p + H(t, x^1, \ldots, x^n, p_1, \ldots, p_n) = 0 \tag{4.5.5}$$

(with $p = \varphi_t$, $p_i = \varphi_{x^i}$); in particular

$$\det \left(G_{x^i \lambda_j} \right) \neq 0. \tag{4.5.6}$$

Jacobi's theorem says that we obtain a $2n$-parameter family of solutions of the canonical equations by solving

$$G_{\lambda_i} = \mu^i$$

$$G_{x^i} = p_i,$$

where the parameters are $\lambda_1, \ldots, \lambda_n, \mu^1, \ldots, \mu^n$. For fixed values of $\lambda_1, \ldots, \lambda_n, \lambda$, G determines a field of solutions of the canonical equations, and by the preceding consideration, it is given by the corresponding invariant integral

$$G(B) - G(A) = \int_\sigma^s \left(G_{x^i} \frac{dx^i}{d\tau} - H \right) d\tau \tag{4.5.7}$$

$$= \int_\sigma^s \left\{ L(\tau, x^i(\tau), \dot{x}^i(\tau)) + \left(\frac{dx^i}{d\tau} - \dot{x}^i(\tau) \right) L_{\dot{x}^i} \right\} d\tau,$$

where $\dot{x}^i(\tau)$ now denotes the derivative in the direction of the solution and not in the direction of the arbitrary curve $x^i(\tau)$ connecting A and B.

We now vary $\lambda_1, \ldots, \lambda_n$, but keep the curve $x^i(\tau)$ fixed. Then the field of solutions varies, and so then does $\dot{x}^i(\tau)$. We also determine λ so that $G(A) = 0$. Differentiating (4.5.7) then yields

$$G_{\lambda_i} = \int_\sigma^s \left(\left(\frac{dx^j}{d\tau} - \dot{x}^j\right) L_{\dot{x}^j \lambda_i}\right) d\tau. \tag{4.5.8}$$

In the same way as $G(B)$, this expression only depends on B (A is kept fixed for the moment) but not on the particular $x^j(\tau)$. For each B, we find B_0 on the surface

$$G(t, x^1, \ldots, x^n, \lambda_1, \ldots, \lambda_n) = 0$$

that can be connected with B by a solution of the canonical equations. Along such a solution, we have

$$\frac{dx^j}{d\tau} = \dot{x}^j,$$

and the integrand in (4.5.8) thus vanishes along this curve. Instead of integrating from A to B, it therefore suffices to integrate from A to B_0, and we obtain

$$G_{\lambda_i} = \mu^i, \tag{4.5.9}$$

with μ^i being the value of the integral from A to B_0. Thus, μ^i can be considered as a constant for the solution passing through B_0.

If, conversely, (4.5.9) defined a family of curves $x^i(t, \lambda_j, \mu^j)$ (the family is locally unique because of (4.5.6)), then, since G_{λ_j} is constant, the integrand in (4.5.8) has to vanish along any curve of the family. Thus

$$\left(\frac{dx^j}{d\tau} - \dot{x}^j\right) L_{\dot{x}^j \lambda_i} = 0 \quad (i = 1, \ldots, n). \tag{4.5.10}$$

In our field we have (cf. (4.2.8))

$$L_{\dot{x}^j} = G_{x^j},$$

hence by assumption (4.5.6)

$$\det L_{\dot{x}^j \lambda_i} = \det G_{x^j \lambda_i} \neq 0.$$

Equation (4.5.10) then implies

$$\frac{dx^j}{d\tau} = \dot{x}^j;$$

this means that the curves defined by (4.5.9) are solutions of the canonical equations contained in the field defined by $G(t, x^1, \ldots, x^n, \lambda_1, \ldots, \lambda_n)$. We also observe that the parameter λ is only used for specifying the surface $G = 0$ and has no geometric meaning beside that.

4.6 Canonical transformations

We want to find transformations, i.e. diffeomorphisms†

$$\psi : \mathbb{R}^{2n} \to \mathbb{R}^{2n}$$
$$(x,p) \mapsto (\xi, \pi),$$

that preserve the canonical equations

$$\begin{aligned} \dot{x} &= H_p \\ \dot{p} &= -H_x. \end{aligned} \qquad (4.6.1)$$

This means that $\xi = \xi(x,p)$, $\pi = \pi(x,p)$ satisfy

$$\begin{aligned} \dot{\xi} &= H^*_\pi \\ \dot{\pi} &= -H^*_\xi. \end{aligned} \qquad (4.6.2)$$

with $H^*(t, \xi(x,p), \pi(x,p)) = H(t, x, p)$.

Equation (4.6.1) constitutes a system of ODE and if the assumptions of the Picard–Lindelöf theorem are satisfied, a solution exists for given initial values $x(t_0) = x_0$, $p(t_0) = p_0$ on some interval $[t_0, t_1]$. For any $\bar{t} \in [t_0, t_1]$, we then obtain such a transformation by letting $\xi(x,p) = x(\bar{t})$, $\pi(x,p) = p(\bar{t})$ where $(x(t), p(t))$ is the solution of (4.6.1) with $x(t_0) = x, p(t_0) = p$. Thus, the evolution of (4.6.1) in time t, the so-called Hamiltonian flow, yields 'canonical transformations'. However, the concept of canonical transformations is more general as we now shall see.

Since

$$\begin{aligned} \dot{\xi}^j &= \frac{\partial \xi^j}{\partial x^i} \dot{x}^i + \frac{\partial \xi^j}{\partial p_i} \dot{p}_i \\ &= \frac{\partial \xi^j}{\partial x^i} H_{p_i} - \frac{\partial \xi^j}{\partial p_i} H_{x^i} \\ \dot{\pi}_j &= \frac{\partial \pi_j}{\partial x^i} \dot{x}^i + \frac{\partial \pi_j}{\partial p_i} \dot{p}_i \\ &= \frac{\partial \pi_j}{\partial x^i} H_{p_i} - \frac{\partial \pi_j}{\partial p_i} H_{x^i} \quad \text{by (4.6.1)} \end{aligned}$$

---

† A diffeomorphism is a bijective map that together with its inverse is everywhere differentiable.

and

$$H^*_{\pi_j} = H_{x^i}\frac{\partial x^i}{\partial \pi_j} + H_{p_i}\frac{\partial p_i}{\partial \pi_j},$$

$$H^*_{\xi^j} = H_{x^i}\frac{\partial x^i}{\partial \xi^j} + H_{p_i}\frac{\partial p_i}{\partial \xi^j},$$

we obtain the conditions

$$\frac{\partial p_i}{\partial \pi_j} = \frac{\partial \xi^j}{\partial x^i},$$

$$\frac{\partial x^i}{\partial \pi^j} = -\frac{\partial \xi^j}{\partial p_i},$$

$$\frac{\partial p_i}{\partial \xi^j} = -\frac{\partial \pi_j}{\partial x^i},$$

$$\frac{\partial x^i}{\partial \xi^j} = \frac{\partial \pi_j}{\partial p_i}, \qquad (4.6.3)$$

or in matrix notation

$$\begin{bmatrix} \frac{\partial \xi}{\partial x} & \frac{\partial \xi}{\partial p} \\ \frac{\partial \pi}{\partial x} & \frac{\partial \pi}{\partial p} \end{bmatrix}^{-1} = \begin{bmatrix} \left(\frac{\partial \pi}{\partial p}\right)^{\mathrm{T}} & -\left(\frac{\partial \xi}{\partial p}\right)^{\mathrm{T}} \\ -\left(\frac{\partial \pi}{\partial x}\right)^{\mathrm{T}} & \left(\frac{\partial \xi}{\partial x}\right)^{\mathrm{T}} \end{bmatrix}, \qquad (4.6.4)$$

where A^{T} denotes the transpose of a matrix A. Obviously, this is a condition that does not depend anymore on the particular Hamiltonian H.

Definition 4.6.1. *A diffeomorphism* $\psi : \mathbb{R}^{2n} \to \mathbb{R}^{2n}$, $(x,p) \mapsto (\xi,\pi)$, *satisfying (4.6.3) (or, equivalently (4.6.4)) is called* canonical transformation.

Canonical transformations can often be used to simplify the canonical equations. Before we return to that topic, however, we interrupt the discussion of the Hamilton–Jacobi theory in order to describe some basic points of *symplectic geometry* (for more information on that subject, we refer to D.Mc Duff, D.Salamon, *Introduction to Symplectic Topology*, Oxford University Press, Oxford, 1995). We denote the $(n \times n)$ unit matrix by $\mathbb{I}_n$ and put

$$J := \begin{pmatrix} 0 & -\mathbb{I}_n \\ \mathbb{I}_n & 0 \end{pmatrix}.$$

Then obviously

$$J^2 = -\mathbb{I}_{2n}. \qquad (4.6.5)$$

4.6 Canonical transformations

Equation (4.6.4) may then be written as

$$(D\psi)^{-1} = -J(D\psi)^T J, \tag{4.6.6}$$

or equivalently

$$(D\psi)^T J D\psi = J. \tag{4.6.7}$$

In this connection, a ψ satisfying (4.6.7), i.e. a canonical transformation, is also called *symplectomorphism*. From these relations, one also easily sees that ψ is a canonical transformation iff ψ^{-1} is.

In terms of J, the canonical equations (4.6.1) can also be written as

$$\dot{z} = -J \nabla^0 H(t, z) \tag{4.6.8}$$

where $z = (x, p)$, $\nabla^0 H(t, z) = (H_x, H_p)$.

For a reader who knows the calculus of exterior differential forms, the following explanation should be useful. We consider the two-form

$$\omega = dx^i \wedge dp_i \quad \text{on } \mathbb{R}^{2n}$$

(here, as always, we use a summation convention: $dx^i \wedge dp_i$ means $\sum_{i=1}^n dx^i \wedge dp_i$). According to the transformation rules for exterior differential forms (i.e. $d\xi^j = \frac{\partial \xi^j}{\partial x^i} dx^i$ etc.), we have, for $\xi = \xi(x,p)$, $\pi = \pi(x,p)$,

$$d\xi^j \wedge d\pi_j = \left(\frac{\partial \xi^j}{\partial x^i} \frac{\partial \pi_j}{\partial p_k} - \frac{\partial \xi^j}{\partial p_k} \frac{\partial \pi_j}{\partial x^i} \right) dx^i \wedge dp_k.$$

Thus, ω remains invariant under the transformation ψ, i.e.

$$d\xi^j \wedge d\pi_j = dx^i \wedge dp_i \tag{4.6.9}$$

precisely if ψ is a canonical transformation. In fact, this is often used as the definition of a canonical transformation. If ω is left invariant under ψ, so is

$$\omega^n := \underbrace{\omega \wedge \cdots \wedge \omega}_{n \text{ times}} = n!(-1)^{\frac{n(n-1)}{2}} dx^1 \wedge \cdots \wedge dx^n \wedge dp_1 \wedge \cdots \wedge dp_n. \tag{4.6.10}$$

Since

$$d\xi^1 \wedge \cdots \wedge d\xi^n \wedge d\pi_1 \wedge \cdots \wedge d\pi_n = (\det D\psi) dx^1 \wedge \cdots \wedge dx^n \wedge dp_1 \wedge \cdots \wedge dp_n,$$

we conclude Liouville's:

Theorem 4.6.1. *Every canonical transformation $\psi : \mathbb{R}^{2n} \to \mathbb{R}^{2n}$ satisfies*

$$\det D\psi \equiv 1. \tag{4.6.11}$$

q.e.d.

One also expresses this result by saying that a canonical transformation is volume preserving in phase space as $dx^1 \wedge \cdots \wedge dx^n \wedge dp_1 \wedge \cdots \wedge dp_n$ can be interpreted as the volume form of $\mathbb{R}^{2n}$. By what was observed in the beginning of this section, this applies in particular to the Hamiltonian flow which constitutes Liouville's original statement.

After this excursion and interruption, we return to our canonical equations (4.6.1) and try to simplify them by suitable canonical transformations. Canonical transformations may be easily obtained from the variational integral

$$I = \int_{t_1}^{t_2} L(t, x, \dot{x}) dt$$

with

$$L(t, x, \dot{x}) = \dot{x} \cdot p - H(t, x, p) \quad (p = L_{\dot{x}}).$$

If W is any differentiable function, then

$$I^* = \int_{t_1}^{t_2} \left(L(t, x, \dot{x}) + \frac{dW}{dt} \right) dt$$

has the same critical points as I, because

$$I^* = I + W(t_2) - W(t_1),$$

so that I^* and I differ only by a constant independent of the particular path $x(t)$. Thus, we may for example take any function $W(t, x, \xi)$ and require that for all choices of $x, \xi, \dot{x}, \dot{\xi}$

$$\dot{x} \cdot p - H(t, x, p) = \dot{\xi} \cdot \pi - H^*(t, \xi, \pi) + \frac{dW}{dt}. \tag{4.6.12}$$

Then, with

$$\Lambda(t, \xi, \dot{\xi}) = \dot{\xi} \cdot \pi - H^*(t, \xi, \pi) \quad (\pi = \Lambda_{\dot{\xi}}),$$

$$I^* = \int_{t_1}^{t_2} \Lambda(t, \xi, \dot{\xi}) dt,$$

differs from I only by a constant. Thus, if $x(t)$ is a critical path for I,

4.6 Canonical transformations

$\xi(x(t), p(t))$ then becomes a critical path for I^*. Since

$$\frac{dW}{dt} = W_t + W_x \cdot \dot{x} + W_\xi \cdot \dot{\xi},$$

(4.6.12) becomes

$$\dot{x} \cdot (p - W_x) - \dot{\xi} \cdot (\pi + W_\xi) - H + H^* - W_t = 0 \qquad (4.6.13)$$

Since (4.6.13) is required to hold for all choices of x, ξ, $\dot{x}$, $\dot{\xi}$, we obtain:

Theorem 4.6.2. *Given an arbitrary (differentiable) function $W(t, x, \xi)$, a canonical transformation (transforming (4.6.1) into (4.6.2)) is obtained through the equations*

$$\begin{aligned} p &= W_x \\ \pi &= -W_\xi \\ H^* &= H \ , \ i.e. \ W_t = 0. \end{aligned} \qquad (4.6.14)$$

$W_t = 0$ *of course means that* $W = W(x, \xi)$.

In the same manner, we may also take a function $W(t, p, \xi)$, $W(t, x, \pi)$ or $W(t, p, \pi)$. In the first case, we obtain for example the equations

$$\begin{aligned} x &= W_p \\ \pi &= -W_\xi \\ H^* &= H \ , \ i.e. \ W_t = 0. \end{aligned}$$

Here and above, of course $H^* = H^*(t, \xi, \pi)$.

We may now easily explain Jacobi's method for solving the canonical equations. We try to find $W(x, \xi)$ satisfying

$$H(t, x, W_x(x, \xi)) = H^*(\xi), \qquad (4.6.15)$$

i.e. reduce the Hamiltonian to a function of the variable ξ alone. We have to require that

$$\det W_{x^i \xi^j} \neq 0. \qquad (4.6.16)$$

This ensures that the equation $\pi = -W_\xi$ determines x, and p then is determined from $p = W_x$. If (4.6.15) holds, (4.6.2) becomes

$$\begin{aligned} \dot{\xi} &= 0 \\ \dot{\pi} &= -H^*_\xi. \end{aligned} \qquad (4.6.17)$$

This implies that $\xi^1, \ldots, \xi^n$ are constants of motion (i.e. independent of t), or so-called integrals of the Hamiltonian flow. A system for which

n independent integrals can be found is called *completely integrable*. Thus, if we can find a so-called generating function $W(x, \xi)$ of the above type reducing the Hamiltonian to a function of ξ alone, the canonical system is completely integrable. Clearly, since in this case $\xi^1, \ldots, \xi^n$ are constant in t, the relation $\dot\pi = -H^*_\xi(\xi)$ can then be used to determine $\pi_1, \ldots, \pi_n$. In other words, a completely integrable canonical system may be solved explicitly through quadratures. Actually, one may show in this case that the sets $T_c = \{\xi^1 = c^1, \ldots, \xi^n = c^n\}$ for a constant vector $c = (c^1, \ldots, c^n)$ are n-dimensional tori, if compact and connected. Thus, the so-called *phase space* $\{(x, p) \in \mathbb{R}^{2n}\}$ is foliated by tori that are invariant under the motion, and on each such torus, the motion is given by straight lines.

It should be pointed out, however, that completely integrable dynamical systems are quite rare, in the sense that the complete integrability usually depends on particular symmetries, and their dynamical behaviour is quite exceptional in the class of all Hamiltonian systems. The invariant tori may disappear under arbitrarily small perturbations. By way of contrast, the Kolmogorov–Arnold–Moser theory asserts that these invariant tori persist under sufficiently small and smooth perturbations if the coordinates of H^*_ξ are rationally independent and satisfy certain Diophantine inequalities, and if the matrix $H^*_{\xi\xi}$ of second derivatives is invertible.

In the older literature, the notion of 'canonical transformation' is usually applied to any transformation $\psi : \mathbb{R}^{2n} \to \mathbb{R}^{2n}$ that preserves the form of the canonical equations, i.e. (4.6.1) is transformed into (4.6.2), but without requiring that

$$H^*(t, \xi, \pi) = H(t, x, p).$$

An example of a canonical transformation in this wider sense is

$$\xi = 2x \,,\, \pi = p$$

with $H^* = 2H$.

If we now take a generating function $W(t, x, \xi)$ as above, the Hamiltonian is transformed into

$$H^* = H + W_t \qquad (4.6.18)$$

while the first two relations of (4.6.14), i.e.

$$p = W_x \,,\, \pi = -W_\xi \qquad (4.6.19)$$

4.6 Canonical transformations

still hold. This may be used to explain Jacobi's theorem once more, as we now shall see.

Let $I(t, x^1, \ldots, x^n, \lambda_1, \ldots, \lambda_n)$ be a solution of the Hamilton–Jacobi equation

$$I_t + H(t, x, I_x) = 0, \qquad (4.6.20)$$

depending on parameters $\lambda_1, \ldots, \lambda_n$ and satisfying as usually

$$\det I_{x^i \lambda_j} \neq 0. \qquad (4.6.21)$$

We now choose

$$W(t, x^1, \ldots, x^n, \xi_1, \ldots, \xi_n) = I(t, x^1, \ldots, x^n, \xi_1, \ldots, \xi_n).$$

The corresponding transformation then is

$$\begin{aligned} p &= I_x \\ \pi &= -I_\xi \\ H^*(t, \xi, \pi) &= H(t, x, p) + I_t. \end{aligned} \qquad (4.6.22)$$

Because of (4.6.20),

$$H^* \equiv 0.$$

Thus, the new canonical equations are just

$$\dot\xi = 0$$

$$\dot\pi = 0.$$

Solutions are of course

$$\xi = \lambda = \text{constant}$$

$$\pi = -I_\lambda = -\mu = \text{constant}.$$

We have thus obtained the statement of Jacobi's Theorem 4.2.1, namely that from a solution of (4.6.20) with (4.6.21), we may obtain solutions of the canonical equations by solving

$$I_\lambda = \mu$$

$$I_x = p$$

with parameters $\lambda = (\lambda_1, \ldots, \lambda_n)$, $\mu = (\mu^1, \ldots, \mu^n)$.

The theory of Hamilton and Jacobi

Classical references for this chapter include:

C.G.J. Jacobi, *Vorlesungen über Analytische Mechanik* (ed. H. Pulte), Vieweg, Braunschweig, Wiesbaden 1996,

C. Caratheodory, *Variationsrechnung und partielle Differentialgleichungen erster Ordnung*, Teubner, Leipzig 1935,

R. Courant, D. Hilbert, *Methoden der Mathematischen Physik II*, Springer, Berlin, 2nd edition, 1968.

The global aspects are developed in

V.I. Arnold, *Mathematical Methods of Classical Mechanics*, GTM60, Springer, New York, 1978.

A recent advanced monograph is

H. Hofer, E. Zehnder, *Symplectic Invariants and Hamiltonian Dynamics*, Birkhäuser, Basel, 1994.

That text will give readers a good perspective on the present research directions in the field.

Exercises

4.1 Discuss the relation between the canonical equations for the energy functional E and the equations for geodesics derived in Chapter 2.

4.2 (Kepler problem) Consider the Lagrangian

$$L(x, \dot{x}) = \frac{1}{2} |\dot{x}|^2 + \frac{1}{|x|} \quad \text{for } x \in \mathbb{R}^3.$$

Compute the corresponding Hamiltonian and write down the canonical equations. Show that the three components of the angular momentum $x \wedge \dot{x}$ are integrals of the Hamiltonian flow.

4.3 For smooth functions $F, G : \mathbb{R}^{2n} \to \mathbb{R}$, define their Poisson bracket as

$$\{G, F\} := \frac{\partial F}{\partial x^j} \frac{\partial G}{\partial p_j} - \frac{\partial F}{\partial p_j} \frac{\partial G}{\partial x^j},$$

where $z = (x, p) = (x^1, \ldots, x^n, p_1, \ldots, p_n)$ are Euclidean coordinates of $\mathbb{R}^{2n}$. Let $z(t) = (x(t), p(t))$ be a solution of a canonical system

$$\dot{x} = H_p$$
$$\dot{p} = -H_x$$

for some Hamiltonian $H(x,p)$ that is independent of t. Show that for any (smooth) $F : \mathbb{R}^{2n} \to \mathbb{R}$

$$\frac{d}{dt}F(z(t)) = \{F, H\}.$$

Show that the Poisson bracket is antisymmetric, i.e.

$$\{F, G\} = -\{G, F\}$$

and satisfies the Jacobi identity

$$\{\{F, G\}, L\} + \{\{G, L\}, F\} + \{\{L, F\}, G\} = 0$$

for all smooth F, G, L.

4.4   Show that a diffeomorphism $\psi : \mathbb{R}^{2n} \to \mathbb{R}^{2n}$ is a canonical transformation if

$$\{F, G\} \circ \psi = \{F \circ \psi, G \circ \psi\}$$

for all smooth F, G.

5
Dynamic optimization

Optimal control theory is concerned with time dependent processes that can be influenced or controlled via the tuning of certain parameters. The aim is to choose these parameters in such a manner that a desired result is achieved and the cost resulting from the intermediate states of the process and from the application or change of the parameters is minimized. In some problems, the control parameter can be applied only at discrete time steps, while other problems can be continuously controlled. As we shall see, however, the discrete and the continuous case can be treated by the same principles. Since the end result may be prescribed, and the value of a parameter at some given time influences the state of the system at subsequent times and therefore typically will also contribute through this influence to the cost of the process at those later times, the determination of the optimal control parameters is best performed in a backward manner. This means in the discrete case that one first selects the best value of the control parameter at the last stage, whatever state the system is in at that time, then the value at the second-to-last stage, so that at this step the contribution of the value of the control parameter at the last stage to the total cost function is already determined and one only needs to optimize the cost function w.r.t. the second-to-last parameter value, and so on.

5.1 Discrete control problems

We consider a process with n states $x_1, \ldots, x_n \in \mathbb{R}^d$. At each state x_i, we may choose a control parameter

$$\lambda_i \in \Lambda_i, \tag{5.1.1}$$

5.1 Discrete control problems

where Λ_i is a given control restriction ($\Lambda_i \subset \mathbb{R}^c$) to determine

$$x_{i+1} = \varphi_i(x_i, \lambda_i) \tag{5.1.2}$$

with cost

$$k_i(x_i, \lambda_i).$$

The total cost of the process starting at the initial state x_ν is

$$K_\nu(x_\nu, \lambda_\nu, \ldots, \lambda_n) := \sum_{i=\nu}^{n} k_i(x_i, \lambda_i), \quad \text{with} \quad x_{i+1} = \varphi_i(x_i, \lambda_i). \tag{5.1.3}$$

We wish to minimize the total cost of the process and define the *Bellman function*

$$I_\nu(x_\nu) := \inf_{\substack{\lambda_i \in \Lambda_i \\ i=\nu,\ldots,n}} K_\nu(x_\nu, \lambda_\nu, \ldots, \lambda_n) \quad (\nu = 1, \ldots, n). \tag{5.1.4}$$

Theorem 5.1.1. *The Bellman function satisfies the Bellman equation*

$$I_\nu(x_\nu) = \inf_{\lambda_\nu \in \Lambda_\nu} \left(k_\nu(x_\nu, \lambda_\nu) + I_{\nu+1}(\varphi_\nu(x_\nu, \lambda_\nu)) \right) \quad \text{for } \nu = 1, \ldots, n \tag{5.1.5}$$

(here, we put $I_{n+1} = 0$). Furthermore, $(\lambda_\nu, \ldots, \lambda_n) \in \Lambda_\nu \times \cdots \times \Lambda_n$, $(x_\nu, \ldots, x_n)$ with (5.1.2) are solutions of (5.1.4) iff

$$I_j(x_j) = k_j(x_j, \lambda_j) + I_{j+1}(x_{j+1}) \quad \text{for } j = \nu, \ldots, n. \tag{5.1.6}$$

Proof. Since

$$K_\nu(x_\nu; \lambda_\nu, \ldots, \lambda_n) = k_\nu(x_\nu, \lambda_\nu) + K_{\nu+1}(\varphi_\nu(x_\nu, \lambda_\nu); \lambda_{\nu+1}, \ldots, \lambda_n),$$

we get

$$I_\nu(x_\nu) = \inf_{\substack{\lambda_i \in \Lambda_i \\ i=\nu,\ldots,n}} K_\nu(x_\nu; \lambda_\nu, \ldots, \lambda_n)$$

$$= \inf_{\lambda_\nu \in \Lambda_\nu} \left(\inf_{\substack{\lambda_j \in \Lambda_j \\ j=\nu+1,\ldots,n}} K_\nu(x_\nu; \lambda_\nu, \ldots, \lambda_n) \right)$$

$$= \inf_{\lambda_\nu \in \Lambda_\nu} \left(k_\nu(x_\nu, \lambda_\nu) + \inf_{\substack{\lambda_j \in \Lambda_j \\ j=\nu+1,\ldots,n}} K_{\nu+1}(\varphi_\nu(x_\nu, \lambda_\nu); \lambda_{\nu+1}, \ldots, \lambda_n) \right)$$

$$= \inf_{\lambda_\nu \in \Lambda_\nu} \left(k_\nu(x_\nu, \lambda_\nu) + I_{\nu+1}(\varphi_\nu(x_\nu, \lambda_\nu)) \right),$$

which is (5.1.5). For $(\lambda_\nu, \ldots, \lambda_n) \in \Lambda_\nu \times \cdots \times \Lambda_n$, $x_{j+1} = \varphi_j(x_j, \lambda_j)$ for

$j = \nu, \ldots, n$,

$$I_\nu(x_\nu) \le k_\nu(x_\nu, \lambda_\nu) + I_{\nu+1}(x_{\nu+1})$$
$$\le k_\nu(x_\nu, \lambda_\nu) + \cdots + k_n(x_n, \lambda_n) = K_\nu(x_\nu, \lambda_n, \ldots, \lambda_n).$$

If the infimum w.r.t. $\lambda_j \in \Lambda_j$ $(j = \nu, \ldots, n)$ is realized, we must have equality, and (5.1.6) follows.

q.e.d.

Corollary 5.1.1. $(\lambda_1, \ldots, \lambda_n) \in \Lambda_1 \times \cdots \times \Lambda_n$, $(x_1, \ldots, x_n)$ with (5.1.2) is a solution of (5.1.4), iff for all $\nu = 1, \ldots, n$, $(\lambda_\nu, \ldots, \lambda_n) \in \Lambda_\nu \times \cdots \times \Lambda_n$, $(x_\nu, \ldots, x_n)$ with (5.1.2) is a solution of (5.1.4).

Corollary 5.1.2. (*Bellman's method*) *An optimal solution of the process can be calculated as follows:*

For any value of x_n, compute $\lambda_n(x_n)$ minimizing (5.1.5) for $\nu = n$. Having computed $\lambda_j(x_j)$ for $j = \nu + 1, \ldots, n$, compute $\lambda_\nu(x_\nu)$ for any value of x_ν as to minimize (5.1.5) and put $x_{\nu+1} = \varphi_\nu(x_\nu, \lambda_\nu(x_\nu))$. For an arbitrary initial value $\bar{x}_1$, an optimal process thus is given by:

$$\bar{\lambda}_1 := \lambda_1(\bar{x}_1)\,, \quad \bar{x}_2 := \varphi_1(\bar{x}_1, \bar{\lambda}_1)\,, \quad \bar{\lambda}_2 = \lambda_2(\bar{x}_2), \ldots \quad .$$

5.2 Continuous control problems

We want to minimize

$$K(t_1, x(t_1)) \quad \text{for a path } x : [t_0, t_1] \to \mathbb{R}^d$$

under the following conditions: We have the initial condition

$$x(t_0) = x_0$$

and the final condition

$$x(t_1) \in B_1$$

with a given set $B_1 \in \mathbb{R}^d$. We have the control equation

$$\dot{x}(t) = f(t, x(t), \lambda(t)) \quad \text{for almost all } t \in (t_0, t_1)$$

for a piecewise continuous control function $\lambda(t)$ satisfying

$$\lambda(t) \in \Lambda$$

5.2 Continuous control problems

for some given $\Lambda \subset \mathbb{R}^c$. Pairs $(\lambda(t), x(t))$ satisfying all these restrictions are called admissible, and the set of admissible pairs is called $P(t_0, x_0)$. We put

$$I(t_0, x_0) := \inf_{(\lambda(t), x(t)) \in P(t_0, x_0)} K(t_1, x(t_1))$$

(Bellman function).

Lemma 5.2.1.

(i) $I(t_1, x_1) = K(t_1, x_1)$ for all $x_1 \in B_1$
(ii) For any path $(\lambda(t), x(t)) \in P(t_0, x_0)$, $I(t, x(t))$ is a monotonically increasing function of $t \in [t_0, t_1]$.

Proof. (i) is obvious. For (ii), if $t_0 \leq \tau_1 \leq \tau_2 \leq t_1$, the set of all admissible paths from $(\tau_2, x(\tau_2))$ to (t_1, B_1) can be considered as a subset of those ones from $(\tau_1, x(\tau_1))$ to $(t_1, x(t_1))$. Namely, if we have any path from $(\tau_2, x(\tau_2))$ to (t_1, x_1) for some $x_1 \in B_1$, we may compose it with $x(t)|_{[\tau_1, \tau_2]}$ to obtain a path from $(\tau_1, x(\tau_1))$ to (t_1, x_1). Thus, every endpoint in B_1 that can be reached from $(\tau_2, x(\tau_2))$ by an admissible path can also be reached from $(\tau_1, x(\tau_1))$ by an admissible path. This implies monotonicity.

q.e.d.

Theorem 5.2.1. $(\bar{\lambda}(t), \bar{x}(t))$ is a solution of the problem, if $I(t, \bar{x}(t))$ is constant in t. Moreover, if there exist a function $J(t, x)$ that satisfies $J(t_1, x_1) = K(t_1, x_1)$ for all $x_1 \in B_1$ and is monotonically increasing along any admissible path, and an admissible path $(\bar{\lambda}(t), \bar{x}(t))$, along which J is constant, then that path is a solution of the problem.

Proof. For a solution,

$$I(t_0, x_0) = K(t_1, \bar{x}(t_1)) = I(t_1, \bar{x}(t_1)) \quad (x_0 = \bar{x}(t_0)), \tag{5.2.1}$$

$I(t, \bar{x}(t))$ then is constant by Lemma 5.2.1 (ii). If $I(t, \bar{x}(t))$ is constant, then (5.2.1) holds, and by Lemma 5.2.1, we have a solution. Given J as described, by the monotonicity of J, for any admissible path $J(t_0, x_0) \leq K(t_1, x(t_1))$ and for the path $(\bar{\lambda}(t), \bar{x}(t))$,

$$J(t_0, x_0) = J(t_1, \bar{x}(t_1)) = K(t_1, \bar{x}(t_1)),$$

and optimality follows.

q.e.d.

Lemma 5.2.1 implies that for those t for which $I(t, x(t))$ is differentiable $((\lambda(t), x(t)) \in P(t_0, x_0))$

$$I_t(t, x(t)) + I_x(t, x(t))f(t, x(t), \lambda(t)) \geq 0.$$

For an optimal $(\bar{\lambda}(t), \bar{x}(t))$, we have by Theorem 5.2.1 then

$$I_t(t, \bar{x}(t)) + I_x(t, \bar{x}(t))f(t, \bar{x}(t), \bar{\lambda}(t)) = 0.$$

Corollary 5.2.1. (*Bellman equation*) *Let $t \in [t_0, t_1]$, $\xi \in \mathbb{R}^d$. Assume that for every $\lambda \in \Lambda$, there exists an admissible pair $(\lambda(t), x(t))$ with $\lambda(\tau) = \lambda$, $x(\tau) = \xi$. Then*

$$\inf_{\lambda \in \Lambda} (I_t(\tau, \xi) + I_x(\tau, \xi)f(\tau, \xi, \lambda)) = 0.$$

Proof. This follows from the proof of Lemma 5.2.1. Namely, the assumption implies that we may select λ such that the path is optimal at the point (τ, ξ) under consideration.

q.e.d.

Example. We want to minimize the integral

$$\int_{t_0}^{t_1} \left(u^2(t) + \lambda^2(t)\right) dt$$

with the initial condition

$$u(t_0) = u_0$$

and the control equation

$$\dot{u}(t) = \alpha u(t) + \beta \lambda(t) \quad \text{with given } \alpha, \beta \in \mathbb{R}. \tag{5.2.2}$$

In order to express this problem as a control problem, we introduce a new dependent variable $v(t)$ as solution of the equation

$$\dot{v}(t) = u^2(t) + \lambda^2(t), \; v(t_0) = 0. \tag{5.2.3}$$

We then want to minimize

$$v(t_1).$$

Given $\rho : [t_0, t_1] \to \mathbb{R}$ with

$$\rho(t_1) = 0$$

and satisfying the Riccati equation

$$\dot{\rho}(t) = -2\alpha\rho(t) + \beta^2 \rho^2(t) - 1, \tag{5.2.4}$$

we put
$$J(t, u, v) = \rho(t)u^2(t) + v(t).$$
Then
$$J(t_1, u(t_1), v(t_1)) = v(t_1)$$
and from (5.2.2), (5.2.3), (5.2.4)
$$\frac{d}{dt} J(t, u(t), v(t)) = \beta^2 \rho^2 u^2 + 2\rho\beta u\lambda + \lambda^2 = (\beta\rho u + \lambda)^2 \geq 0,$$
and this expression vanishes precisely if
$$\lambda(t) = -\beta\rho(t)u(t). \tag{5.2.5}$$

By Theorem 5.2.1, $\bar{x}(t) = (u(t), v(t))$ and $\bar{\lambda}(t) = -\beta\rho(t)u(t)$ yield an optimal solution.

If we substitute $\lambda(t)$ through the control equation (5.2.2) in the variational integral, we obtain the integral
$$\int_{t_0}^{t_1} \left(\frac{1}{\beta^2} \dot{u}(t)^2 + \left(1 + \frac{\alpha^2}{\beta^2}\right) u(t)^2 - \frac{2\alpha}{\beta^2} u(t)\dot{u}(t) \right) dt,$$
which is essentially the same as the one considered at the end of 4.2 with integrand given by (4.2.28). We recall that the latter one had also been reduced to a Riccati equation.

Equation (5.2.5) expresses the control parameter as a function of the state of the system. We just have a *feedback control*: knowing the state at a given time determines the control needed to reach an optimal state at the next time.

5.3 The Pontryagin maximum principle

We consider the control problem
$$\int_{t_0}^{t_1} F(t, x(t), \lambda(t)) dt \to \min \quad (x : [t_0, t_1] \to \mathbb{R}^d, \lambda : [t_0, t_1] \to \mathbb{R}^c) \tag{5.3.1}$$
with the control conditions
$$x(t_0) = x_0 \tag{5.3.2}$$
$$\dot{x}(t) = f(t, x(t), \lambda(t))$$

with controls

$$\lambda(t) \in \Lambda \subset \mathbb{R}^c$$

and the end condition

$$g(t_1, x(t_1)) = 0. \qquad (5.3.3)$$

Here, $\lambda(t)$ is required to be piecewise continuous, and $x(t)$ to be continuous. (Equation (5.3.2) then has to be interpreted as an integral equation $x(t) = x_0 + \int_{t_0}^{t} f(\tau, x(\tau), \lambda(\tau)) d\tau$.) F, f, and g are required to be of class C^1. Also, t_0 is fixed, whereas $t_1 \geq t_0$ is variable subject to the restriction (5.3.3). We define the *Pontryagin function*

$$\mathcal{H}(x, \lambda, p, t, \mu_0) := p \cdot f(t, x, \lambda) - \mu_0 F(t, x, \lambda).$$

We now state the *Pontryagin maximum principle*

Theorem 5.3.1. *If $(x(t), \lambda(t))$ is a solution of the control problem, there exist $\mu_0 \geq 0$, $\alpha = (\alpha_1, \ldots, \alpha_d) \in \mathbb{R}^d$ ($\alpha \neq 0$ if $\mu_0 = 0$) and a continuous $p = (p_1, \ldots, p_d)$ on $[t_0, t_1]$ such that at all points where $\lambda(t)$ is continuous, we have*

$$\mathcal{H}(x(t), \lambda(t), p(t), t, \mu_0) = \max_{\lambda \in \Lambda} \mathcal{H}(x(t), \lambda, p(t), t, \mu_0) \qquad (5.3.4)$$

and

$$\dot{p} = -\mathcal{H}_x \; , \; \dot{x} = \mathcal{H}_p \qquad (5.3.5)$$

and at the end point t_1, we have the transversality condition

$$p(t_1) = -\frac{\partial g^j}{\partial x}(t_1, x(t_1)) \cdot \alpha_j. \qquad (5.3.6)$$

There also exists a continuous function $\eta : [t_0, t_1] \to \mathbb{R}$ such that at all points where $\lambda(t)$ is continuous

$$\eta(t) = \mathcal{H}(x(t), \lambda(t), p(t), t, \mu_0) \qquad (5.3.7)$$

and

$$\dot{\eta}(t) = \mathcal{H}_t \qquad (5.3.8)$$

$$\eta(t_1) = \frac{\partial g^j}{\partial t}(t_1, x(t_1))\alpha_j. \qquad (5.3.9)$$

Also, one may always achieve $\mu_0 = 0$ or 1.

5.3 The Pontryagin maximum principle

Remarks:

(1) The equation $\dot{x} = \mathcal{H}_p$ is just the control equation

$$\dot{x} = f(t, x(t), \lambda(t)).$$

(2) If $\Lambda = \mathbb{R}^c$, then (5.3.4) becomes

$$\mathcal{H}_\lambda(x(t), \lambda(t), p(t), t, \mu_0) = 0.$$

(3) If we want to guarantee a fixed end time $\bar{t}_1$, we simply introduce an additional variable

$$x^{d+1} \equiv t$$

with control conditions

$$\dot{x}^{d+1} = 1$$
$$x^{d+1}(t_0) = t_0$$

and end condition

$$x^{d+1}(t_1) = \bar{t}_1.$$

We now want to exhibit the Hamilton–Jacobi theory as a special case of optimal control theory. Concretely, we want to derive the Euler–Lagrange equations which are equivalent to the canonical equations of Chapter 4 from the Pontryagin maximum principle. We thus consider the variational problem

$$\int_{t_0}^{t_1} L(t, x(t), \dot{x}(t)) dt \to \min$$

with $x(t_0) = x_0$, $x(t_1) = x_1$, $x : [t_0, t_1] \to \mathbb{R}^d$ and where $x(t)$ is required to have piecewise continuous first derivatives. We introduce the control variable through the control equation

$$\lambda(t) - \dot{x}(t)$$

with $\Lambda = \mathbb{R}^d$, i.e. no constraint imposed. We have $g(t_1, x(t_1)) = x_1 - x(t_1)$. The Pontryagin function of this problem is

$$\mathcal{H}(x, \lambda, p, t, \mu_0) = p \cdot \lambda - \mu_0 L(t, x, \lambda).$$

By Theorem 5.3.1 there exists $\mu_0 = 0$ or 1, $\alpha \in \mathbb{R}^d$ ($\alpha \neq 0$ for $\mu_0 = 0$)

and $p \in C^0([t_0, t_1], \mathbb{R}^d)$ with
$$\dot{p} = -\mathcal{H}_x$$
$$p(t_1) = \alpha$$
$$\mathcal{H}(t, x(t), \lambda(t), p(t), \mu_0) = \max_{\lambda \in \mathbb{R}^d} \mathcal{H}(t, x(t), \lambda, p(t), \mu_0)$$

and $\eta \in C^0([t_0, t_1], \mathbb{R})$ with
$$\eta(t) = \mathcal{H}(t, x(t), \lambda(t), p(t), \mu_0)$$
$$\dot{\eta} = \mathcal{H}_t ,$$
$$\eta(t_1) = 0.$$

We now want to exclude that $\mu_0 = 0$. In that case, we would have
$$\dot{\eta} = \mathcal{H}_t = 0 \quad , \text{ hence } \quad \eta \equiv 0 \quad \text{since} \quad \eta(t_1) = 0$$
and
$$\dot{p} = -\mathcal{H}_x = 0 \quad , \text{ hence } \quad p \equiv \alpha \quad \text{since} \quad p(t_1) = \alpha.$$

Thus
$$\mathcal{H} = \alpha \cdot \lambda,$$
and since $\mathcal{H}_t = 0$, $\mathcal{H}(x(t), \lambda(t), p(t), t, 0) \equiv 0$, and thus
$$\alpha = 0,$$
contradicting the statement of the theorem that $\alpha \neq 0$ in case $\mu_0 = 0$. We may thus assume $\mu_0 = 1$.

The Pontryagin maximum principle then gives the Weierstraß condition
$$L(t, x(t), \lambda) - L(t, x(t), \dot{x}(t)) \geq p \cdot (\lambda - \dot{x}(t)) \quad \text{for all } \lambda \in \mathbb{R}^d \quad (5.3.10)$$
and
$$\mathcal{H}_\lambda(t, x(t), \lambda(t), p(t), 1) = 0 \quad (5.3.11)$$
and the Legendre condition
$$L_{\dot{x}\dot{x}}(t, x(t), \dot{x}(t)) \quad \text{is positive semidefinite.} \quad (5.3.12)$$

Equation (5.3.11) implies
$$p = L_{\dot{x}},$$
and together with
$$\dot{p} = -\mathcal{H}_x = L_x,$$

5.3 The Pontryagin maximum principle

we obtain the Euler–Lagrange equations

$$\frac{d}{dt}L_{\dot{x}} = L_x. \tag{5.3.13}$$

A basic reference for the variational aspects of optimization and control theory where also a detailed proof of the Pontryagin maximum principle together with many applications is given is

E. Zeidler, *Nonlinear Functional Analysis and its Applications*, III, Springer, New York, 1984, pp. 93–6, 422–40.

Part two
Multiple integrals in the calculus of variations

1
Lebesgue measure and integration theory

1.1 The Lebesgue measure and the Lebesgue integral

In this section, we recall the basic notions and results about the Lebesgue measure and the Lebesgue integral that will be used in the sequel. Most proofs are omitted as they can be readily found in standard textbooks, e.g. J. Jost, *Postmodern Analysis*, Springer, Berlin, 1998, pp. 151–97 and 209–15.

Definition 1.1.1. *A collection Σ of subsets of $\mathbb{R}^d$ is called a σ-algebra (on $\mathbb{R}^d$) if*

(i) $\mathbb{R}^d \in \Sigma$
(ii) *If* $A \in \Sigma$, *then also* $\mathbb{R}^d \setminus A \in \Sigma$
(iii) *If* $A_n \in \Sigma$, $n = 1, 2, 3 \ldots$, *then also* $\bigcup_{n=1}^{\infty} A_n \in \Sigma$.

The Borel σ-algebra *is the smallest σ-algebra containing all open subsets of $\mathbb{R}^d$. The elements of the Borel σ-algebra are called* Borel sets.

Easy consequences of (i)–(iii) are

(iv) $\emptyset \in \Sigma$
(v) If $\Lambda_n \in \Sigma$, $n = 1, 2, 3 \ldots$, then also $\bigcap_{n=1}^{\infty} \in \Sigma$.
(vi) If $A, B \in \Sigma$, then also $A - B := A \setminus (A \cap B) \in \Sigma$.

Definition 1.1.2. *Let Σ be a σ-algebra. A* measure *μ on Σ is a countably additive function*

$$\mu : \Sigma \to \mathbb{R}^+ \cap \{\infty\}.$$

'Countably additive' here means that

$$\mu\left(\bigcup_{n=1}^{\infty} A_n\right) = \sum_{m=1}^{\infty} \mu(A_n)$$

for any collection of mutually disjoint ($A_m \cap A_n = \emptyset$ for $m \neq n$) elements of Σ. A measure defined on the Borel σ-algebra is called a Borel measure. A Borel measure μ is called a Radon measure if $\mu(K) < \infty$ for every compact $K \subset \mathbb{R}^d$ and $\mu(B) = \sup\{\mu(K) \mid K \subset B, K \text{ compact}\}$ for every Borel set B.

A measure μ on Σ enjoys the following properties:

(vii) $\mu(\emptyset) = 0$
(viii) If $A, B \in \Sigma$, $A \subset B$, then $\mu(A) \leq \mu(B)$
(ix) If $A_n \in \Sigma$, $n = 1, 2, 3, \ldots$ and $A_n \subset A_{n+1}$ for all n, then

$$\mu\left(\bigcup_{n=1}^{\infty} A_n\right) = \lim_{n \to \infty} \mu(A_n).$$

Theorem 1.1.1. *There exist a (unique) σ-algebra Σ on $\mathbb{R}^d$ and a (unique) measure μ in Σ satisfying*

(x) *Any open subset of $\mathbb{R}^d$ is contained in Σ (i.e. Σ contains the Borel σ-algebra)*
(xi) *For*

$$Q := \{x = (x^1, \ldots, x^d) \in \mathbb{R}^d \mid a_j < x^j < b_j, \, j = 1, \ldots, d\},$$

for numbers $a_1, \ldots, a_d, b_1, \ldots, b_d$, we have

$$\mu(Q) = \prod_{j=1}^{d} (b_j - a_j)$$

(xii) *(translation invariance) For $x \in \mathbb{R}^d$, $A \in \Sigma$ we have*

$$x + A := \{x + y \mid y \in A\} \in \Sigma \text{ and } \mu(x + A) = \mu(A)$$

(xiii) *If $A \subset B$, $B \in \Sigma$, $\mu(B) = 0$, then $A \in \Sigma$ (and, consequently, $\mu(A) = 0$).*

This μ is called **Lebesgue measure**, and the elements of Σ are called *(Lebesgue) measurable*.
In later chapters, we shall however write meas in place of μ for Lebesgue measure.

1.1 The Lebesgue measure and the Lebesgue integral

One should note that the σ-algebra of (Lebesgue) measurable sets is larger than the Borel σ-algebra.

We say that a property holds almost everywhere in $A \subset \mathbb{R}^d$ if it holds on $A \setminus B$ for some $B \subset A$ with $\mu(B) = 0$. We say that two functions $f, g : A \to \mathbb{R} \cup \{\pm\infty\}$ are equivalent if $f(x) = g(x)$ for almost all $x \in A$. A set contained in a set of measure 0 is called a null set.

We usually write meas A instead of $\mu(A)$ for a measurable set A.

Definition 1.1.3. Let $A \subset \mathbb{R}^d$ be measurable. A function
$$f : A \to \mathbb{R} \cup \{\pm\infty\}$$
is called measurable if
$$\{x \in A \mid f(x) < \lambda\}$$
is measurable for every $\lambda \in \mathbb{R}$.

If f_n, $n \in \mathbb{N}$, are measurable, $c \in \mathbb{R}$, then f_1+f_2, cf_1, $f_1 f_2$, $\max(f_1, f_2)$, $\min(f_1, f_2)$, $\limsup_{n\to\infty} f_n$, $\liminf_{n\to\infty} f_n$ are likewise measurable. Any continuous function f is measurable, because in that case $\{f(x) < \lambda\}$ is open in its domain of definition. We have the following important composition property:

Theorem 1.1.2. Let $g : A \to \mathbb{R}^c$ be measurable (i.e. $g = (g^1, \ldots, g^c)$, and each component g^j is measurable), $y : \mathbb{R}^c \to \mathbb{R}$ continuous. Then $y \circ g$ is measurable.

The characteristic function χ_A of $A \subset \mathbb{R}^d$ is defined as
$$\chi_A(x) := \begin{cases} 1 & \text{if } x \in A \\ 0 & \text{otherwise.} \end{cases}$$
Thus, A is measurable if and only if its characteristic function χ_A is measurable.

More generally, $s : A \to \mathbb{R}$ is called a *simple function* or a *step function* if it assumes only finitely many values, say $s(A) = \{\lambda_1, \ldots, \lambda_k\}$, and if all the sets $\{s(x) = \lambda_i\}$ are measurable. Thus
$$s = \sum_{i=1}^{k} \lambda_i \chi_{\{s(x)=\lambda_i\}}.$$

Theorem 1.1.3. $f : A \to \mathbb{R}$ is measurable if and only if it is the pointwise limit of a sequence of simple functions. If $f : A \to \mathbb{R}$ is measurable and bounded, then it is the uniform limit of a sequence of simple functions.

Definition 1.1.4.

(1) Let $A \subset \mathbb{R}^d$ be measurable with $\mu(A) < \infty$,

$$s = \sum_{i=1}^{k} \lambda_i \chi_{\{s(x)=\lambda_i\}}$$

a simple function on A. The Lebesgue integral of s is

$$\int_A s(x) dx := \sum_{i=1}^{k} \lambda_i \mu(\{s(x) = \lambda_i\}).$$

(2) Let A be as in *(1)*, $f : A \to \mathbb{R}$ measurable and bounded. Let $s_n : A \to \mathbb{R}$ be a sequence of simple functions converging uniformly to f according to Theorem 1.1.3. The Lebesgue integral of f then is

$$\int_A f(x) dx := \lim_{n \to \infty} \int_A s_n(x) dx$$

(this integral is independent of the choice of the sequence $(s_n)_{n \in \mathbb{N}}$).

(3) A as in *(1)*, $f : A \to \mathbb{R} \cup \{\pm\infty\}$ measurable. Put

$$f_{m,n} := \begin{cases} m & \text{if } f(x) \leq m \\ n & \text{if } f(x) \geq n \\ f(x) & \text{if } m < f(x) < n. \end{cases}$$

We say that f is integrable if

$$\lim_{\substack{m \to -\infty \\ n \to \infty}} \int_Y f_{m,n}(x) dx$$

exists. That limit then is called the Lebesgue integral $\int_A f(x) dx$ of f.

(4) $A \subset \mathbb{R}^d$ measurable, $f : A \to \mathbb{R} \cup \{\pm\infty\}$ measurable. f is called integrable if for any increasing sequence $A_1 \subset A_2 \subset \cdots \subset A$ of measurable subsets of A with $\mu(A_n) < \infty$ for all n, $f_{\chi A_n}$ is integrable on A_n and

$$\lim_{n \to \infty} \int_{A_n} f(x) \chi_{A_n}(x) dx$$

exists. That limit then is independent of the choice of (A_n) and called the Lebesgue integral $\int_A f(x) dx$ of f.

Theorem 1.1.4. *The Lebesgue integral is a linear nonnegative functional on $\mathcal{L}^1(A)$, the vector space of Lebesgue integrable functions on a measurable set A, and it satisfies:*

1.1 The Lebesgue measure and the Lebesgue integral

(1) If $f \in \mathcal{L}^1(A)$, and if $f = g$ almost everywhere on A, i.e.
$$\mu\{x \in A \mid f(x) \neq g(x)\} = 0,$$
then $g \in \mathcal{L}^1(A)$, and
$$\int_A f(x)dx = \int_A g(x)dx.$$
In particular,
$$\int_A f(x)dx = 0 \text{ if } \mu(A) = 0.$$

(2) If $f \in \mathcal{L}^1(A)$, then $|f| \in \mathcal{L}^1(A)$, and
$$\left|\int_A f(x)dx\right| \leq \int_A |f(x)|\,dx.$$

(3) If $f \in \mathcal{L}^1(A)$, $h: A \to \mathbb{R} \cup \{\pm\infty\}$ measurable with $|h| \leq f$, then $h \in \mathcal{L}^1(A)$ and
$$\left|\int_A h(x)dx\right| \leq \int_A f(x)dx.$$

(4) If $\mu(A) < \infty$, $f: A \to \mathbb{R}$ measurable with $m \leq f \leq M$, then $f \in \mathcal{L}^1(A)$, and
$$m\mu(A) \leq \int_A f(x)dx \leq M\mu(A).$$

(5) If $(A_n)_{n \in \mathbb{N}}$ is a sequence of mutually disjoint ($A_m \cap A_n = \emptyset$ for $m \neq n$) measurable sets, $A := \bigcup_{n=1}^\infty A_n$, $f \in \mathcal{L}^1(A_n)$ for every n, and if
$$\sum_{n=1}^\infty \int_{A_n} |f(x)|\,dx < \infty,$$
then $f \in \mathcal{L}^1(A)$, and
$$\int_A f(x)dx = \sum_{n=1}^\infty \int_{A_n} f(x)dx.$$
Conversely, if $f \in \mathcal{L}^1(A)$, then this equation holds for any such sequence $(A_n)_{n \in \mathbb{N}}$.

(6) If $f \in \mathcal{L}^1(A)$, then for every $\epsilon > 0$, there exists
$$\varphi \in C_0^0(\mathbb{R}^d) := \{g : \mathbb{R}^d \to \mathbb{R} \text{ continuous} \mid \{x \mid g(x) \neq 0\} \text{ bounded}\}$$
with
$$\int_A |f(x) - \varphi(x)| \, dx < \epsilon.$$

Theorem 1.1.5 (Fubini). *Let $A \subset \mathbb{R}^c$, $B \subset \mathbb{R}^d$ be measurable, and write $x = (\xi, \eta) \in A \times B$. If $f : A \times B \to \mathbb{R} \cup \{\pm\infty\}$ is integrable, then*
$$\int_{A \times B} f(x) dx = \int_A \left(\int_B f(\xi, \eta) d\eta \right) d\xi$$
$$= \int_B \left(\int_A f(\xi, \eta) d\xi \right) d\eta.$$

(Here, for example $\int_B f(\xi, \eta) d\eta$ exists for almost all $\xi \in A$.)

For $f \in \mathcal{L}^1(A)$, we put
$$\fint f := \frac{1}{\mu(A)} \int_A f(x) dx.$$

We then have Jensen's inequality:

Theorem 1.1.6. *Let $A \subset \mathbb{R}^d$ be bounded and measurable, f a convex function. Then for all $\psi \in \mathcal{L}^1(A)$*
$$f\left(\fint_A \psi\right) \le \fint_A f \circ \psi.$$

1.2 Convergence theorems

In this section, again no proofs are given, and the reader is referred to J. Jost, loc. cit., pp. 199–208.

Theorem 1.2.1 (B. Levi). *Let $A \subset \mathbb{R}^d$ be measurable, and let $f_n : A \to \mathbb{R} \cup \{\pm\infty\}$ be a monotonically increasing sequence (i.e. $f_n(x) \le f_{n+1}(x)$ for all $x \in A$, $n \in \mathbb{N}$) of integrable functions. If*
$$\lim_{n \to \infty} \int_A f_n(x) dx < \infty,$$
then $f := \lim_{n \to \infty} f_n$ (pointwise limit) is integrable, and
$$\int_A f(x) dx = \lim_{n \to \infty} \int_A f_n(x) dx.$$

1.2 Convergence theorems

Corollary 1.2.1. *Let $A \subset \mathbb{R}^d$ be measurable, $f_n : A \to \mathbb{R}^+ \cup \{\pm\infty\}$ (nonnegative and) integrable. If*

$$\sum_{n=1}^{\infty} \int_A f_n(x) dx < \infty,$$

then $\sum_{n=1}^{\infty} f_n$ is integrable, and

$$\int_A \sum_{n=1}^{\infty} f_n(x) dx = \sum_{n=1}^{\infty} \int_A f_n(x) dx.$$

Theorem 1.2.2 (Fatou). *Let $A \subset \mathbb{R}^d$ be measurable, $f_n : A \to \mathbb{R} \cup \{\pm\infty\}$ integrable for $n \in \mathbb{N}$. Assume that there exists some integrable $F : A \to \mathbb{R} \cup \{\pm\infty\}$ with*

$$f_n \geq F \quad \text{for all } n \in \mathbb{N},$$

$$\int_A f_n(x) dx \leq K < \infty \text{ for some } K \text{ independent of } n.$$

Then $\liminf_{n \to \infty} f_n$ is integrable, and

$$\int_A \liminf_{n \to \infty} f_n(x) dx \leq \liminf_{n \to \infty} \int_A f_n(x) dx.$$

Theorem 1.2.3 (Lebesgue). *Let $A \subset \mathbb{R}^d$ be measurable, $f_n : A \to \mathbb{R} \cup \{\pm\infty\}$ a sequence of integrable functions converging pointwise almost everywhere on A to some function $f : A \to \mathbb{R} \cup \{\pm\infty\}$. Suppose there exists some integrable $F : A \to \mathbb{R} \cup \{\pm\infty\}$ with*

$$|f_n| \leq F \quad \text{for all } n.$$

Then f is integrable, and

$$\int_A f(x) dx = \lim_{n \to \infty} \int_A f_n(x) dx.$$

Thoerem 1.2.3 is called the theorem on dominated convergence.

Let us consider an example that shows the necessity of the hypotheses in the previous results:
$f_n : [0, 1] \to \mathbb{R}$ is defined as

$$f_n(x) := \begin{cases} n & \text{for } 1/n \leq x \leq 2/n \\ 0 & \text{otherwise.} \end{cases} \quad (n \geq 2)$$

Then
$$f := \lim_{n\to\infty} f_n \equiv 0,$$
and
$$\lim_{n\to\infty} \int_0^1 f_n(x)dx = 1 \neq 0 = \int_0^1 f(x)dx.$$

The f_n do not form a monotonically increasing sequence so that B.Levi's theorem does not apply, and they are not bounded by some integrable function that is independent of n so that Lebesgue's theorem does not apply either. Considering $-f_n$ instead of f_n, we finally obtain a sequence for which Fatou's theorem does not hold.

As a corollary of Theorem 1.2.3 one has (approximate the derivative by difference quotients):

Corollary 1.2.2 (Differentiation under the integral). *Let $I \subset \mathbb{R}$ be an open interval, $A \subset \mathbb{R}^d$ measurable, and suppose $f : A \times I \to \mathbb{R} \cup \{\pm\infty\}$ satisfies*
(i) for any $t \in I$, $f(\cdot, t)$ is integrable on A
(ii) for almost all $x \in A$, $f(x, \cdot)$ is differentiable on I
(iii) there exists an integrable $\phi : A \to \mathbb{R} \cup \{\pm\infty\}$ with the property that for all $t \in I$ and almost all $x \in A$
$$\left|\frac{\partial}{\partial t}f(x,t)\right| \leq \phi(x).$$

Then
$$\varphi(t) := \int_A f(x,t)dx$$
is a differentiable function of $t \in I$, with
$$\frac{d}{dt}\varphi(t) = \int_A \frac{\partial}{\partial t}f(x,t)dx.$$

q.e.d.

2
Banach spaces

In this chapter, we present some results from functional analysis that will be needed in the sequel, in particular in the next chapter. All proofs are supplied. As a reference, one may use any good book on functional analysis, e.g. K. Yosida, *Functional Analysis*, Springer, Berlin, 5th edition, 1978, pp. 52–5, 81–3, 90–92, 102–28, 139–45 or F. Hirzebruch, W. Scharlau, *Einführung in die Funktionalanalysis*, Bibliograph. Inst., Mannheim, 1971, pp. 60–88, 107–12. (These were also our main sources when compiling this chapter.)

2.1 Definition and basic properties of Banach and Hilbert spaces

Definition 2.1.1. *A vector space V over $\mathbb{R}$ is called a normed space if there exists a map*

$$\|\cdot\| : V \to \mathbb{R}, \text{ called norm,}$$

satisfying

   (i) $\|v\| > 0$ *for all* $v \in V$, $v \neq 0$
   (ii) $\|\lambda v\| = |\lambda|\, \|v\|$ *for all* $\lambda \in \mathbb{R}$, $v \in V$
   (iii) $\|v + w\| \leq \|v\| + \|w\|$ *for all* $v, w \in V$ *(triangle inequality)*

A sequence $(v_n)_{n \in \mathbb{N}} \subset V$ *is said to converge to* $v \in V$ *if*

$$\lim_{n \to \infty} \|v_n - v\| = 0.$$

(In order to distinguish the notion of convergence just defined from the notion of weak convergence to be defined in the next section, we sometimes call it norm convergence or strong convergence.)

A sequence $(v_n)_{n \in \mathbb{N}} \subset V$ is called a *Cauchy sequence* if for every $\epsilon > 0$ we may find $N \in \mathbb{N}$ such that for all $n, m \geq N$

$$||v_n - v_m|| < \epsilon.$$

A normed space $(V, ||\cdot||)$ is called a *Banach space* if it is complete w.r.t. the notion of convergence just defined, i.e. if every Cauchy sequence converges to some $v \in V$.

Examples

(1) Every finite dimensional normed vector space is a Banach space, for example $\mathbb{R}^d$ with its Euclidean norm $|\cdot|$.

(2) Let $K \subset \mathbb{R}^d$ be compact. $C^0(K) := \{f : K \to \mathbb{R} \text{ continuous}\}$, $||f||_\infty := \sup_{x \in K} |f(x)|$ for $f \in C^0(K)$, defines a Banach space. If we equip $C^m(K) := \{f : K \to \mathbb{R} \text{ } m\text{-times continuously differentiable}\}$, $m \in \mathbb{N}$, with the norm $||\cdot||_\infty$, it is not a Banach space, because it is not complete. Namely the convergence w.r.t. $||\cdot||_\infty$ is uniform convergence, and while the uniform limit of continuous functions is continuous, in general the uniform limit of differentiable functions is not necessarily differentiable.

(3) Let $(V, ||\cdot||)$ be a Banach space, $W \subset V$ a linear subspace that is closed w.r.t. $||\cdot||$ i.e. if $(w_n)_{n \in \mathbb{N}} \subset W$ converges to $v \in V (\lim_{n \to \infty} ||w_n - v|| = 0)$, then $v \in W$. Then $(W, ||\cdot||)$ is a Banach space itself.

Definition 2.1.2. *A Hilbert space is a vector space H over $\mathbb{R}$ equipped with a scalar product, i.e. a map $(\cdot, \cdot) : H \times H \to \mathbb{R}$ satisfying*

(i) $(v, w) = (w, v)$ for all $v, w \in H$
(ii) $(\lambda_1 v_1 + \lambda_2 v_2, w) = \lambda_1(v_1, w) + \lambda_2(v_2, w)$ for all $\lambda_1, \lambda_2 \in \mathbb{R}$, $v_1, v_2, w \in H$
(iii) $(v, v) > 0$ for all $v \in H \setminus \{0\}$.

In addition, we require

(iv) H is complete w.r.t. the norm $||v|| := (v, v)^{\frac{1}{2}}$, i.e. a Banach space.

In order to justify the preceding definition, we need to verify that $||v|| = (v, v)^{\frac{1}{2}}$ defines indeed a norm in the sense of Definition 2.1.1. Since the properties (i), (ii) of Definition 2.1.1 are clearly satisfied, we only need to check the triangle inequality:

2.1 Basic properties of Banach and Hilbert spaces

Lemma 2.1.1. *Let* $(\cdot,\cdot) : H \times H \to \mathbb{R}$ *satisfy* (i)–(iii) *of Definition 2.1.2. Then we have the Schwarz inequality:* $|(v,w)| \leq ||v|| \cdot ||w||$ *for all* $v, w \in H$, *with equality if and only if* v *and* w *are linearly dependent.*

Proof. We have for $v, w \in H$, $\lambda \in \mathbb{R}$

$$(v + \lambda w, v + \lambda w) \geq 0 \quad \text{by (iii)}.$$

Inserting $\lambda = -\frac{(v,w)}{(w,w)}$ and expanding with the help of (i), (ii) yields the Schwarz inequality.
Since $||v + w||^2 = (v + w, v + w) = ||v||^2 + ||w||^2 + 2(v, w)$, the Schwarz inequality in turn implies the triangle inequality.

q.e.d.

Definition 2.1.3. *Let V be a vector space (over $\mathbb{R}$, as always). $M \subset V$ is called convex if whenever $x, y \in M$, then also*

$$tx + (1-t)y \in M \quad \text{for all } 0 \leq t \leq 1.$$

Example 2.1.1. Let $(V, ||\cdot||)$ be a normed space. Then for every $\mu < 0$, $B_\mu := \{x \in V \mid ||x|| \leq \mu\}$ is convex. Namely if $x, y \in B_\mu$, i.e. $|x| \leq \mu$, $|y| \leq \mu$, then for $0 \leq t \leq 1$

$$|tx + (1-t)y| \leq t|x| + (1-t)|y| \leq \mu,$$

hence $tx + (1-t)y \in B_\mu$.

The following definition contains a sharpening of the convexity of the balls B_μ. It will be formulated only for $\mu = 1$, but by homogeneity ((ii) of Definition 2.1.1), it implies an analogous condition for any $\mu > 0$.

Definition 2.1.4. *A normed space $(V, ||\cdot||)$ is called uniformly convex if for all $\epsilon > 0$ there exists $\delta > 0$ with the property that for all $x, y \in V$ with $||x|| = ||y|| = 1$, we have*

$$\left|\left|\frac{1}{2}(x+y)\right|\right| > 1 - \delta \Rightarrow ||x - y|| < \epsilon. \tag{2.1.1}$$

Remark 2.1.1. An equivalent form of the implication (2.1.1) is

$$||x - y|| \geq 1 \Rightarrow \left|\left|\frac{1}{2}(x+y)\right|\right| \leq 1 - \delta \tag{2.1.2}$$

(again for $||x|| = ||y|| = 1$).

Example 2.1.2. In a Hilbert space $(H, (\cdot, \cdot))$, we have the parallelogram identity

$$\left\|\frac{1}{2}(x+y)\right\|^2 = \frac{1}{2}\|x\|^2 + \frac{1}{2}\|y\|^2 - \frac{1}{4}\|x-y\|^2 \tag{2.1.3}$$

which follows by expanding the norms in terms of the scalar product. Therefore, any Hilbert space is uniformly convex.

Lemma 2.1.2. *In Definition 2.1.3, the condition $\|x\| = \|y\| = 1$ may be replaced by*

$$\|x\| \leq 1, \ \|y\| \leq 1.$$

Proof. In the situation of Definition 2.1.3, for $\epsilon_0 > 0$, we may find $\delta_0 > 0$ such that for all z, w with $\|z\| = \|w\| = 1$, we have

$$\left\|\frac{1}{2}(z+w)\right\| > 1 - \delta_0 \Rightarrow \|z - w\| < \epsilon_0. \tag{2.1.4}$$

Let now $\epsilon > 0$, $\|x\| \leq 1$, $\|y\| \leq 1$. If for $\delta \leq \frac{1}{2}$

$$1 - \delta < \left\|\frac{1}{2}(x+y)\right\|$$

then

$$\|x\| > 1 - 2\delta, \ \|y\| > 1 - 2\delta.$$

In particular, $x, y \neq 0$, and by the triangle inequality,

$$\left\|\frac{1}{2}\left(\frac{x}{\|x\|} + \frac{y}{\|y\|}\right)\right\| \geq \left\|\frac{1}{2}(x+y)\right\| - \frac{1}{2}\left\|x - \frac{x}{\|x\|}\right\| - \frac{1}{2}\left\|y - \frac{y}{\|y\|}\right\|$$
$$> 1 - 3\delta.$$

We apply (2.1.4) with $z = \frac{x}{\|x\|}$, $w = \frac{y}{\|y\|}$, $\epsilon_0 = \frac{\epsilon}{2}$. If $3\delta \leq \delta_0$, we then get

$$\left\|\frac{x}{\|x\|} - \frac{y}{\|y\|}\right\| < \frac{\epsilon}{2}.$$

Now

$$\|x - y\| \leq \left\|x - \frac{x}{\|x\|}\right\| + \left\|y - \frac{y}{\|y\|}\right\| + \left\|\frac{x}{\|x\|} - \frac{y}{\|y\|}\right\|$$
by the triangle inequality
$$< 4\delta + \frac{\epsilon}{2}.$$

2.1 Basic properties of Banach and Hilbert spaces

Choosing $\delta = \min(3\delta_0, \epsilon/8)$, we have shown the implication

$$\left\|\frac{1}{2}(x+y)\right\| > 1 - \delta \Rightarrow \|x - y\| < \epsilon$$

for $\|x\| \leq 1$, $\|y\| \leq 1$.

q.e.d.

Lemma 2.1.3. *Let $(V, \|\cdot\|)$ be a uniformly convex Banach space. Let $(x_n)_{n \in \mathbb{N}} \subset V$ be a sequence with*

$$\limsup_{n \to \infty} \|x_n\| \leq 1 \quad \text{for all } n \in \mathbb{N} \tag{2.1.5}$$

and

$$\lim_{n,m \to \infty} \left\|\frac{1}{2}(x_n + x_m)\right\| = 1. \tag{2.1.6}$$

Then (x_n) converges to some $x \in V$ with $\|x\| = 1$.

Proof. Let $\epsilon > 0$. (2.1.5) and (2.1.6) imply $\lim \|x_n\| = 1$. Therefore, by replacing x_n by $\frac{x_n}{\|x_n\|}$, we may assume w.l.o.g $\|x_n\| = 1$. Because of (2.1.5), we may apply Lemma 2.1.2. By (2.1.6), we may find $N \in \mathbb{N}$ such that for $n, m \geq N$

$$\left\|\frac{1}{2}(x_n + x_m)\right\| \geq 1 - \delta,$$

with δ determined by Lemma 2.1.2. We obtain

$$\|x_n - x_m\| < \epsilon,$$

i.e. $(x_n)_{n \in \mathbb{N}}$ is a Cauchy sequence. Since $(V, \|\cdot\|)$ is a Banach space, it has a limit x, and

$$\|x\| = \lim_{n \to \infty} \|x_n\| = 1.$$

q.e.d.

In order to formulate the Hahn–Banach theorem, a fundamental extension result for linear functionals from a linear space to the whole space, we need:

Definition 2.1.5. *Let V be a (real) vector space.*

$$p : V \to \mathbb{R}^+ \quad (\mathbb{R}^+ := \{t \in \mathbb{R} \mid t \geq 0\})$$

is called *convex* if

(i) $p(x+y) \leq p(x) + p(y)$ for all $x, y \in V$
(ii) $p(\lambda x) = \lambda p(x)$ for all $x \in V$, $\lambda > 0$

Example 2.1.3. The norm on a normed vector space.

Let V_0 be a linear subspace of the vector space V, $f_0 : V_0 \to \mathbb{R}$ linear. A linear $f : V \to \mathbb{R}$ is called an *extension* of f_0 if

$$f|_{V_0} = f_0.$$

Theorem 2.1.1 (Hahn–Banach). *Let V_0 be a linear subspace of the vector space V, $p : V \to \mathbb{R}^+$ convex. Suppose that $f_0 : V_0 \to \mathbb{R}$ is linear and satisfies*

$$f_0(x) \leq p(x) \quad \text{for all } x \in V_0. \tag{2.1.7}$$

Then there exists an extension $f : V \to \mathbb{R}$ of f_0 with

$$f(x) \leq p(x) \quad \text{for all } x \in V. \tag{2.1.8}$$

Remark 2.1.2. We shall need the Hahn–Banach theorem only in the case where V possesses a countable basis, i.e. is separable (see p. 130).

Proof. We may assume $V_0 \neq V$. Let $v \in V \setminus V_0$, V_1 be the linear subspace of V spanned by V_0 and v, i.e.

$$V_1 := \{x + tv \mid x \in V_0, \, t \in \mathbb{R}\}.$$

We shall now investigate how f_0 can be extended to $f_1 : V_1 \to \mathbb{R}$ with

$$f_1(x) \leq p(x) \quad \text{for all } x \in V_1. \tag{2.1.9}$$

We put $f_1(v) =: \alpha$. Then as an extension of f_0, f_1 satisfies

$$f_1(x + tv) = f_0(x) + t\alpha.$$

Equation (2.1.9) requires

$$f_0(x) + t\alpha \leq p(x + tv). \tag{2.1.10}$$

For $t > 0$, this is equivalent to

$$\alpha \leq p\left(\frac{x}{t} + v\right) - f_0\left(\frac{x}{t}\right), \tag{2.1.11}$$

and for $t < 0$ to

$$\alpha \geq -p\left(-\frac{x}{t} - v\right) - f_0\left(\frac{x}{t}\right). \tag{2.1.12}$$

For $x_1, x_2 \in V$, we have

$$f_0(x_2) - f_0(x_1) \leq p(x_2 - x_1)$$
$$= p((x_2 + v) - (x_1 + v))$$
$$\leq p(x_2 + v) + p(-x_1 - v),$$

hence

$$-f_0(x_2) + p(x_2 + v) \geq -f_0(x_1) - p(-x_1 - v). \tag{2.1.13}$$

Thus

$$\alpha_2 := \inf_{x_2} \big(- f_0(x_2) + p(x_2 + tv) \big)$$
$$\geq \alpha_1 := \sup_{x_1} \big(- f_0(x_1) - p(-x_1 - v) \big).$$

Therefore, any α with

$$\alpha_1 \leq \alpha \leq \alpha_2$$

satisfies (2.1.11) and (2.1.12), hence (2.1.10). Thus, the desired extension f_1 exists. If V possesses a countable basis, we may use the preceding construction to extend f_0 inductively to all of V.

If V does not possess a countable basis, we need to use Zorn's lemma to complete the proof. For that purpose, let

$$\Phi := \{\varphi : W \to \mathbb{R} \text{ extension of } f_0 \text{ to some}$$
$$\text{linear subspace } W, V_0 \subset W \subset V,$$
$$\text{satisfying } \varphi(x) \leq p(x) \text{ for all } x \in W\}$$

On Φ, we have an obvious ordering relation (namely, for $\varphi_i : W_i \to \mathbb{R}$, $i = 1, 2$, we have $\varphi_1 \leq \varphi_2$ if $W_1 \subset W_2$ and $\varphi_2|_{W_1} = \varphi_1$), and every totally ordered subset Φ_0 of Φ possesses a maximal element, namely φ_0 defined on the union of the domains of all $\varphi \in \phi_0$ and coinciding with each such φ on its domain of definition. By Zorn's lemma, Φ then contains a maximal element f. Let W be the domain of definition of f. f then extends f_0 to W. If W were not the whole space V, we could use the preceding construction to extend f to a larger subspace of V, contradicting the maximality of f. Therefore, f furnishes the desired extension of f_0.

q.e.d.

Corollary 2.1.1. *Let V_0 be a linear subspace of the normed vector space $(V, \|\cdot\|)$, $\lambda > 0$, $f_0 : V_0 \to \mathbb{R}$ linear with*

$$|f_0(x)| \leq \lambda \|x\| \quad \text{for all } x \in V_0.$$

Then there exists an extension $f : V \to \mathbb{R}$ of f_0 with
$$|f(x)| \leq \lambda \, ||x|| \quad \text{for all } x \in V.$$

Theorem 2.1.2 (Helly). Let $(V, ||\cdot||)$ be a Banach space, $f_1, \ldots, f_n$ linear functionals $V \to \mathbb{R}$ that are continuous w.r.t. the norm convergence, $\mu, \alpha_1, \ldots, \alpha_n \in \mathbb{R}$. Suppose that for any $\lambda_1, \ldots, \lambda_n \in \mathbb{R}$

$$\left| \sum_{i=1}^n \lambda_i \alpha_i \right| \leq \mu \left\| \sum_{i=1}^n \lambda_i f_i \right\|. \tag{2.1.14}$$

Then for each $\epsilon > 0$, there exists $x_\epsilon \in V$ with

$$f_i(x_\epsilon) = \alpha_i \quad \text{for } i = 1, 2, \ldots, n \tag{2.1.15}$$

and

$$||x_\epsilon|| \leq \mu + \epsilon.$$

Proof. Let $m \leq n$ be the maximal number of linearly independent f_i, $i = 1, \ldots, n$. It suffices to consider m linearly independent f_i, w.l.o.g. $f_1, \ldots, f_m$, since the remaining ones are easily seen to be taken care of by (2.1.14). $F(x) := (f_1(x), \ldots, f_m(x))$ may then be considered as a linear map onto $\mathbb{R}^m$. We equip $\mathbb{R}^m$ with its Euclidean structure. Let

$$B_{\mu+\epsilon} := \{x \in V \mid ||x|| \leq \mu + \epsilon\}.$$

Then $F(B_{\mu+\epsilon})$ is a convex set containing 0 as an interior point. Also, $F(B_{\mu+\epsilon})$ is balanced in the sense that with $p \in \mathbb{R}^m$ it also contains $-p$.

We now assume that $\alpha_1, \ldots, \alpha_m$ is not contained in $F(B_{\mu+\epsilon})$. Because of the properties of $F(B_{\mu+\epsilon})$ just noted, we may then find $\lambda = (\lambda_1, \ldots, \lambda_m)$ with

$$\sum_{i=1}^m \lambda_i \alpha_i \geq \sup_{x \in B_{\mu+\epsilon}} \left| \sum_{i=1}^m \lambda_i f_i(x) \right|$$

$$= (\mu + \epsilon) \left\| \sum_{i=1}^m \lambda_i f_i \right\|,$$

contradicting (2.1.14). Thus $(\alpha_1, \ldots, \alpha_m) \in F(B_{\mu+\epsilon})$, implying the claim. q.e.d.

2.2 Dual spaces and weak convergence

Let V be a vector space. The linear functionals

$$f : V \to \mathbb{R}$$

2.2 Dual spaces and weak convergence

then also form a vector space. If $(V, ||\cdot||)$ is a normed vector space, we define the norm of a linear functional $f : V \to \mathbb{R}$ as

$$||f||_* := \sup_{x \neq 0} \frac{|f(x)|}{||x||} \in \mathbb{R}^+ \cup \{\infty\}. \tag{2.2.1}$$

Lemma 2.2.1. *A linear functional $f : V \to \mathbb{R}$ is continuous if and only if $||f||_* < \infty$.*

The easy *proof* is left to the reader. (See also Lemma 2.3.1 below.)

q.e.d.

Definition 2.2.1. $V^* := \{f : V \to \mathbb{R} \text{ linear with } ||f||_* < \infty\}$ *equipped with the norm (2.2.1) is called the dual space of $(V, ||\cdot||)$. (It is easy to verify that (2.2.1) defines a norm on V^* in the sense of Definition 2.1.1.)*

Lemma 2.2.2. $(V^*, ||\cdot||_*)$ *is a Banach space.*

Proof. Let $(f_n)_{n \in \mathbb{N}} \subset V^*$ be a Cauchy sequence. For every $\epsilon > 0$ we may then find $N \in \mathbb{N}$ such that for $n, m \in \mathbb{N}$

$$||f_n - f_m||_* < \epsilon.$$

By (2.2.1), this implies that for every $x \in V$

$$|f_n(x) - f_m(x)| < \epsilon.$$

Therefore, since $\mathbb{R}$ is complete, $(f_n(x))_{n \in \mathbb{N}}$ converges for every $x \in X$. We denote the limit by $f(x)$. $f : V \to \mathbb{R}$ then is a linear functional. It is an easy consequence of the triangle inequality that $||f||_* < \infty$ and that $\lim_{n \to \infty} ||f_n - f||_* = 0$. This implies that $(f_n)_{n \in \mathbb{N}}$ converges to $f \in V^*$, and $(V^*, ||\cdot||_*)$ therefore is complete, hence a Banach space.

q.e.d.

Remark 2.2.1. We did not assume that V itself is a Banach space.

We now consider

$$(V^*)^* =: V^{**},$$

the dual space of V^*, with norm denoted by $||\cdot||_{**}$. Any $x \in V$ defines a linear functional

$$i(x) : V^* \to \mathbb{R}$$
$$i(x)(f) := (f, x) := f(x).$$

Lemma 2.2.3. $||i(x)||_{**} = ||x||$. Thus, the linear functional $i(x) : V^* \to \mathbb{R}$ is contained in V^{**}, i.e. we have a linear isometric map $i : V \to V^{**}$.

Proof. We have
$$|(f, x)| \leq ||f||_* \, ||x||,$$
and therefore
$$||x|| \geq \sup_{f \in V^*} \frac{|f(x)|}{||f||_*} = ||i(x)||_{**}. \tag{2.2.2}$$

Conversely, let $x \in V$. Let
$$f_0(tx) := t\,||x|| \quad \text{for } t \in \mathbb{R}.$$

By the Hahn–Banach theorem (Corollary 2.1.1), we may extend f_0 from $\{tx \mid t \in \mathbb{R}\}$ to V as a linear functional f with
$$||f||_* = 1$$
and
$$|(f, x)| = ||x||.$$

Therefore
$$||i(x)||_{**} = \sup_{f \in V^*} \frac{|(f, x)|}{||f||_*} \geq ||x||. \tag{2.2.3}$$

Equations (2.2.2) and (2.2.3) imply the result.

q.e.d.

Definition 2.2.2. *A normed linear space $(V, ||\cdot||)$ is called reflexive if*
$$i : V \to V^{**}$$
is a bijective isometry (i.e. $||x|| = ||i(x)||_{**}$ for all $x \in V$).

Remark 2.2.2.

(1) Since $(V^{**}, ||\cdot||_{**})$ is a Banach space by Lemma 2.2.2, any reflexive space is complete, i.e. a Banach space.
(2) By the remark before Definition 2.2.2, the crucial condition in that definition is the surjectivity of i.

2.2 Dual spaces and weak convergence

Definition 2.2.3.

(i) Let $(V, \|\cdot\|)$ be a normed linear space. $(x_n)_{n \in \mathbb{N}} \subset V$ is said to be *weakly convergent* to $x \in V$ if $f(x_n)$ converges to $f(x)$ for all $f \in V^*$, in symbols:

$$x_n \rightharpoonup x.$$

(ii) Let $(V^*, \|\cdot\|_*)$ be the dual of a normed linear space. $(f_n)_{n \in \mathbb{N}} \subset V^*$ is said to be *weak* convergent* to $f \in V^*$ if $f_n(x)$ converges to $f(x)$ for all $x \in V$.

Theorem 2.2.1. *Let V be a separable† normed linear space. Let $(f_n)_{n \in \mathbb{N}} \subset V^*$ be bounded, i.e. $\|f_n\|_* \le$ constant (independent of n). Then (f_n) contains a weak* convergent subsequence.*

Proof. Let $(y_\nu)_{\nu \in \mathbb{N}}$ by a dense subset of V. Since $(f_n(y_1))_{n \in \mathbb{N}}$ is bounded, a subsequence $(f_n^1(y_1))$ of $(f_n(y_1))$ converges. Having iteratively found a subsequence (f_n^m) of (f_n) for which $(f_n^m(y_\nu))_{n \in \mathbb{N}}$ converges for $1 \le \nu \le m$, we may find a subsequence (f_n^{m+1}) of (f_n^m) for which also $(f_n^{m+1}(y_{m+1}))_{n \in \mathbb{N}}$ converges. The diagonal sequence $(f_n^n)_{n \in \mathbb{N}}$ then converges at every y_ν, $\nu \in \mathbb{N}$, and since $(y_\nu)_{\nu \in \mathbb{N}}$ is dense in V, $(f_n^n(x))_{n \in \mathbb{N}}$ has to converge for every $x \in V$. Thus, we have found a weak* convergent subsequence of $(f_n)_{n \in \mathbb{N}}$.

q.e.d.

Remark 2.2.3.

(1) The argument employed in the preceding proof is called Cantor diagonalization.
(2) Theorem 2.2.1 remains true without the assumption that V is separable, and so does the following:

Corollary 2.2.1. *Let $(V, \|\cdot\|)$ be a separable reflexive Banach space. Then every bounded sequence $(x_n)_{n \in \mathbb{N}}$ contains a weakly convergent subsequence.*

Proof. By (2.2.2) or reflexivity, $(i(x_n))_{n \in \mathbb{N}}$ is a bounded sequence in V^{**} and therefore contains a weak* convergent subsequence. Since V is

---

† Separable means that V contains a countable subset $(y_\nu)_{\nu \in \mathbb{N}}$ that is dense w.r.t. $\|\cdot\|$, i.e. for every $y \in V$, $\epsilon > 0$ there exists y_ν with $\|y - y_\nu\| < \epsilon$.

reflexive, the limit is of the form $i(x)$ for some $x \in V$. Thus
$$f(x_n) = (f, x_n) \to (f, x) = f(x) \quad \text{for every } f \in V^*$$
so that $(x_n)_{n \in \mathbb{N}}$ converges weakly to x.

q.e.d.

Theorem 2.2.2. *Any weakly convergent sequence $(x_n)_{n \in \mathbb{N}}$ in a Banach space is bounded.*

Proof. We shall show that $i(x_n)$ is uniformly bounded on $\{f \in V^* \mid \|f\|_* \leq 1\}$. Then also
$$\|x_n\| = \|i(x_n)\|_{**} = \sup_{f \in V^*} \frac{|(f, x_n)|}{\|f\|_*} \tag{2.2.4}$$
is bounded (see Lemma 2.2.3 for the first equality). Since $i(x_n)$ is linear, it suffices to show uniform boundedness on some ball in V^*. Otherwise, we find a sequence B_j of closed balls,
$$B_j = \{f \in V^* \mid \|f - f_j\| \leq \varrho_i\} \quad \text{for some } f_j \in V^*, \varrho_j > 0$$
with
$$B_{j+1} \subset B_j \quad \text{and} \quad \lim_{j \to \infty} \varrho_j = 0$$
and a subsequence (x'_n) of (x_n) with
$$|(f, x'_n)| > j \quad \text{for all } f \in B_j. \tag{2.2.5}$$
By construction, $(f_j)_{j \in \mathbb{N}}$ forms a Cauchy sequence and therefore converges to some $f_0 \in V^*$, with
$$f_0 \in \bigcap_{j=1}^{\infty} B_j.$$
Because of (2.2.5), we have
$$|(f_0, x'_n)| > j \quad \text{for all } j \in \mathbb{N}.$$
This is impossible since (f_0, x'_n) converges because $(x'_n)_{n \in \mathbb{N}}$ converges weakly.

q.e.d.

Example 2.2.1.

(1) In a finite dimensional normed vector space (which automatically is complete, i.e. a Banach space), weak convergence is just componentwise convergence and therefore equivalent to the usual convergence w.r.t. the norm.

2.2 Dual spaces and weak convergence

(2) In an infinite dimensional reflexive Banach space $(V, \|\cdot\|)$, this is no longer so, because one may always find a sequence $(e_n)_{n \in \mathbb{N}} \subset V$ with $\|e_i\| \leq 1$ for all i and $\|e_i - e_j\| \geq 1$ for $i \neq j$. Such a sequence cannot converge w.r.t. $\|\cdot\|$, because it is not a Cauchy sequence, but it always contains a weakly convergent subsequence according to Corollary 2.2.1 (we have shown Corollary 2.2.1 only under the assumption of separability, but it holds true in general).

Lemma 2.2.4. *Let $(V, \|\cdot\|)$ be a separable normed space. Then V^* satisfies the first axiom of countability w.r.t. the weak* topology, i.e. for each $f \in V^*$, there exists a sequence $(U_\nu)_{\nu \in \mathbb{N}}$ of subsets of V^* that are open in the weak* topology such that every U that is open in this topology and contains x is contained in some U_n. Consequently, if $(V, \|\cdot\|)$ is also reflexive, then V^* satisfies the first axiom of countability w.r.t. the weak topology.*

Proof. Let $f \in V^*$. Every neighbourhood of f w.r.t. the weak* topology contains a neighbourhood of the form

$$U_{\epsilon, v_1, \ldots, v_k}(f) := \{g \in V^* \mid |g(v_i) - f(v_i)| < \epsilon \quad \text{for } i = 1, \ldots, k\}.$$

Since V is separable, there exists a sequence $(w_n)_{n \in \mathbb{N}} \subset V$ that is dense w.r.t the $\|\cdot\|$ topology. We claim that the neighbourhoods of the form

$$U_{\frac{1}{n}, w_{i_1}, \ldots, w_{i_k}}(f)$$

form a basis of the neighbourhood system of f of the required type, i.e. every $U_{\epsilon; v_1, \ldots, v_k}(f)$ contains some such $U_{\frac{1}{n}; w_{i_1}, \ldots, w_{i_k}}(f)$. For that purpose, we choose n with $\frac{3}{n} < \epsilon$ and $w_{i_1}, \ldots, w_{i_k}$ with $|v_j - w_{i_j}| < \frac{1}{n}$ for $j = 1, \ldots, k$. For $g \in U_{\frac{1}{n}; w_{i_1}, \ldots, w_{i_k}}(f)$, we then have

$$|g(v_j) - f(v_j)| \leq |g(w_{i_j}) - f(w_{i_j})| + |(g-f)(v_j - w_{ij})| \leq \frac{1}{n} + \frac{2}{n} < \epsilon,$$

i.e. $g \in v_{\epsilon; v_1, \ldots, v_k}(f)$ as required.

Finally, if V is reflexive, then the weak* and the weak topology of V^* coincide.

q.e.d.

We now present some further applications of the Hahn–Banach theorem that will be used in Chapter 3.

Lemma 2.2.5. *Let $(V, \|\cdot\|)$ be a normed space, V_0 a closed linear subspace. Then V_0 is also closed w.r.t. weak convergence.*

Proof. By the Hahn–Banach theorem (Corollary 2.1.1), for every $x_0 \in V \setminus V_0$, we may find a continuous linear functional $f_0 : V \to \mathbb{R}$ with

$$f_0(x_0) = 1$$
$$f_0|_{V_0} = 0.$$

Thus, x_0 cannot be a weak limit of a sequence in V_0.

q.e.d.

Lemma 2.2.6. *Let $(V, ||\cdot||)$ be a reflexive Banach space, V_0 a closed linear subspace. Then V_0 is reflexive.*

Proof. We may identify V_0^{**} with a subspace of V^{**}, by putting $v(f) = v(f|_{V_0})$ for $f \in V^*$, $v \in V_0^{**}$. Let $v \in V_0^{**}$. Since V is reflexive, there exists $x \in V$ with

$$v(f) = f(x) \quad \text{for all } f \in V^*.$$

We claim $x \in V_0$. Otherwise, by the Hahn–Banach theorem (Corollary 2.1.1), there exists $f \in V^*$ with

$$f(x) \neq 0$$
$$f|_{V_0} = 0.$$

Since $f(x) = v(f|_{V_0})$ by the above, this is impossible. Since every $f_0 \in V_0^*$ can be extended to $f \in V^*$, again by Hahn–Banach, we conclude

$$v(f_0) = f_0(x) \quad \text{for all } f \in V_0^*.$$

Thus, $v = i(x)$. This implies $V_0^{**} = i(V_0)$, i.e. reflexivity of V_0.

q.e.d.

Corollary 2.2.2. *A Banach space $(V, ||\cdot||)$ is reflexive if and only if its dual $(V^*, ||\cdot||_*)$ is reflexive.*

Proof. If $V = V^{**}$, then also $V^* = V^{***}$. Thus, if V is reflexive, so is V^*. Consequently, if conversely V^* is reflexive, so then is V^{**}. Since V can be identified with a closed subspace of V^{**} by Lemma 2.2.2, Lemma 2.2.6 then yields reflexivity of V.

q.e.d.

Lemma 2.2.7. *Let $(V, ||\cdot||)$ be a normed space, and suppose that $(x_n)_{n \in \mathbb{N}} \subset V$ converges weakly to $x \in V$. Then*

$$||x|| \leq \liminf_{n \to \infty} ||x_n||.$$

2.2 Dual spaces and weak convergence

Proof. After selection of a subsequence, we may assume that $||x_n||$ converges (see Theorem 2.2.2). Assume
$$||x|| > \lim_{n \to \infty} ||x_n||.$$
As in the proof of Lemma 2.2.3, we may find $f \in V^*$ with
$$||f||_* = 1$$
$$|f(x)| = ||x||.$$
But then
$$|f(x)| > \lim_{n \to \infty} ||x_n|| \geq \limsup_{n \to \infty} |f(x_n)|,$$
while the weak convergence of $(x_n)_{n \in \mathbb{N}}$ to x implies
$$f(x) = \lim_{n \to \infty} f(x_n).$$
This contradiction establishes the claim.

q.e.d.

Theorem 2.2.3 (Milman). *Any uniformly convex Banach space is reflexive.*

Proof (Kakutani). Let $(V, ||\cdot||)$ be a uniformly convex Banach space, and let $x_0^{**} \in V^{**}$. We need to show that there exists some $x_0 \in V$ with
$$i(x_0) = x_0^{**} \tag{2.2.6}$$
(see Remark 2 after Definition 2.2.2). We may assume w.l.o.g. that
$$||x_0^{**}|| = 1. \tag{2.2.7}$$
For every $n \in \mathbb{N}$, we may then find $f_n \in V^*$ with $||f_n|| = 1$ and
$$1 - \frac{1}{n} < x_0^{**}(f_n) \leq 1. \tag{2.2.8}$$
We now claim that for every $n \in \mathbb{N}$, we may find $x_n \in V$ with
$$f_i(x_n) = x_0^{**}(f_i) \quad \text{for } i = 1, \ldots, n \tag{2.2.9}$$
and
$$||x_n|| \leq ||x_0^{**}|| + \frac{1}{n} = 1 + \frac{1}{n}. \tag{2.2.10}$$
For any $\lambda_1, \ldots, \lambda_n \in \mathbb{R}$, we have
$$\left| \sum_{i=1}^n \lambda_i x_0^{**}(f_i) \right| = \left| x_0^{**}\left(\sum_{i=1}^n \lambda_i f_i\right) \right| \leq ||x_0||^{**} \left\| \sum_{i=1}^n \lambda_i f_i \right\|$$

and so the claim follows from Helly's Theorem 2.1.2. Since in addition to (2.2.10) also

$$||x_n|| = ||f_n||\,||x_n|| \geq f_n(x_n) = x_0^{**}(f_n) \geq 1 - \frac{1}{n},$$

we must have

$$\lim_{n\to\infty} ||x_n|| = 1.$$

For $m \geq n$, we have

$$2 - \frac{2}{n} \leq f_n(x_n) + f_n(x_m) \leq ||x_n + x_m|| \leq ||x_m|| + ||x_m|| \leq 2 + \frac{2}{n}.$$

By Lemma 2.1.3, $(x_n)_{n\in\mathbb{N}}$ is a Cauchy sequence and converges to some $x_0 \in V$, satisfying

$$||x_0|| = 1 \qquad (2.2.11)$$

and

$$f_i(x_0) = x_0^{**}(f_i) \quad \text{for } i = 1, 2, 3, \ldots \qquad (2.2.12)$$

The solution x_0 of (2.2.11), (2.2.12) is unique. Namely, if there were another solution x_0', on one hand, we would have

$$||x_0 + x_0'|| < 2 \qquad (2.2.13)$$

by uniform convexity. On the other hand

$$f_i(x_0 + x_0') = 2x_0^{**}(f_i) \quad \text{for all } i,$$

hence

$$2 - \frac{2}{i} \leq 2x_0^{**}(f_i) = f_i(x_0 + x_0') \leq ||x_0 + x_0'||,$$

hence

$$||x_0 + x_0'|| \geq 2.$$

This contradicts (2.2.13), and so x_0 is unique. We now claim that

$$f_0(x_0) = x_0^{**}(f_0) \quad \text{for any } f_0 \in V^*, \qquad (2.2.14)$$

so that $x_0^{**} = i(x_0)$, proving the theorem. Let this $f_0 \in V^*$ be given. In the above reasoning, we replace the sequence $f_1, f_2, f_3, \ldots$ by $f_0, f_1, f_2, f_3, \ldots$. We then obtain $x_0' \in V$ with

$$||x_0'|| = 1$$

and

$$f_i(x_0') = x_0^{**}(f_i) \quad \text{for } i = 0, 1, 2, 3, \ldots \qquad (2.2.15)$$

2.2 Dual spaces and weak convergence

Since the solution x_0 of (2.2.11), (2.2.12) was shown to be unique, however, we must have $x_0' = x_0$. Equation (2.2.15) for $i = 0$ then is (2.2.14).

q.e.d.

Corollary 2.2.3 (Riesz). *Any Hilbert space $(H, (\cdot, \cdot))$ can be identified with its dual H^*.*

Proof. Since a Hilbert space is uniformly convex, Therem 2.2.3 implies $H = H^{**}$. On the other hand, any $x \in H$ induces an $f_x \in H^*$ by

$$f_x(y) := (x, y) \quad \text{for } y \in H.$$

We have

$$\|f_x\| = \sup_{\|y\|=1} (x, y) \leq \|x\|$$

and $f_x(x) = (x, x) = \|x\|^2$, hence

$$\|f_x\| = \|x\|.$$

Thus, H is isometrically embedded into H^*. For the same reason, H^* is isometrically embedded into H^{**}, and since $H = H^{**}$, one readily verifies that these embeddings must be surjective, hence $H = H^* = H^{**}$.

q.e.d.

Let M be a linear subspace of a Hilbert space H. The orthogonal complement $M^\perp$ of M is defined as

$$M^\perp := \{x \in H : (x, y) = 0 \quad \text{for all } y \in M\}.$$

It is clear that $M^\perp$ is a closed linear subspace of H. M need not be closed here, but the orthogonal complement of M is the same as the one of its closure $\overline{M}$ in H.

Corollary 2.2.4. *Let M be a closed linear subspace of the Hilbert space H. Then every $x \in H$ can be uniquely decomposed as*

$$x = x_1 + x_2 \text{ with } x_1 \in M, x_2 \in M^\perp.$$

Proof. By the proof of Corollary 2.2.3, $x \in H$ corresponds to $f_x \in H^*$ with

$$f_x(y) = (x, y) \text{ for all } y \in H.$$

We let f_x^M be the restriction of f_x to M. M, since closed, is a Hilbert space itself, and f_x^M is an element of the dual M^*. By Corollary 2.2.3, it corresponds to some $x_1 \in M$, i.e.

$$f_x^M(y) = (x_1, y) \text{ for all } y \in M.$$

We put $x_2 := x - x_1$. Then for all $y \in M$,

$$(x - x_1, y) = f_x(y) - f_x^M(y) = 0 \text{ since } f_x = f_x^M \text{ on } M.$$

Therefore, $x_2 \in M^\perp$. Thus, we have constructed the required decomposition. Concerning uniqueness, if

$$x = x_1 + x_2 = x_1' + x_2' \text{ with } x_1, x_1' \in M, x_2, x_2' \in M^\perp,$$

then for all $y \in M$

$$(x, y) = (x_1, y) = (x_1', y),$$

and by Corollary 2.2.3 applied to M, $x_1 = x_1'$, and therefore also $x_2 = x_2'$.
q.e.d.

Of course, the reader knows the preceding result in the case where H is finite dimensional, i.e. a Euclidean space. x_1 is interpreted as the orthogonal projection of x onto the subspace M, and therefore Corollary 2.2.4 is called the **projection theorem**.

The next result will be needed for Sections 4.2 and 4.3 when we establish the existence of minimizers for lower semicontinuous, convex functionals.

Theorem 2.2.4 (Mazur). *Suppose $(x_n)_{n \in \mathbb{N}}$ converges weakly to x in some Banach space V. For every $\varepsilon > 0$, we may then find a convex combination*

$$\sum_{n=1}^{N} \lambda_n x_n \quad (\lambda_n > 0, \sum_{n=1}^{N} \lambda_n = 1)$$

with

$$\left\| \sum_{n=1}^{N} \lambda_n x_n - x \right\| \leq \varepsilon. \tag{2.2.16}$$

Proof. We consider the set C_0 of all convex combinations of the x_n, i.e.

$$C_0 := \left\{ \sum_{n=1}^{N} \lambda_n x_n \text{ with } \lambda_n \geq 0, \sum_{n=1}^{N} \lambda_n = 1 \right\}.$$

2.2 Dual spaces and weak convergence

Replacing all x_n by $x_n - x_1$ and x by $x - x_1$, we may assume $0 \in C_0$. If (2.2.16) is not true, then there exists $\varepsilon > 0$ with

$$||x - y|| > \varepsilon \text{ for all } y \in C_0. \qquad (2.2.17)$$

$$C_1 := \{z \in V : ||z - y|| \leq \frac{\varepsilon}{2} \text{ for some } y \in C_0\}$$

is convex and contains the ball with radius $\frac{\varepsilon}{2}$ and center 0. We consider the Minkowski functional p of C_1 defined by

$$p(z) := \inf\{\lambda > 0 : \lambda^{-1} z \in C_1\}.$$

p is convex in the sense of Definition 2.1.5 since C_1 is convex, and continuous since C_1 contains the ball of radius $\frac{\varepsilon}{2} > 0$ about 0. Since, because of (2.2.17),

$$||x - z|| > \frac{\varepsilon}{2} \text{ for every } z \in C_1,$$

we have

$$p(x) > 1.$$

More precisely, there exists y_0 with

$$x = \lambda^{-1} y_0, \quad 0 < \lambda < 1$$
$$p(y_0) = 1.$$

We consider the linear subspace

$$V_0 = \{\mu y_0, \mu \in \mathbb{R}\} \subset V$$

and the linear functional

$$f_0(\mu y_0) = \mu \text{ on } V_0.$$

Then

$$f_0 \leq p \text{ on } V_0,$$

and by the Hahn–Banach Theorem 2.1.1, there exists an extension f of f_0 to all V with

$$f \leq p.$$

Since p is continuous, f is also continuous (see Lemma 2.2.1). We have

$$\sup_{y \in C_0} f(y) \leq \sup_{y \in C_1} f(y) \leq \sup_{y \in C_1} p(y) = 1$$
$$< \lambda^{-1} = f(\lambda^{-1} y_0) = f(x).$$

This, however, contradicts the fact that $(x_n)_{n \in \mathbb{N}} \subset C_0$ converges weakly to x. Thus, (2.2.17) cannot hold, and (2.2.16) is established.

q.e.d.

2.3 Linear operators between Banach spaces

The results of this section will be used in Chapter 8. In Section 2.2, we considered linear functionals

$$f : V \to \mathbb{R};$$

in the beginning, V was a normed linear space, with norm denoted by $\|\cdot\|$, and later, we also assumed that V was complete, i.e. a Banach space. In the present section, we replace the target $\mathbb{R}$ by a general Banach space W, with norm also denoted by $\|\cdot\|$. We thus consider linear operators

$$T : V \to W,$$

and we put

$$\|T\| := \sup_{x \neq 0} \frac{\|Tx\|}{\|x\|} \in \mathbb{R}^+ \cup \{\infty\}. \tag{2.3.1}$$

Lemma 2.3.1. *The linear operator $T : V \to W$ is continuous if and only if $\|T\| < \infty$.*

Proof. If $\|T\| < \infty$, then the inequality

$$\|Tx\| \leq \|T\| \, \|x\| \tag{2.3.2}$$

implies that T is continuous. (Of course, this uses the linearity of T.) Conversely, if T is continuous, we recall the usual $\varepsilon - \delta$ criterion for continuity, and so for $\varepsilon = 1$, we find some $\delta > 0$ with the property that

$$\|Ty\| \leq 1 \text{ if } \|y\| \leq \delta.$$

For $x \in V \setminus \{0\}$, we then have with $y = \delta \frac{x}{\|x\|}$ ($\|y\| \leq \delta$)

$$\|Tx\| = \left\| \frac{\|x\|}{\delta} Ty \right\| \leq \frac{1}{\delta} \|x\|.$$

Thus

$$\|T\| \leq \frac{1}{\delta} < \infty.$$

q.e.d.

2.3 Linear operators between Banach spaces

The space of continuous linear operators $T : V \to W$ between the normed spaces $(V, ||\cdot||)$ and $(W, ||\cdot||)$ is denoted by $L(V, W)$. It becomes a normed space with norm $||T||$.

Lemma 2.3.2. *If $(W, ||\cdot||)$ is a Banach space, then so is $(L(V, W), ||\cdot||)$.*

The *proof* is the same as the one of Lemma 2.2.2, simply replacing $(\mathbb{R}, |\cdot|)$ by $(W, ||\cdot||)$.

Remark 2.3.1. Again, $(V, ||\cdot||)$ need not be a Banach space here.

Lemma 2.3.3. *Let $T \in L(V, W)$. Then*
$$\ker T := \{x \in V : Tx = 0\}$$
is a closed linear subspace of V.

Proof. $\ker T = T^{-1}(0)$ is the pre-image of a closed set under a continuous map, hence closed.
q.e.d.

In the sequel, we shall encounter bijective continuous linear operators
$$T : V \to W$$
between Banach spaces. It is a general theorem in functional analysis, the inverse operator theorem, that the inverse of T, denoted by T^{-1}, is then continuous as well. Here, however, we do not want to prove that result, and we shall therefore frequently assume that T^{-1} is continuous although that assumption is automatically fulfilled in the light of that theorem.

Lemma 2.3.4. *Let*
$$T : V \to W$$
be a bijective continuous linear map between Banach spaces, with a continuous inverse T^{-1}. If $S \in L(V, W)$ satisfies
$$||T - S|| < \frac{1}{||T^{-1}||}, \qquad (2.3.3)$$
then S is bijective, and S^{-1} is continuous, too.

Proof. We have
$$S = T(Id - T^{-1}(T - S)).$$

As with the geometric series, the inverse of S then is given by

$$\left(\sum_{\nu=0}^{\infty}(T^{-1}(T-S))^{\nu}\right)T^{-1}, \tag{2.3.4}$$

provided that series converges. However,

$$\left\|\sum_{\nu=m}^{n}(T^{-1}(T-S))^{\nu}\right\| \leq \sum_{\nu=m}^{n}\|(T^{-1}(T-S))^{\nu}\|$$

$$\leq \sum_{\nu=m}^{n}(\|T^{-1}\|\,\|T-S\|)^{\nu},$$

and since $\|T^{-1}\|\,\|T-S\| < 1$ by assumption, the series satisfies the Cauchy property and hence converges to a linear operator with finite norm.

q.e.d.

If V is a vector space, we say that V is the direct sum of the subspaces V_1, V_2,

$$V = V_1 \oplus V_2$$

if for every $x \in V$, we can find unique elements $x_1 \in V_1$, $x_2 \in V_2$, with

$$x = x_1 + x_2.$$

We then also call V_1 and V_2 complementary subspaces of V. Easy linear algebra also shows that if V_1 possesses a complementary subspace of finite dimension, then the dimension of that space is uniquely determined, i.e. if $V_1 \oplus V_2 = V_1 \oplus V_2'$, then $\dim V_2 = \dim V_2'$.

We now consider a normed vector space $(V, \|\cdot\|)$. Then every finite dimensional subspace V_0 is complete, hence closed. We also have:

Lemma 2.3.5. *Let $V_0 \subset V$ be a finite dimensional subspace of the normed vector space $(V, \|\cdot\|)$. Then V_0 possesses a closed complementary subspace V_1, i.e. $V = V_0 \oplus V_1$.*

Proof. Let $e_1, ..., e_n$ be a basis of V_0, $f_0^j : V_0 \to \mathbb{R}$ be the linear functionals with

$$f_0^j(e_i) = \delta_{ij} \quad (i, j = 1, ..., n).$$

By Corollary 2.1.1, we may find extensions $f^j : V \to \mathbb{R}$ with $f^j_{|V_0} = f_0^j$.

2.3 Linear operators between Banach spaces

We define $\pi : V \to V$ as

$$\pi x := \sum_{j=1}^{n} f^j(x) e_j.$$

π is continuous, with $\pi(V) = V_0$.

$$V_1 := \ker \pi$$

then is closed as the kernel of a continuous linear operator (Lemma 2.3.3), and every $x \in V$ admits the unique decomposition

$$x = \pi(x) + (x - \pi(x))$$

with $\pi(x) \in V_0$, $x - \pi(x) \in V_1$, because $\pi \circ \pi = \pi$.

q.e.d.

Definition 2.3.1. *Let $T : V \to W$ be a continuous linear operator between Banach spaces $(V, \|\cdot\|)$ and $(W, \|\cdot\|)$. T is called a Fredholm operator if the following conditions hold:*

(i) *$V_0 = \ker T$ is finite dimensional. Consequently, according to Lemma 2.3.5, there exists a closed subspace V_1 of V with*

$$V = V_0 \oplus V_1. \tag{2.3.5}$$

(ii) *There exists a finite dimensional subspace W_0 of W, called the cokernel of T (coker T) giving rise to a decomposition of W into closed subspaces*

$$W = W_0 \oplus W_1 \tag{2.3.6}$$

with

$$W_1 = T(V) =: R(T) \quad (\text{range of } T).$$

Thus, T yields bijective continuous linear operator $T_1 : V_1 \to W_1$. We finally require

(iii) *$T^{-1} : W_1 \to V_1$ is continuous.*

For a Fredholm operator T, we call

$$\operatorname{ind} T := \dim V_0 - \dim W_0 \quad (= \dim \ker T - \dim \operatorname{coker} T)$$

the index of T. The set of all Fredholm operators $T : V \to W$ is denoted by $F(V, W)$.

Remark 2.3.2. Question to the reader: Why is $F(V, W)$ not a vector space?

Remark 2.3.3. As mentioned, condition (iii) is automatically satisfied as a consequence of the inverse operator theorem.

Remark 2.3.4. In our conventions, the cokernel of T is only determined up to isomorphism, i.e. any W_0 satisfying (2.3.6) with $W_1 = T(V)$ is a cokernel. Usually, one defines the cokernel as the quotient space W/W_1, but here we do not want to introduce quotient spaces of Banach spaces.

Theorem 2.3.1. *Let V, W be Banach spaces. $F(V, W)$ is open in $L(V, W)$, and*
$$\text{ind} : F(V, W) \to \mathbb{Z}$$
is continuous, hence constant on each connected component of $F(V, W)$.

Proof. Let $T : V \to W$ be a Fredholm operator. We use the decompositions
$$V = V_0 \oplus V_1 \quad \text{with} \quad V_0 = \ker T$$
$$W = W_0 \oplus W_1 \quad \text{with} \quad W_0 = \operatorname{coker} T$$
of Definition 2.3.1. For $S \in L(V, W)$, we define a continuous linear operator
$$S' : V_1 \times W_0 \to W$$
$$(x, z) \mapsto Sx + z,$$
and we obtain a continuous linear operator
$$L(V, W) \to L(V_1 \times W_0, W)$$
$$S \mapsto S'.$$

Since $T_1 : V_1 \to W_1$ is bijective with a continuous inverse, T' is also bijective with a continuous inverse, and by Lemma 2.3.4 this then also holds for all S in some neighbourhood of T. For such S, $S'(V_1)$ is closed as V_1 is closed and S' is continuous, and we have the decomposition
$$W = S'(V_1) \oplus S'(W_0),$$
and since $S'(V_1) = S(V_1)$ also
$$W = S(V_1) \oplus S'(W_0), \tag{2.3.7}$$
and since W_0 is finite dimensional, so is $S'(W_0)$. Then $S(V) \supset S(V_1)$ is also closed since $S(V_1)$ is closed and possesses a complementary subspace of finite dimension.

Finally, the dimension of the kernel of S is upper semicontinuous.

2.3 Linear operators between Banach spaces

Namely, if S is in our above neighbourhood of T, then since S is bijective, S is injective on V_1, and hence the kernel of S is contained in some complementary subspace of V_1, and as observed above, the dimension of such a subspace equals the one of V_0. Thus

$$\dim \ker S \leq \dim \ker T \tag{2.3.8}$$

if S is in a suitable neighbourhood of T in $L(V, W)$.

Altogether, we have verified that S is a Fredholm operator if it is sufficiently close to T.

From the preceding, we see that there exist finite dimensional subspaces $V_0' = \ker S$ and V_0'' of V with

$$V = V_0' \oplus V_0'' \oplus V_1,$$

and thus

$$\dim V_0' + \dim V_0'' = \dim V_0 \quad (V_0 = \ker T). \tag{2.3.9}$$

S thus is injective on $V_0'' \oplus V_1$, and since S coincides with S' on V_1, we get a decomposition

$$W = S(V_1) \oplus S(V_0'') \oplus W_0',$$

with $W_0' = \operatorname{coker} S$ and from (2.3.7)

$$\dim S(V_0'') + \dim W_0' = \dim S'(W_0) = \dim W_0 \tag{2.3.10}$$

since S' is bijective.

Consequently

$$\begin{aligned}
\operatorname{ind} S &= \dim \ker S - \dim \operatorname{coker} S \\
&= \dim V_0' - \dim W_0' \\
&= (\dim V_0 - \dim V_0'') - (\dim W_0 - \dim S(V_0'')) \text{ by (2.3.9), (2.3.10)} \\
&= \dim V_0 - \dim W_0 \quad \text{since } S \text{ is injective on } V_0'' \\
&= \operatorname{ind} T.
\end{aligned}$$

for S in some neighborhood of T.

q.e.d.

The following result motivates the definition of a Fredholm operator:

Theorem 2.3.2 (Fredholm alternative). *Let V be a Banach space, $T: V \to V$ a Fredholm operator of index 0. We consider the equation*

$$Tx = y. \tag{2.3.11}$$

Either

(i) *Either $Tx = y$ is solvable for all y, and thus T is surjective, hence also injective as $\operatorname{ind} T = 0$, and so the solution x is uniquely determined by y,*

or

(ii) *$Tx = y$ is only solvable if y is contained in some proper subspace of V (with a finite dimensional complementary subspace), and for each such y, the solutions x constitute a finite dimensional affine subspace.*

Proof. A direct consequence of the definition.

q.e.d.

2.4 Calculus in Banach spaces

In this section, we collect some material that will only be used in Chapters 8 and 9.

Definition 2.4.1. *Let $(V, \|\cdot\|_V)$, $(W, \|\cdot\|_W)$ be Banach spaces, $F : V \to W$ a map. F is called differentiable (in the sense of Fréchet) at $u \in V$ if there exists a bounded linear map*

$$DF(u) : V \to W$$

with

$$\lim_{\substack{v \to 0 \\ (v \neq 0)}} \frac{\|F(u+v) - F(u) - DF(u)(v)\|_W}{\|v\|_V} = 0. \quad (2.4.1)$$

f is called differentiable in $U \subset V$ if it is differentiable at every $u \in U$. f is said to be of class C^1 if $DF(u)$ depends continuously on u. f is said to be of class C^2 if $DF(u)$ is differentiable in u and the derivative $D^2F(u) := D(DF)(u)$ depends continuously on u.

It is easy to show that a differentiable map is continuous.

We now wish to derive the implicit and inverse function theorems in Banach spaces that will be used in Chapter 8. We shall need a technical tool, the Banach fixed point theorem:

Lemma 2.4.1. *Let A be a closed subset of some Banach space $(V, \|\cdot\|)$. Let $0 \leq q < 1$, and suppose $G : A \to A$ satisfies*

$$\|Gy_1 - Gy_2\| \leq q \|y_1 - y_2\| \quad \text{for all } y_1, y_2 \in A. \quad (2.4.2)$$

Then there exists a unique $y \in A$ with
$$Gy = y. \qquad (2.4.3)$$

If we have a continuous family $G(x)$ where all the $G(x)$ satisfy (2.4.2) (with q not depending on x), then the solution $y = y(x)$ of (2.4.3) depends continuously on X.

Proof. We choose $y_0 \in A$ and put iteratively
$$y_n := Gy_{n-1}.$$

We have
$$y_n = \sum_{i=1}^n (y_i - y_{i-1}) + y_0 = \sum_{i=1}^n \left(G^{i-1}y_1 - G^{i-1}y_0\right) + y_0. \qquad (2.4.4)$$

We obtain from (2.4.2)
$$\sum_{i=1}^n \|G^{i-1}y_1 - G^{i-1}y_0\| \leq \sum_{i=1}^n q^{i-1} \|y_1 - y_0\| \leq \frac{1}{1-q} \|y_1 - y_0\|.$$

Consequently, the series y_n in (2.4.4) converges absolutely and uniformly to some $y \in A$, noting that A is assumed to be closed and the limit function $y = y(x)$ is continuous. We have
$$y = \lim_{n \to \infty} Gy_n = G\left(\lim_{n \to \infty} y_n\right) = Gy,$$

hence (2.4.3). The uniqueness of a solution of (2.4.3) follows from (2.4.2), since $q < 1$.

q.e.d.

Theorem 2.4.1 (Implicit Function Theorem). *Let V_1, V_2, W be Banach spaces with all norms denoted by $\|\cdot\|$, $U \subset V_1 \times V_2$ open, $(x_0, y_0) \in U$, $F \in C^1(U, W)$, i.e. F is continuously differentiable. For purposes of normalization solely, we assume*
$$F(x_0, y_0) = 0. \qquad (2.4.5)$$

We also suppose that
$$D_2 F(x_0, y_0) : V_2 \to W,$$

the derivative of $F(x_0, \cdot) : V_2 \to W$ at $y = y_0$, is invertible. By our differentiability assumption, $D_2 F(x_0, y_0)$ is continuous, and we assume that its inverse is likewise continuous. Then there exist open neighbourhoods U_1 of x_0, U_2 of y_0 with $U_1 \times U_2 \in U$, and a differentiable map

$$\varphi : U_1 \to U_2$$

with

$$F(x, \varphi(x)) = 0 \tag{2.4.6}$$

and

$$D\varphi(x) = -(D_2 F(x, \varphi(x)))^{-1} \circ D_1 F(x, \varphi(x)) \quad \text{for all } x \in U_1 \tag{2.4.7}$$

$(D_1 F(\cdot, y) : V_1 \to W$ is the derivative of $F(\cdot, y) : V_1 \to W)$. In fact, for every $x \in U_1$, $\varphi(x)$ is the only solution of (2.4.6) in U_2.

The content of the implicit function theorem is that the equation

$$F(x, y) = 0$$

can be solved locally uniquely for y as a function of x, if the derivative of F w.r.t. y is continuously invertible.

Proof. The idea is to transform the problem into a fixed point problem for which the Banach fixed point theorem is applicable. We put

$$l := D_2 F(x_0, y_0).$$

With this notation, our fixed point equation is

$$\Phi(x, y) := y - l^{-1} F(x, y) = y \tag{2.4.8}$$

which clearly is equivalent to our orginal equation $F(x, y) = 0$. For every x, we thus want to find a fixed point of

$$y \mapsto \Phi(x, y).$$

Using $l^{-1} \circ l = id$ (note that l is invertible by assumption), we get

$$\Phi(x, y_1) - \Phi(x, y_2) = l^{-1} (D_2 F(x_0, y_0)(y_1 - y_2) - (F(x, y_1) - F(x, y_2))).$$

In Lemma 2.4.1, we take $q = \frac{1}{2}$, and by the differentiability of F at (x_0, y_0) and the continuity of l^{-1}, we may find $\delta' > 0, \varepsilon > 0$ with the property that for

$$||x - x_0|| \leq \delta'$$

and

$$||y_1 - y_0|| \leq \varepsilon, ||y_2 - y_0|| \leq \varepsilon \quad (\text{ hence also } ||y_1 - y_2|| < 2\varepsilon),$$

we have
$$\|\Phi(x,y_1) - \Phi(x,y_2)\| \leq \frac{1}{2}\|y_1 - y_2\|.$$

Furthermore, we may find $\delta'' > 0$ with the property that for
$$\|x - x_0\| \leq \delta'',$$

we have
$$\|\Phi(x,y_0) - \Phi(x_0,y_0)\| < \frac{\varepsilon}{2}.$$

Since $\Phi(x_0, y_0) = y_0$ by assumption, we then have for $\|y - y_0\| \leq \varepsilon$
$$\|\Phi(x,y) - y_0\| \leq \|\Phi(x,y) - \Phi(x,y_0)\| + \|\Phi(x,y_0) - \Phi(x_0,y_0)\|$$
$$\leq \frac{1}{2}\|y - y_0\| + \frac{\varepsilon}{2} < \varepsilon$$

whenever $\|x - x_0\| \leq \delta := \min(\delta', \delta'')$. This means that if $\|x - x_0\| \leq \delta$, $\Phi(x, y)$ maps the closed ball
$$A := \{y \in V_2 : \|y - y_0\| \leq \varepsilon\}$$

onto itself. By Lemma 2.4.1, for every x with $\|x - x_0\| \leq \delta$, there exists a unique $y =: \varphi(x)$ with $\|y - y_0\| \leq \varepsilon$ and $y = \Phi(x, y)$, i.e. $F(x, y) = 0$. Moreover, y depends continuously on x. We consider the open balls
$$U_1 := \{x : \|x - x_0\| < \delta\}, U_2 := \{y : \|y - y_0\| < \varepsilon\}.$$

($\Phi(x, \cdot)$ also maps the open ball U_2 onto itself.) By choosing $\delta, \varepsilon > 0$ smaller, if necessary, we may assume
$$U_1 \times U_2 \subset U.$$

It remains to show that $\varphi(x)$ is differentiable and that its derivative is given by (2.3.7). We consider
$$(x_1, \varphi(x_1)) \in U_1 \times U_2,$$

and abbreviate $y_1 := \varphi(x_1)$. We put
$$l_1 := D_1 F(x_1, y_1), l_2 := D_2 F(x_1, y_1).$$

Since F is differentiable, we may write
$$F(x,y) = l_1(x - x_1) + l_2(x - x_2) + r(x,y)$$

where the remainder term satisfies
$$\lim_{\substack{x \to x_1 \\ y \to y_1}} \frac{r(x,y)}{\|x - x_1\| + \|y - y_1\|} = 0. \tag{2.4.9}$$

Since $F(x, \varphi(x)) = 0$ for $x \in U_1$ by construction of φ, we obtain
$$\varphi(x) = -l_2^{-1}l_1(x - x_1) + y_1 - l_2^{-1}r(x, \varphi(x)). \tag{2.4.10}$$

By (2.4.9), we may find $\eta, \rho > 0$ such that for
$$\|x - x_1\| \leq \eta, \ \|y - y_1\| \leq \rho$$
$$\|r(x, y)\| \leq \frac{1}{2\|l_2^{-1}\|}(\|x - x_1\| + \|y - y_1\|).$$

Thus
$$\|r(x, \varphi(x))\| \leq \frac{1}{2\|l_2^{-1}\|}(\|x - x_1\| + \|\varphi(x) - \varphi(x_1)\|). \tag{2.4.11}$$

By (2.4.10), (2.4.11),
$$\|\varphi(x) - \varphi(x_1)\| \leq \|l_2^{-1}l_1\|\,\|x - x_1\| + \frac{1}{2}\|x - x_1\| + \frac{1}{2}\|\varphi(x) - \varphi(x_1)\|,$$
hence
$$\|\varphi(x) - \varphi(x_1)\| \leq c\|x - x_1\| \text{ for a constant } c.$$

We abbreviate $r_0(x) := -l_2^{-1}r(x, \varphi(x))$ and rewrite (2.4.10) as
$$\varphi(x) - \varphi(x_1) = -l_2^{-1}l_1(x - x_1) + r_0(x), \tag{2.4.12}$$

with
$$\lim_{x \to x_1} \frac{r_0(x)}{\|x - x_1\|} = 0 \quad \text{from (2.4.9)}. \tag{2.4.13}$$

(2.4.12) and (2.4.13) yields the differentiability of φ and (2.4.7).
q.e.d.

Corollary 2.4.1 (Inverse Function Theorem). *Let V, W be Banach spaces, $U \subset V$ open, $y_0 \in U$. Let $f : U \to W$ be continuously differentiable, and assume that the derivative $Df(y_0)$ is invertible with a continuous inverse. Then there exist open neighbourhoods $U_2 \subset U$ of y_0, U_1 of $f(y_0) =: x_0$ so that f maps U_2 bijective onto U_1, and the inverse $\varphi := f^{-1} : U_1 \to U_2$ is differentiable with*
$$D\varphi(x_0) = (Df(y_0))^{-1}. \tag{2.4.14}$$

Proof. We shall apply Theorem 2.4.1 to $F(x, y) := f(y) - x$, and find an open neighbourhood U_1 of x_0 and a differentiable function
$$\varphi : U_1 \to V$$

2.4 Calculus in Banach spaces

with $\varphi(U_1) \subset U_2$ for a neighbourhood U_2 of y_0, with $\varphi(x_0) = y_0$ and
$$F(x, \varphi(x)) = 0, \text{ i.e. } f(\varphi(x)) = x \text{ for } x \in U_1.$$

As $\varphi(U_1) = f^{-1}(U_1)$ is open, we may redefine U_2 as $\varphi(U_1)$, and φ then yields a bijection between U_1 and U_2. As $f(\varphi(x)) = x$, the chain rule implies
$$Df(\varphi(x_0)) \cdot D\varphi(x_0)) = \text{id}, \text{ i.e. } (2.4.14).$$

q.e.d.

The next topic concerns ordinary differential equations in Banach spaces. In Chapter 9, we shall use the Picard–Lindelöf theorem in a Banach space that we shall now derive.

We need the integral of a continuous function
$$x : I \to V$$
from some interval $I = [a,b] \subset \mathbb{R}$ into some Banach space V,
$$\int_a^b x(t)dt.$$
This can be defined as a Riemann integral as in the case of real-valued functions through approximation by step functions.

Given a continouous
$$\Phi : \mathbb{R} \times V \to V,$$
we say that $x(t)$ solves the ODE (ordinary differential equation) on I,
$$\frac{d}{dt}x(t) = \dot{x}(t) = \Phi(t, x(t)) \quad \text{with } x(a) = x_0 \tag{2.4.15}$$
if for all $t \in I$
$$x(t) = x_0 + \int_a^b \Phi(\tau, x(\tau))d\tau. \tag{2.4.16}$$

Theorem 2.4.2 (Picard–Lindelöf). *Suppose that Φ is uniformly Lipschitz continuous, i.e. suppose there exists some $L < \infty$ with*
$$\|\Phi(t_1, x_1) - \Phi(t_2, x_2)\| \leq L\left(|t_1 - t_2| + \|x_1 - x_2\|\right)$$
$$\text{for all } t \subset I, x_1, x_2 \in V. \tag{2.4.17}$$
Then for any $x_0 \in V$, there exists a unique solution of (2.4.15).

Proof. We shall solve (2.4.16) with the help of Lemma 2.4.1. For a continuous $y : I \to V$, we define $Gy \in C^0(I, V)$,

$$(Gy)(t) := x_0 + \int_a^b \Phi(\tau, y(\tau)) d\tau.$$

We note that $C^0(I, V)$, the space of continuous functions from I with values in V, is a Banach space w.r.t. the norm

$$\|y\|_{C^0} := \sup_{t \in I} \|y(t)\|.$$

(To verify this, one just needs to observe that any sequence $(y_n)_{n \in \mathbb{N}} \subset C^0(I, V)$ with

$$\lim_{n,m \to \infty} \|y_n - y_m\|_{C^0} \left(= \lim_{n,m \to \infty} \sup_{t \in I} |y_n(t) - y_m(t)| \right) = 0$$

converges uniformly to some continuous function $y : I \to V$.)
We have

$$\|Gy_1 - Gy_2\|_{C^0} = \sup_{t \in I} \left| \int_a^t (\Phi(\tau, y_1(\tau)) - \Phi(\tau, y_2(\tau))) d\tau \right|$$
$$\leq |t - a| L \|y_1 - y_2\|_{C_0} \quad \text{because of (2.4.17).}$$

We choose $\epsilon > 0$ so small that

$$L\epsilon \leq \frac{1}{2}.$$

Lemma 2.4.1 with V replaced by $C^0([a, a + \epsilon], V)$ and with $q = \frac{1}{2}$ then implies that there exists a unique $y \in C^0([a, a + \epsilon], V)$ with

$$Gy(t) = x_0 + \int_a^b \Phi(\tau, y(\tau)) d\tau \quad \text{for } a \leq t \leq a + \epsilon.$$

Repeating the construction with $a + \epsilon$ in place of a and $y(t + \epsilon)$ in place of x_0 yields the solution on $[a, a + 2\epsilon]$, and so on.

q.e.d.

Remark 2.4.1. If I is an infinite or semi-infinite interval, e.g. $I = [a, \infty)$, and if (2.4.17) holds on I, we obtain a solution of (2.4.15) on I, since Theorem 2.4.2 yields a solution on every interval $[a, b]$ with $b < \infty$.

Corollary 2.4.2. *Let the assumptions of Theorem 2.4.2 be satisfied on the interval $I = [0, \infty)$, and suppose that Φ does not depend explicitly on t, i.e. $\Phi : V \to V$, $\Phi = \Phi(x)$. For $x_0 \in V$ we thus consider the ODE*

$$\dot{x}(t) = \Phi(x(t)), \quad x(0) = x_0. \tag{2.4.18}$$

($x(0)$, the value at 'time' 0, is called initial value). We denote the solution by $x(x_0, t)$. Then for $s, t \geq 0$,

$$x(x_0, t+s) = x(x(t), s) \quad \text{(semigroup property)}.$$

Thus, the solution with initial value x_0 at 'time' $t + s$ is the same as the solution with initial value $x(t)$ computed at 'time' s.

Proof. This follows from the uniqueness statement in Theorem 2.4.2, as both sides of (2.4.18) are solutions.

q.e.d.

Exercises

2.1 Let $(V, \|\cdot\|_V)$ $(W, \|\cdot\|_W)$ be normed linear spaces. For a linear functional

$$f : V \to W,$$

put

$$\|f\| := \sup_{x \in V \setminus \{0\}} \frac{\|f(x)\|_W}{\|x\|_V}.$$

Show that f is continuous iff $\|f\| < \infty$. Let $L(V, W) := \{f : V \to W \text{ linear with } \|f\| < \infty\}$. Show that if $(W, \|\cdot\|_W)$ is a Banach space then so is $(L(V, W), \|\cdot\|)$.

2.2 Show that a normed space $(V, \|\cdot\|)$ is uniformly convex if the following condition holds:
Whenever $(x_n)_{n \in \mathbb{N}}, (y_n)_{n \in \mathbb{N}} \subset V$ satisfy

$$\limsup_{n \to \infty} \|x_n\| \leq 1, \quad \limsup_{n \to \infty} \|y_n\| \leq 1$$

and

$$\lim_{n \to \infty} \|x_n + y_n\| = 2,$$

then

$$\lim_{n \to \infty} (x_n - y_n) = 0.$$

2.3 A normed space $(V, \|\cdot\|)$ is called strictly normed if the following condition holds: Whenever $x, y \in V$, $x, y \neq 0$ satisfy

$$\|x + y\| = \|x\| + \|y\|,$$

then there exists $\alpha > 0$ with

$$x = \alpha y.$$

Show that any uniformly convex normed space is strictly normed.

2.4 Does the Banach fixed point theorem (Lemma 2.4.1) continue to hold if we replace (2.4.2) by the condition

$$\|Gy_1 - Gy_2\| < \|y_1 - y_2\| \quad \text{for all } y_1, y_2 \in A?$$

3
L^p and Sobolev spaces

3.1 L^p spaces

In the sequel, instead of functions $f : A \to \mathbb{R} \cup \{\pm\infty\}$ (A measurable), we shall consider equivalence classes of functions, where f and g are equivalent if $f(x) = g(x)$ for almost all $x \in A$. We shall be lax with the notation, however, not distinguishing between a function and its equivalence class. The equivalence class of the zero function is called the null class, and a function in that class is called a null function.

Definition 3.1.1. Let $A \subset \mathbb{R}^d$ be measurable, $p \in \mathbb{R} \setminus \{0\}$.

$$L^p(A) = \{(\text{equivalence classes of}) \text{ measurable} \\ \text{functions } f : A \to \mathbb{R} \cup \{\pm\infty\} \text{ with} \\ |f(x)|^p \in \mathcal{L}^1(A)\}.$$

For $f \in L^p(A)$, we put

$$\|f\|_p := \|f\|_{L^p(A)} := \left(\int_A |f(x)|^p \, dx\right)^{\frac{1}{p}}. \tag{3.1.1}$$

The notation suggests that $\|\cdot\|_p$ is a norm, and we now proceed to verify this for $p \geq 1$. First of all,

$$\|f\|_p = 0 \Leftrightarrow f \text{ is a null function}. \tag{3.1.2}$$

Thus, $\|\cdot\|_p$ is positive definite (on the set of equivalence classes). Next, for $c \in \mathbb{R}$,

$$\|cf\|_p = |c| \, \|f\|_p. \tag{3.1.3}$$

It remains to verify the triangle inequality. This is obvious for $p = 1$:

$$\|f_1 + f_2\|_{L^1(A)} \leq \|f_1\|_{L^1(A)} + \|f_2\|_{L^1(A)}. \tag{3.1.4}$$

For $p > 1$, we need

Lemma 3.1.1 (Hölder's inequality). *Let $p, q > 1$ satisfy $\frac{1}{p} + \frac{1}{q} = 1$, $f_1 \in L^p(A)$, $f_2 \in L^q(A)$. Then $f_1, f_2 \in L^1(A)$, and*

$$\|f_1 f_2\|_1 \leq \|f_1\|_p \|f_2\|_q. \tag{3.1.5}$$

Proof. By homogeneity, we may assume w.l.o.g.

$$\|f_1\|_p = 1, \quad \|f_2\|_q = 1. \tag{3.1.6}$$

Recalling Young's inequality, namely

$$ab \leq \frac{a^p}{p} + \frac{b^q}{q} \quad \text{for } a, b \geq 0, \ p, q > 1, \ \frac{1}{p} + \frac{1}{q} = 1, \tag{3.1.7}$$

we have for $x \in A$

$$|f_1(x) f_2(x)| \leq \frac{|f_1(x))|^p}{p} + \frac{|f_2(x)|^q}{q},$$

hence by our normalization (3.1.6)

$$\int_A |f_1(x) f_2(x)| \, dx \leq \frac{1}{p} + \frac{1}{q} = 1 = \|f_1\|_p \|f_2\|_q.$$

q.e.d.

We now obtain the triangle inequality:

Lemma 3.1.2 (Minkowski's inequality). *Let $f_1, f_2 \in L^p(A)$, $p \geq 1$. Then*

$$\|f_1 + f_2\|_p \leq \|f_1\|_p + \|f_2\|_p. \tag{3.1.8}$$

Proof. The case $p = 1$ is given by (3.1.4). We now consider $p > 1$ and put $q := \frac{p}{p-1}$ (so that $\frac{1}{q} + \frac{1}{p} = 1$). For $\psi(x) := |f_1(x) + f_2(x)|^{p-1}$, we have

$$\psi^q = |f_1 + f_2|^p,$$

i.e. $\psi \in L^q(A)$. Since

$$|f_1(x) + f_2(x)|^p \leq |f_1(x)\psi(x)| + |f_2(x)\psi(x)|,$$

we get

$$\|f_1 + f_2\|_p^p \leq \|f_1 \psi\|_1 + \|f_2 \psi\|_1$$
$$\leq \|f_1\|_p \|\psi\|_q + \|f_2\|_p \|\psi\|_q$$
$$\text{by Hölder's inequality}$$
$$= \left(\|f_1\|_p + \|f_2\|_p\right) \|f_1 + f_2\|_p^{\frac{p}{q}}.$$

3.1 L^p spaces

Since $p - \frac{p}{q} = 1$, (3.1.8) follows.

q.e.d.

We have thus verified that $\|\cdot\|_p$ is a norm on $L^p(A)$. In fact, we have:

Theorem 3.1.1 (Riesz-Fischer). *Let A be measurable, $p \geq 1$. Then $L^p(A)$ is a Banach space.*

Proof. Let $(f_n)_{n \in \mathbb{N}} \subset L^p(A)$ be a Cauchy sequence. For every $\nu \in \mathbb{N}$, we may then find $n_\nu \in \mathbb{N}$ with

$$\|f_n - f_{n_\nu}\|_p \leq \frac{1}{2^\nu} \quad \text{for all } n \geq n_\nu.$$

This implies that the series

$$\|f_{n_1}\|_p + \sum_{\nu=1}^{\infty} \|f_{n_{\nu+1}} - f_{n_\nu}\|_p \tag{3.1.9}$$

converges. We claim that the series

$$f_{n_1} + \sum_{\nu=1}^{\infty} \left(f_{n_{\nu+1}} - f_{n_\nu} \right) \tag{3.1.10}$$

then converges in $L^p(A)$. Since all elements of the series are nonnegative, $(g_m)_{m \in \mathbb{N}}$ converges to some $g : A \to \mathbb{R}^+ \cup \{\infty\}$ pointwise in A, and Corollary 1.2.1 implies that (g_m) also converges to g in $L^p(A)$. In particular, $g(x) < \infty$ for almost all $x \in A$. Thus, our original sequence (3.1.10) is absolutely convergent for almost all $x \in A$, towards some f with $|f| \leq g + |f_{n_1}|$; in particular $f \in L^p(\Omega)$. We interrupt the proof to record:

Lemma 3.1.3. *Let $(f_n)_{n \in \mathbb{N}}$ converge to f in $L^p(A)$. Then some subsequence converges pointwise almost everywhere to f.*

In order to complete the proofs of Lemma 3.1.3 and Theorem 3.1.1, it remains to show that the series (3.1.10) converges to f in $L^p(A)$. (Then a subsequence of (f_n) converges to f in $L^p(A)$. Since (f_n) was assumed to be a Cauchy sequence in $L^p(A)$, the whole sequence has to converge in $L^p(A)$. It is in general not true, however, that the whole sequence also converges pointwise almost everywhere to f.)
This is easy:

$$f_{n_1}(x) + \sum_{\nu=1}^{\infty} \left(f_{n_{\nu+1}}(x) - f_{n_\nu}(x) \right) - f(x)$$

converges to 0 almost everywhere in A, and since

$$\left| f_{n_1}(x) + \sum_{\nu=1}^{\infty} \left(f_{n_{\nu+1}}(x) - f_{n_\nu}(x) \right) - f(x) \right| \leq 2\,|g(x)| + 2\,|f_{n_1}(x)|,$$

we may apply Lebesgue's Theorem 1.2.3 on dominated convergence to conclude that we get convergence also w.r.t. $\|\cdot\|_p$.

q.e.d.

Corollary 3.1.1. $L^2(A)$ *is a Hilbert space with scalar product*

$$(f_1, f_2) := \int_A f_1(x) f_2(x)\, dx.$$

Proof. It follows from Hölder's inequality (Lemma 3.1.1) that

$$|(f_1, f_2)| \leq \|f_1\|_2 \, \|f_2\|_2.$$

Thus $(\cdot, \cdot)$ is finite on $L^2(A) \times L^2(A)$. All the other properties are obvious or follow from Theorem 3.1.1.

q.e.d.

Definition 3.1.2. *Let $A \subset \mathbb{R}^d$ be measurable, $f : A \to \mathbb{R} \cup \{\pm\infty\}$ measurable.*

$$\operatorname*{ess\,sup}_{x \in A} f(x) := \inf \{\lambda \in \mathbb{R} \mid f(x) \leq \lambda \text{ for almost all } x \in A\}$$

(essential supremum), and

$$L^\infty(A) := \big\{ \textit{(equivalence classes of) measurable} \\ \textit{functions } f : A \to \mathbb{R} \cup \{\pm\infty\} \textit{ with} \\ \|f\|_\infty := \|f\|_{L^\infty(A)} := \operatorname*{ess\,sup}_{x \in A} |f(x)| < \infty \big\}$$

Theorem 3.1.2. $L^\infty(A)$ *is a Banach space.*

Proof. If is straightforward to verify that $\|\cdot\|_\infty$ is a norm. It remains to show completeness. Thus, let $(f_n)_{n \in \mathbb{N}}$ be a Cauchy sequence in L^∞. For $\nu \in \mathbb{N}$, we find $n_\nu \in \mathbb{N}$ such that for $m, n \geq n_\nu$

$$\|f_n - f_m\|_\infty < \frac{1}{2^\nu}.$$

Thus

$$\left\{ x \in A \mid |f_n(x) - f_m(x)| \geq \frac{1}{2^\nu} \right\}$$

is a null set for $m, n \geq n_\nu$, and so then is

$$N := \bigcup_{\substack{\nu \in \mathbb{N} \\ m,n \geq n_\nu}} \left\{ x \in A \mid |f_n(x) - f_m(x)| \geq \frac{1}{2^\nu} \right\}$$

as the countable union of null sets. Since

$$|f_n(x) - f_m(x)| < \frac{1}{2^\nu}$$

for $m, n \geq n_\nu$ and $x \in A \setminus N$, f_n converges uniformly on $A \setminus N$ towards some f. We simply put $f(x) = 0$ for $x \in N$. Then

$$\operatorname*{ess\,sup}_{x \in A} |f_n(x) - f(x)| = \operatorname*{ess\,sup}_{x \in A \setminus N} |f_n(x) - f(x)|,$$

since the essential supremum is not affected by null sets,

$$< \frac{1}{2^\nu},$$

and f_n converges to f in $L^\infty(A)$.

q.e.d.

We also note that Hölder's inequality admits the following extension to the case $p = 1$, $q = \infty$:

Lemma 3.1.4. *Let $f_1 \in L^1(A)$, $f_2 \in L^\infty(A)$. Then $f_1 f_2 \in L^1(A)$, and*

$$\|f_1 f_2\|_1 \leq \|f_1\|_1 \|f_2\|_\infty. \tag{3.1.11}$$

Proof.

$$\int_A |f_1(x) f_2(x))| \, dx \leq \operatorname*{ess\,sup}_{x \in A} |f_2(x)| \int_A |f_1(x)| \, dx$$
$$= \|f_2\|_\infty \|f_1\|_1.$$

q.e.d.

Theorem 3.1.3. *Let $A \subset \mathbb{R}^d$ be measurable. Let $1 < p < \infty$, $q = \frac{p}{p-1}$, i.e. $\frac{1}{p} + \frac{1}{q} = 1$. Then $L^q(A)$ is the dual space of $L^p(A)$. In particular, $L^p(A)$ is reflexive.*

Remark 3.1.1. The dual space of $L^1(A)$ is given by $L^\infty(A)$ while the dual space of $L^\infty(A)$ is larger than $L^1(A)$. Therefore, neither $L^1(A)$ nor $L^\infty(A)$ is reflexive.

In order to prepare the proof of Theorem 3.1.3, we first derive:

Theorem 3.1.4 (Clarkson). Let $A \subset \mathbb{R}^d$ be measurable, $2 \leq q < \infty$. Then $L^q(A)$ is uniformly convex.

Remark 3.1.2. Clarkson's theorem holds more generally for $1 < q < \infty$. The proof for $1 < q < 2$ is a little more complicated than the one for $2 \leq q < \infty$.

The proof of Theorem 3.1.4 is based on:

Lemma 3.1.5. Let $2 \leq q < \infty$, $f, g \in L^q(A)$. Then

$$\|f+g\|_q^q + \|f-g\|_q^q \leq 2^{q-1}\left(\|f\|_q^q + \|g\|_q^q\right). \tag{3.1.12}$$

Proof. For $x, y \geq 0$, we have

$$(x^q + y^q)^{\frac{1}{q}} \leq (x^2 + y^2)^{\frac{1}{2}} \leq 2^{\frac{q-2}{2q}}(x^q + y^q)^{\frac{1}{q}}. \tag{3.1.13}$$

(In order to verify the left inequality in (3.1.13), we may assume w.l.o.g. $x^2 + y^2 = 1$. Then $x^q \leq x^2$, $y^q \leq y^2$ since $q \leq 2$, and the desired inequality easily follows. The right inequality follows for example from Hölder's inequality (Lemma 3.1.1) applied to the following functions

$$f_1, f_2 : (-1, 1) \to \mathbb{R}$$
$$f_1 \equiv 1,$$
$$f_2(t) = \begin{cases} a^2 & \text{for } -1 < t \leq 0 \\ b^2 & \text{for } 0 < t \leq 1. \end{cases})$$

The left hand side of (3.1.13) implies

$$(|a+b|^q + |a-b|^q)^{\frac{1}{q}} \leq \left(|a+b|^2 + |a-b|^2\right)^{\frac{1}{2}}$$
$$\leq \sqrt{2}(a^2 + b^2)^{\frac{1}{2}} \tag{3.1.14}$$

for $a, b \in \mathbb{R}$, and by the right-hand-side of (3.1.13), we have

$$\sqrt{2}(a^2+b^2)^{\frac{1}{2}} \leq 2^{\frac{q-1}{q}}(|a|^q + |b|^q)^{\frac{1}{q}}. \tag{3.1.15}$$

Equations (3.1.14) and (3.1.15) imply

$$|f(x)+g(x)|^q + |f(x)-g(x)|^q \leq 2^{q-1}(|f(x)|^q + |g(x)|^q), \tag{3.1.16}$$

and (3.1.12) follows by integrating (3.1.16).

q.e.d.

Proof (Theorem 3.1.4). Let $f, g \in L^q(A)$ with

$$\|f\|_q = \|g\|_q = 1.$$

By (3.1.12),
$$\|f+g\|_q^q + \|f-g\|_q^q \le 2^q.$$
Therefore, for $\epsilon > 0$, we may find $\delta > 0$ such that
$$\|f-g\|_q < \epsilon$$
whenever $\left\|\frac{1}{2}(f+g)\right\|_q > 1 - \delta$. This shows uniform convexity.

q.e.d.

Proof (Theorem 3.1.3). We consider the map
$$i : L^p(A) \to L^q(A)$$
with
$$i(f)(g) := \int_A f(x)g(x)dx.$$
By Hölder's inequality (Lemma 3.1.1)
$$\|i(f)\| = \sup_{\substack{g \in L^q(A) \\ \|g\|_q \le 1}} |i(f)(g)| \le \|f\|_p. \qquad (3.1.17)$$

Thus $i(f)$ is indeed an element of $L^q(A)^*$. We claim that we have equality in (3.1.17). This means that there exists some $g \in L^q$ with
$$\left|\int_A f(x)g(x)dx\right| = \|f\|_p \|g\|_q. \qquad (3.1.18)$$

We put $g(x) := \text{sign}\, f(x)\, |f(x)|^{p-1}$. Then $|g|^q = |f|^p$, hence $g \in L^q(A)$, and
$$\left|\int_A f(x)g(x)dx\right| = \int_A |f(x)g(x)|\, dx$$
$$= \int_A |f(x)|^p\, dx$$
$$= \left(\int_A |f(x)|^p\, dx\right)^{\frac{1}{p}} \left(\int_A |f(x)|^p\, dx\right)^{\frac{1}{q}}$$
$$= \|f\|_p \|g\|_q.$$

This verifies (3.1.18), hence equality in (3.1.17). Equality in (3.1.17) implies that i is an isometry, in particular injective. In order to complete the proof we need to show that i is surjective. Suppose on the contrary that
$$L^q(A)^* \setminus i(L^p(A)) \ne \emptyset.$$

Since $L^p(A)$ is complete and i is continuous, $i(L^p(A))$ is complete, hence closed. By the Hahn–Banach theorem (Corollary 2.1.1), there then exists $v \in L^q(A)^{**}$, $v \neq 0$, with

$$v|_{i(L^p(A))} = 0.$$

We now suppose for a moment that $1 < p \leq 2$. Then $2 \leq q < \infty$, and $L^q(A)$ is reflexive by Theorems 3.1.4 and 2.2.3. We may therefore find a g in $L^q(A)$ with

$$F(g) = v(F) \quad \text{for all } F \in L^q(A)^*.$$

We then have for any $\varphi \in L^p(A)$

$$0 = v(i(\varphi)) = i(\varphi)(g) = \int_A \varphi(x)g(x)dx,$$

hence $g = 0$ (by a reasoning as in the derivation of (3.1.18)), hence also $v = 0$, a contradiction. We have shown that i furnishes an isomorphism between $L^p(A)$ and $L^q(A)^*$. Since $L^q(A)$ is reflexive, so is $L^q(A)^*$ by Corollary 2.2.2, hence $L^p(A)$. In conclusion, $L^p(A)$ has to be reflexive for any $1 < p < \infty$, and its dual space is given by $L^q(A)$.

q.e.d.

3.2 Approximation of L^p functions by smooth functions (mollification)

In this section, we shall smooth out L^p functions by integrating them against smooth kernels. As these kernels approach the Dirac distribution, these regularizations will tend towards the original function. For that purpose, we need some $\varrho \in C_0^\infty(\mathbb{R}^d)$† with

$$\varrho(x) \geq 0 \quad \text{for all } x \in \mathbb{R}^d \quad (3.2.1)$$

$$\varrho(x) = 0 \quad \text{for } |x| \geq 1 \quad (3.2.2)$$

$$\int_{\mathbb{R}^d} \varrho(x)dx \left(= \int_{B(0,1)} \varrho(x)dx \right) = 1. \quad (3.2.3)$$

Such a ϱ is called a Friedrichs mollifier. In this §, Ω will always denote an open subset of $\mathbb{R}^d$. Let $f \in L^1(\Omega)$. We extend f to all of $\mathbb{R}^d$ by putting

---

† For $\Omega \subset \mathbb{R}^d$ open, $C_0^\infty(\Omega)$ is the space of all C^∞ functions φ on Ω for which the closure of $\{x \in \Omega \mid \varphi(x) \neq 0\}$, the support of φ (supp φ), is a compact subset of Ω. Elements of $C_0^\infty(\Omega)$ are often called test functions.

3.2 Approximation of L^p functions by smooth functions

$f(x) = 0$ for $x \in \mathbb{R}^d \setminus \Omega$. Let $h > 0$.

$$f_h(x) := \frac{1}{h^d} \int_{\mathbb{R}^d} \varrho\left(\frac{x-y}{h}\right) f(y) dy. \tag{3.2.4}$$

f_h is called the mollification of f with parameter h. In order to appreciate this definition, we first observe

$$\operatorname{supp} \varrho\left(\frac{x-y}{h}\right) \subset B(y, h) := \{z \in \mathbb{R}^d \mid |z - y| \leq h\}, \tag{3.2.5}$$

where $\varrho\left(\frac{x-y}{h}\right)$ is considered as a function of x, and

$$\frac{1}{h^d} \int_{\mathbb{R}^d} \varrho\left(\frac{x-y}{h}\right) dx = 1. \tag{3.2.6}$$

For these reasons, one expects that f_h tends towards f as h tends to 0. It remains to clarify the type of convergence, however. The advantage of approximating f by f_h comes from:

Lemma 3.2.1. Let $\Omega' \subset\subset \Omega$†, $h < \operatorname{dist}(\Omega', \partial\Omega)$. Then

$$f_h \in C^\infty(\Omega').$$

Proof. By Corollary 1.2.2, we may differentiate w.r.t. x under the integral sign in (3.2.4), and since $\varrho \in C^\infty$ so then is f_h.

q.e.d.

We now start investigating the convergence of f_h towards f.

Lemma 3.2.2. If $f \in C^0(\Omega)$, then for each $\Omega' \subset\subset \Omega$, f_h converges uniformly to f on Ω' as $h \to 0$. In symbols: $f_h \rightrightarrows f$ on Ω' as $h \to 0$.

Proof. We have

$$f(x) = \int_{|w|\leq 1} \varrho(w) f(x) dw \quad \text{by (3.2.3)} \tag{3.2.7}$$

and

$$f_h(x) = \int_{|w|\leq 1} \varrho(w) f(x - hw) dw \tag{3.2.8}$$

by using the substitution $w = \frac{x-y}{h}$ in (3.2.4). For $\Omega' \subset\subset \Omega$ and $h <$

---

† '$\Omega' \subset\subset \Omega$' means that the closure of Ω' is compact and contained in Ω. We say that Ω' is relatively compact in Ω.

$\frac{1}{2}\operatorname{dist}(\Omega', \partial\Omega)$, we then have

$$\sup_{x \in \Omega'} |f(x) - f_h(x)| \leq \sup_{x \in \Omega'} \int_{|w| \leq 1} \varrho(w) |f(x) - f(x - hw)| \, dw$$

$$\leq \sup_{\substack{x \in \Omega' \\ |w| \leq 1}} |f(x) - f(x - hw)|$$

using (3.2.3) once more.

Since Ω' is bounded, $\{x \in \Omega \mid \operatorname{dist}(x, \Omega') \leq h\}$ is compact (recall the choice of h). Therefore, f is uniformly continuous on that set, and we conclude that

$$\sup_{x \in \Omega'} |f(x) - f_h(x)| \to 0 \quad \text{as } h \to 0,$$

i.e. uniform convergence.

q.e.d.

Theorem 3.2.1. *Let $f \in L^p(\Omega)$, $1 \leq p \leq \infty$. Then f_h converges to f in $L^p(\Omega)$ as $h \to 0$.*

Proof. We have for $g \in L^p(\Omega)$

$$\int_\Omega |g_h(x)|^p \, dx$$

$$= \int_\Omega \int_{|w| \leq 1} \varrho(w) g(x - hw) \, dw \, dx$$

$$\leq \int_\Omega \left(\int_{|w| \leq 1} \varrho(w) \, dw \right)^{p-1} \left(\int_{|w| \leq 1} \varrho(w) |g(x - hw)|^p \, dw \right).$$

by Hölder's inequality

$$= \int_{|w| \leq 1} \varrho(w) \int_\Omega |g(x - hw)|^p \, dx \, dw,$$

using (3.2.3) and Fubini's theorem,

$$= \int_{|w| \leq 1} \varrho(w) \int_{\mathbb{R}^d} |g(y)|^p \, dy \, dw$$

$$= \int_\Omega |g(y)|^p \, dy,$$

using (3.2.3) again.

Thus

$$\|g_h\|_{L^p(\Omega)} \leq \|g\|_{L^p(\Omega)}. \tag{3.2.9}$$

3.2 Approximation of L^p functions by smooth functions

Let $\epsilon > 0$. By Theorem 1.1.4, (6), we may find $\varphi \in C_0^0(\mathbb{R}^d)$ with

$$\|f - \varphi\|_{L^p(\mathbb{R}^d)} < \frac{\epsilon}{3}. \tag{3.2.10}$$

Since φ has compact support, we may apply Lemma 3.2.2 to conclude that for sufficiently small $h > 0$,

$$\|\varphi - \varphi_h\|_{L^p(\mathbb{R}^d)} \leq \frac{\epsilon}{3}. \tag{3.2.11}$$

Applying (3.2.9) to $f - \varphi$, we obtain

$$\|f_n - \varphi_n\|_{L^p(\mathbb{R}^d)} \leq \|f - \varphi\|_{L^p(\mathbb{R}^d)}. \tag{3.2.12}$$

(3.2.10)–(3.2.12) yield

$$\|f - f_h\|_{L^p(\Omega)} \leq \|f - f_n\|_{L^p(\mathbb{R}^d)} < \epsilon. \tag{3.2.13}$$

q.e.d.

Corollary 3.2.1. *For $1 \leq p < \infty$, $C_0^\infty(\Omega)$ is dense in $L^p(\Omega)$.*

Proof. Let $f \in L^p(\Omega)$, $\epsilon > 0$. We may then find $\Omega' \subset\subset \Omega$ with

$$\|f\|_{L^p(\Omega \setminus \Omega')} < \frac{\epsilon}{2}.$$

We put $f' := f\chi_{L^p(\Omega')}$. Then

$$\|f - f'\|_{L^p(\Omega)} < \frac{\epsilon}{2}. \tag{3.2.14}$$

By Theorem 3.2.1, for sufficiently small h,

$$\|f' - f'_h\|_{L^p(\Omega)} < \frac{\epsilon}{2}. \tag{3.2.15}$$

By (3.2.13), (3.2.14)

$$\|f - f'_h\|_{L^p(\Omega)} < \frac{\epsilon}{2}.$$

Since $f'_h \in C_0^\infty(\Omega)$ for $h < \text{dist}(\Omega', \partial\Omega)$, the claim follows.

q.e.d.

Corollary 3.2.2. *$L^p(\Omega)$ is separable for $1 \leq p < \infty$. Every $f \in L^p(\Omega)$ can be approximated by piecewise constant functions.*

Proof. By Corollary 3.2.1, it suffices to find a countable subset B_Ω of $L^p(\Omega)$ with the property that for every $\varphi \in C_0^\infty(\Omega)$ and every $\epsilon > 0$, there exists some $a \in B_\Omega$ with

$$\|\varphi - a\|_{L^p(\Omega)} < \epsilon. \tag{3.2.16}$$

Let B the set of all functions a on $\mathbb{R}^d$ of the following form: There exist some $k, N \in \mathbb{N}$ and rational numbers $\alpha_1, \ldots, \alpha_k$ and cubes $Q_1, \ldots, Q_k \in \mathbb{R}^d$ with corners having all their coordinates in $\frac{1}{N}\mathbb{Z}$ and of edge length $\frac{1}{N}$ such that

$$a(x) = \begin{cases} \alpha_i & \text{for } x \in Q_i \\ 0 & \text{otherwise.} \end{cases}$$

Clearly, B is countable. Since a continuous function φ with compact support is uniformly continuous, we may easily find some $a \in B$ with

$$\|a - \varphi\|_{L^p(\Omega)} \leq \|a - \varphi\|_{L^p(\mathbb{R}^d)} < \epsilon. \tag{3.2.17}$$

We put $B_\Omega := \{a\chi_\Omega \mid a \in B\}$. B_Ω is likewise countable, and from (3.2.15), (3.2.16), we conclude that B_Ω is dense in $L^p(\Omega)$.

q.e.d.

Remark 3.2.1. The separability of $L^p(\Omega)$ can also be seen by using Corollary 3.2.1 and the Weierstrass approximation theorem that allows the approximation of continuous function with compact support by polynomials with rational coefficients.

The preceding results do not hold for $L^\infty(\Omega)$. Namely, if a sequence of continuous functions converges w.r.t. $\|\cdot\|_{L^\infty(\Omega)}$, then it converges uniformly, and therefore, the limit is again continuous. Therefore, noncontinuous elements of $L^\infty(\Omega)$ cannot be approximated by continuous functions in the L^∞-norm. Also, $L^\infty(\Omega)$ is not separable. To see this, let $(a_n)_{n \in \mathbb{N}}$ be any subsequence of $\{0, 1\}$, i.e. $a_n \in \{0, 1\}$ for all n. To (a_n), we associate the function $f_{(a_n)}$ on $(0, 1)$ defined by

$$f_{(a_n)} := \begin{cases} 1 & \text{for } \frac{1}{2^k} < x \leq \frac{1}{2^{k-1}} \text{ if } a_k = 1 \\ 0 & \text{for } \frac{1}{2^k} < x \leq \frac{1}{2^{k-1}} \text{ if } a_k = 0 \end{cases} \quad \text{for } k \in \mathbb{N}.$$

Then for any two different sequences $(a_n), (b_n)$,

$$\|f_{(a_n)} - f_{(b_n)}\|_{L^\infty((0,1))} = 1.$$

Since the set of subsequences of $\{0, 1\}$ is uncountable, this implies that $L^\infty((0, 1))$ is not separable. Of course, a similar construction is possible for $L^\infty(\Omega)$, Ω any open subset of $\mathbb{R}^d$.

We finally note:

Lemma 3.2.3. *Let $f \in L^2(\Omega)$, and suppose that for all $\varphi \in C_0^\infty(\Omega)$*

$$\int_\Omega f(x)\varphi(x)dx = 0.$$

Then $f = 0$.

Proof. Since $C_0^\infty(\Omega)$ is dense in $L^2(\Omega)$, and since

$$g \mapsto \int_\Omega f(x)g(x)dx$$

is a continuous linear functional on $L^2(\Omega)$, we obtain that

$$\int_\Omega f(x)g(x)dx = 0 \quad \text{for all } g \in L^2(\Omega).$$

Putting $g = f$ yields the result.

q.e.d.

3.3 Sobolev spaces

In this section, we wish to introduce certain extensions of the L^p spaces, the so-called Sobolev spaces. They will play a fundamental rôle in subsequent chapters because they constitute function spaces that are complete w.r.t. norms naturally occurring in variational problems. In this section, Ω will always denote an open subset of $\mathbb{R}^d$. We shall use the following notation: For a d-tuple $\boldsymbol{\alpha} := (\alpha_1, \ldots, \alpha_d)$ of nonnegative integers,

$$|\boldsymbol{\alpha}| := \sum_{i=1}^d \alpha_i, \quad D_\alpha := \left(\frac{\partial}{\partial x^1}\right)^{\alpha_1} \cdots \left(\frac{\partial}{\partial x^d}\right)^{\alpha_d}.$$

Definition 3.3.1. *Let $u, v \in L^1(\Omega)$. Then v is said to be the $\boldsymbol{\alpha}$-th weak derivative of u, $v := D_\alpha u$, if*

$$\int \varphi v \, dx = (-1)^{|\alpha|} \int_\Omega u D_\alpha \varphi \, dx \tag{3.3.1}$$

for every $\varphi \in C_0^{|\alpha|}$.

We can now define, for $k \in \mathbb{N}$ and $1 \leq p < \infty$, the Sobolev space

$$W^{k,p}(\Omega) := \{ u \in L^p(\Omega) \mid D_\alpha u \text{ exists and lies in } L^p(\Omega) \text{ for all } |\alpha| \leq k \},$$

$$\|u\|_{W^{k,p}(\Omega)} := \left(\sum_{|\alpha| \leq k} \int_\Omega |D_\alpha u|^p \right)^{\frac{1}{p}}.$$

Finally, let $H^{k,p}(\Omega)$ and $H_0^{k,p}(\Omega)$ be the closures of $C^\infty(\Omega) \cap W^{k,p}(\Omega)$ and $C_0^\infty \cap W^{k,p}(\Omega)$, respectively in $W^{k,p}(\Omega)$.

We shall use the following abbreviations for $u \in W^{1,1}(\Omega)$, $1 \le i \le d$. $D_i u$ is the weak derivative for the multiindex $(0, \ldots, 0, 1, 0, \ldots, 0)$, 1 at the ith position, and Du is the vector $(D_1 u, \ldots, D_d u)$ of all first weak derivatives.

The following result is obvious.

Lemma 3.3.1. *Let $u \in C^k(\Omega)$, and suppose all derivatives of u of order $\le k$ are in $L^p(\Omega)$. Then $u \in W^{k,p}(\Omega)$, and the weak derivatives are given by the ordinary derivatives.*

q.e.d.

Thus, the $W^{k,p}$ spaces constitute a generalization of the spaces of k times differentiable functions. The $W^{k,p}$ norm is considerably weaker than the C^k-norm, and so the $W^{k,p}$ spaces are larger than the C^k spaces.

Before investigating the properties of these spaces, it should be useful to consider an example: Let $\Omega = (-1, 1) \subset \mathbb{R}$, $u(x) := |x|$. We claim that $u \in W^{1,p}(\Omega)$ for $1 \le p < \infty$. In order to see this it suffices that the first weak derivative of u is given by

$$Du(x) = v(x) := \begin{cases} 1 & \text{for } 0 < x < 1 \\ -1 & \text{for } -1 < x < 0. \end{cases}$$

Indeed, we have for $\varphi \in C_0^1((-1, 1))$

$$\int_{-1}^1 \varphi(x) v(x) dx = - \int_{-1}^1 \varphi'(x) |x|\, dx.$$

We claim, however, that u is not contained in $W^{2,p}(\Omega)$. Namely if $w(x)$ were the second weak derivative of u, it would have to be the first weak derivative of v, and consequently, we would have $w(x) = 0$ for $x \ne 0$. The rule for integration by parts (3.3.1) would then require that for all $\varphi \in C_0^1((-1, 1))$

$$\begin{aligned} 0 &= \int_{-1}^1 \varphi(x) w(x) dx \\ &= - \int_{-1}^1 \varphi'(x) v(x) dx \\ &= \int_{-1}^0 \varphi'(x) dx - \int_0^1 \varphi'(x) dx \\ &= 2\varphi(0) \end{aligned}$$

which is not the case. Thus, v does not have a first weak derivative.

3.3 Sobolev spaces

Remark 3.3.1. Some readers may have encountered the notion of a distributional derivative. It is important to distinguish between weak and distributional derivatives. Any $L^1(\Omega)$ function possesses distributional derivatives of any order, but as the preceding example shows, not necessarily weak derivatives. In the example, of course, the second distributional derivative of u is $2\delta_0$, where δ_0 is the Dirac delta distribution at 0. u does not possess a second weak derivative because the delta distribution cannot be represented by an L^1 function.

Theorem 3.3.1. *The Sobolev spaces $W^{k,p}(\Omega)$ are separable Banach spaces w.r.t. $\|\cdot\|_{W^{k,p}(\Omega)}$.*

Proof. That $\|\cdot\|_{W^{k,p}(\Omega)}$ is a norm follows from the fact that $\|\cdot\|_{L^p(\Omega)}$ is a norm (see section 3.1). Similarly, we shall now derive completeness of $W^{k,p}(\Omega)$ from the completeness of the $L^p(\Omega)$ spaces (Theorem 3.1.1). Thus, let $(v_n)_{n\in\mathbb{N}} \subset W^{k,p}(\Omega)$ be a Cauchy sequence w.r.t. $\|\cdot\|_{W^{k,p}(\Omega)}$. This implies that $(D_\alpha u_n)_{n\in\mathbb{N}}$ is a Cauchy sequence w.r.t. $\|\cdot\|_{L^p(\Omega)}$ for all $|\alpha| \le k$. By Theorem 3.1.1, $(D_\alpha u_n)$ therefore converges in $L^p(\Omega)$ towards some v_α. For $\varphi \in C_0^{|\alpha|}(\Omega)$

$$\int_\Omega D_\alpha u_n \cdot \varphi = (-1)^{|\alpha|} \int_\Omega u_n D_\alpha \varphi. \qquad (3.3.2)$$

Therefore, v_α is the α-th weak derivative of v_0, the L^p-limit of $(u_n)_{n\in\mathbb{N}}$, and consequently $v_0 \in W^{k,p}(\Omega)$. The separability again follows from the corresponding property for $L^p(\Omega)$ (Corollary 3.2.2).

q.e.d.

Theorem 3.3.2. $W^{k,p}(\Omega) = H^{k,p}(\Omega)$.

This result says that elements of $W^{k,p}(\Omega)$ can be approximated by $C^\infty(\Omega)$ functions w.r.t. $\|\cdot\|_{W^{k,p}(\Omega)}$. In general, however, for $k \ge 1$ one has $H_0^{k,p}(\Omega) \ne W^{k,p}(\Omega)$ so that $W^{k,p}(\Omega)$ functions cannot be approximated by $C_0^\infty(\Omega)$ functions, in contrast to $L^p(\Omega)$ functions where this is possible (Corollary 3.2.1). This is seen from the following simple example:
$\Omega = (-1,1) \subset \mathbb{R}$, $u(x) \equiv 1$. If $(\varphi_n)_{n\in\mathbb{N}} \subset C_0^\infty(\Omega)$ converges to u in $L^1(\Omega)$, then after selection of a subsequence, it converges pointwise almost everywhere (Lemma 3.1.3), and therefore, for sufficiently large n, there exists $x_n \in (-1,1)$ with $\varphi_n(x_n) \ge \frac{1}{2}$. Since $\varphi_n(-1) = 0 = \varphi_n(1)$, this implies that

$$\int_{-1}^1 |\varphi_n'(x)| \, dx \ge 1.$$

Therefore, φ'_n cannot converge to $u' \equiv 0$ in $L^p((-1,1))$, and therefore φ_n cannot converge to u in $W^{1,p}((-1,1))$.

Proof (Theorem 3.3.2). We have to show that any $u \in W^{k,p}(\Omega)$ can be approximated by $C^\infty(\Omega)$ functions. As in §3.2, we extend u to be 0 outside Ω and consider the mollifications $u_h \in C^\infty(\Omega)$. We compute

$$D_\alpha(u_h(x)) = \frac{1}{h^d} \int_{\mathbb{R}^d} D_{\alpha,x} \varrho\left(\frac{x-y}{h}\right) \cdot u(y) dy \quad \text{(using Corollary 1.2.2)}$$

where $D_{\alpha,x}$ is the derivative w.r.t. x,

$$= (-1)^\alpha \int_{\mathbb{R}^d} D_{\alpha,x} \varrho\left(\frac{x-y}{h}\right) \cdot u(y) dy$$

$$= \int_{\mathbb{R}^d} \varrho\left(\frac{x-y}{h}\right) D_\alpha u(y) dy$$

by definition of $D_\alpha u$

$$= (D_\alpha u)_h(x). \tag{3.3.3}$$

Thus, the derivative of the mollification is the mollification of the derivative. Since $D_\alpha u \in L^p(\Omega)$, by Theorem 3.2.1, $(D_\alpha u)_h$ converges to $D_\alpha u$ in $L^p(\Omega)$ for $h \to 0$. By (3.3.4), we conclude that $D_\alpha(u_h)$ converges to $D_\alpha u$ in $L^p(\Omega)$, for all $|\alpha| \leq k$, and this means that u_h converges to u in $W^{k,p}(\Omega)$.

q.e.d.

Theorem 3.3.3. $W^{k,p}(\Omega)$ *is reflexive for* $k \in \mathbb{N}$, $1 < p < \infty$.

Proof. It follows from Theorem 3.1.3 that the dual space of $W^{k,p}(\Omega)$ is given by $W^{k,q}(\Omega)$, with $\frac{1}{p} + \frac{1}{q} = 1$. This implies reflexivity.

q.e.d.

Theorem 3.3.4. $H_0^{k,p}(\Omega)$ *is closed under weak convergence in* $W^{k,p}(\Omega)$.

Proof. This follows from Lemma 2.2.5, since $H_0^{k,p}(\Omega)$ by its definition is a closed subspace (w.r.t. strong convergence) of $W^{k,p}(\Omega)$.

q.e.d.

Theorem 3.3.5. *For* $1 < p < \infty$, $k \in \mathbb{N}$, *any sequence in* $W^{k,p}(\Omega)$ *that is bounded w.r.t.* $\|\cdot\|_{W^{k,p}(\Omega)}$ *contains a weakly convergent subsequence.*

3.4 Rellich's theorem, Poincaré and Sobolev inequalities   175

Proof. By Theorems 3.3.1 and 3.3.3, $W^{k,p}(\Omega)$ is separable and reflexive. Therefore, the result follows from Corollary 2.2.1.

q.e.d.

3.4 Rellich's theorem and the Poincaré and Sobolev inequalities

The compactness theorem of Rellich is:

Theorem 3.4.1. *Let $\Omega \subset \mathbb{R}^d$ be open and bounded. Let $(u_n)_{n \in \mathbb{N}} \subset H_0^{1,p}(\Omega)$ be bounded, i.e. $\|u_n\|_{W^{1,p}(\Omega)} \leq c$ (independent of n). Then a subsequence of $(u_n)_{n \in \mathbb{N}}$ converges in $L^p(\Omega)$.*

Remark 3.4.1. Rellich originally proved the theorem for $p = 2$. Kondrachev proved the stronger result that some subsequence converges in $L^q(\Omega)$ for $1 \leq q < \frac{dp}{d-p}$ if $p < d$ and for $1 \leq q < \infty$ if $p \geq d$. Of course, these exponents come from the Sobolev Embedding Theorem (see (3.4.12)). See Corollary 3.4.1 below.

Proof. Since $u_n \in H_0^{1,p}(\Omega)$, for every $n \in \mathbb{N}$ and $\epsilon > 0$, there exists some $v_n \in C_0^1(\Omega)$ with

$$\|u_n - v_n\|_{W^{1,p}(\Omega)} < \frac{\epsilon}{3}. \tag{3.4.1}$$

Therefore

$$\|v_n\|_{W^{1,p}(\Omega)} \leq c' \quad (= c + \frac{\epsilon}{3}). \tag{3.4.2}$$

We consider the mollification

$$v_{n,h}(x) = \frac{1}{h^d} \int_\Omega \varrho\left(\frac{x-y}{h}\right) v_n(y) dy$$

of v_n and estimate

$$|v_n(x) - v_{n,h}(x)|$$

$$= \left| \int_{|w| \leq 1} \varrho(w)(v_n(x) - v_n(x - hw)) dw \right| \quad \text{by (3.2.7), (3.2.8)}$$

$$\leq \int_{|w| \leq 1} \varrho(w) \int_0^{h|w|} \left| \frac{\partial}{\partial r} v_n(x - r\vartheta) \right| dr dw \quad \text{with } \vartheta = \frac{w}{|w|}. \tag{3.4.3}$$

This implies

$$\int_\Omega |v_n(x) - v_{n,h}(x)|^p \, dx$$
$$\leq \int_\Omega \left(\int_{|w|\leq 1} \varrho(w) \int_0^{h|w|} \left| \frac{\partial}{\partial r} v_n(x - r\vartheta) \right| dr dw \right)^p dx$$
$$= \int_\Omega \left(\int_{|w|\leq 1} \left(\varrho(w)^{1-\frac{1}{p}} \right) \varrho(w)^{\frac{1}{p}} \int_0^{h|w|} \left| \frac{\partial}{\partial r} v_n(x - r\vartheta) \right| dr dw \right)^p dx$$
$$\leq \left(\int_{|w|\leq 1} \varrho(w) dw \right)^{p-1} \left(\int_{|w|\leq 1} \varrho(w) h^p |w|^p \int |Dv_n(x)|^p \, dx dw \right),$$

using Hölder's inequality, Fubini's theorem and the notation

$$Dv_n = \left(\frac{\partial}{\partial x^1} v_n, \ldots, \frac{\partial}{\partial x^d} v_n \right).$$

Since $\int_{|w|\leq 1} \varrho(w) dw = 1$ (by (3.2.3)), we obtain

$$\|v_n - v_{n,h}\|_{L^p(\Omega)} \leq h \|Dv_n\|_{L^p(\Omega)}$$
$$\leq hc' \quad \text{by (3.4.2)}$$
$$\leq \frac{\epsilon}{3} \quad \text{if } h \text{ is sufficiently small.} \quad (3.4.4)$$

Next,

$$|v_{n,h}(x)| \leq \frac{1}{h^d} c_0 \|v_n\|_{L^1(\Omega)}$$

with $c_0 := \sup_z \varrho(z)$ by definition of $v_{n,h}$

$$\leq \frac{1}{h^d} c_0 (\operatorname{meas} \Omega)^{1-\frac{1}{p}} \|v_n\|_{L^p(\Omega)} \quad (3.4.5)$$

by Hölder's inequality,

and similarly

$$\left| \frac{\partial}{\partial x^i} v_{n,h}(x) \right| \leq \frac{1}{h^{d+1}} c_i (\operatorname{meas} \Omega)^{1-\frac{1}{p}} \|v_n\|_{L^p(\Omega)} \quad (3.4.6)$$

with $c_i := \sup_z \left| \frac{\partial}{\partial z^i} \varrho(z) \right|$. From (3.4.2), (3.4.5), (3.4.6), we see that for fixed $h > 0$,

$$\|v_{n,h}\|_{C^1(\Omega)} \leq \text{constant} \quad (3.4.7)$$

(where the constant depends on h). Therefore, $(v_{n,h})_{n\in\mathbb{N}}$ contains a uniformly convergent subsequence by the Arzela–Ascoli theorem. Since uni-

3.4 Rellich's theorem, Poincaré and Sobolev inequalities

form convergence implies L^p-convergence (e.g. by Theorem 1.2.3), the closure of $v_{n,h}$ is compact in $L^p(\Omega)$. Since a compact subset of a metric space (e.g. a Banach space) is totally bounded, there exist finitely many $w_1, \ldots, w_N \in L^p(\Omega)$ such that for every $n \in \mathbb{N}$ there exists $1 \leq j \leq N$ with

$$\|v_{n,h} - w_j\|_{L^p(\Omega)} < \frac{\epsilon}{3}. \tag{3.4.8}$$

By (3.4.1), (3.4.4), (3.4.8), for every $n \in \mathbb{N}$ we find $1 \leq j \leq N$ with

$$\|u_n - w_j\|_{L^p(\Omega)} < \epsilon.$$

Thus, $(u_n)_{n \in \mathbb{N}}$ is totally bounded in $L^p(\Omega)$. Therefore, the closure of $(u_n)_{n \in \mathbb{N}}$ in $L^p(\Omega)$ is compact (again, a general result for metric spaces), and it thus contains a convergent subsequence in $L^p(\Omega)$.

q.e.d.

We now come to the Poincaré inequality:

Theorem 3.4.2. *Let $\Omega \subset \mathbb{R}^d$ be open and bounded. For any $u \in H_0^{1,p}(\Omega)$*

$$\|u\|_{L^p(\Omega)} \leq \left(\frac{\operatorname{meas}\Omega}{\omega_d}\right)^{\frac{1}{d}} \|Du\|_{L^p(\Omega)} \tag{3.4.9}$$

where ω_d is the Lebesgue measure of the unit ball in $\mathbb{R}^d$.

Proof. Since $C_0^1(\Omega)$ is dense in $H_0^{1,p}(\Omega)$, we may assume $u \in C_0^1(\Omega)$. We put $u(x) = 0$ for all $x \in \mathbb{R}^d \setminus \Omega$. For $\vartheta \in \mathbb{R}^d$ with $|\vartheta| = 1$, we have

$$u(x) = -\int_0^\infty \frac{\partial}{\partial r} u(x + r\vartheta) dr.$$

Integration w.r.t. ϑ yields

$$|u(x)| = \left| -\frac{1}{d\omega_d} \int_0^\infty \int_{|\vartheta|=1} \frac{\partial}{\partial r} u(x + r\vartheta) d\vartheta dr \right|$$

$$\leq \frac{1}{d\omega_d} \int_\Omega \frac{1}{|x-y|^{d-1}} |Du(y)| \, dy.$$

Therefore

$$\left(\int_\Omega |u(x)|^p \, dx\right)^{\frac{1}{p}}$$

$$\leq \left(\int_\Omega \left(\int_\Omega \frac{1}{|x-y|^{d-1}} |Du(y)| \, dy\right)^p dx\right)^{\frac{1}{p}}$$

$$\leq \frac{1}{d\omega_d} \left(\int_\Omega \left(\int_\Omega \frac{1}{|x-y|^{d-1}} |Du(y)|^p \, dy\right) \left(\int_\Omega \frac{1}{|x-y|^{d-1}} dy\right)^{p-1} dx\right)^{\frac{1}{p}}$$

by Hölder's inequality

$$= \frac{1}{d\omega_d} \left(\int |Du(y)|^p \, dy\right)^{\frac{1}{p}} \left(\int_\Omega \frac{1}{|x-y|^{d-1}} dx\right) \qquad (3.4.10)$$

using Fubini's theorem to exchange the order of integration in the first factor.

In order to control

$$\int_\Omega \frac{1}{|x-y|^{d-1}} dx,$$

we choose R with

$$\operatorname{meas} \Omega = \operatorname{meas} B(y, R) = \omega_d R^d$$

$(B(y, R) := \{z \in \mathbb{R}^d \mid |z| \leq R\})$.

Since

$$\frac{1}{|x-y|^{d-1}} \leq \frac{1}{R^{d-1}} \quad \text{for } |x-y| \geq R$$

$$\frac{1}{|x-y|^{d-1}} \geq \frac{1}{R^{d-1}} \quad \text{for } |x-y| \leq R,$$

we have

$$\int_\Omega \frac{1}{|x-y|^{d-1}} dx \leq \int_{B(y,R)} \frac{1}{|x-y|^{d-1}} dx$$
$$= d\omega_d R \qquad (3.4.11)$$
$$= d\omega_d^{1-\frac{1}{d}} (\operatorname{meas} \Omega)^{\frac{1}{d}}.$$

Equations (3.4.10) and (3.4.11) yield (3.4.9).

q.e.d.

3.4 Rellich's theorem, Poincaré and Sobolev inequalities

We now come to somewhat stronger results that will however only be needed in Chapter 9. Namely, we have the Sobolev inequalities.

Theorem 3.4.3. Let $u \in H_0^{1,p}(\Omega)$.

(i) If $p < d$, then $u \in L^{\frac{dp}{d-p}}(\Omega)$, and

$$\|u\|_{\frac{dp}{d-p}} \leq c \|Du\|_p. \tag{3.4.12}$$

(ii) If $p > d$, then $u \in C^0(\bar{\Omega})$, and

$$\sup_{\Omega} |u| \leq c (\operatorname{meas} \Omega)^{\frac{1}{d} - \frac{1}{p}} \|Du\|_p \tag{3.4.13}$$

with constants c depending only on p and d. (Actually, by a Theorem of Morrey, for $p > d$, $u \in H_0^{1,p}(\Omega)$ is even Hölder continuous with exponent $1 - \frac{d}{p}$.)

We only prove (i) as (ii) will not be used in the present book:

Proof. We first assume $u \in C_0^1(\Omega)$. Since u has compact support, we have

$$|u(y)| \leq \int_{-\infty}^{\infty} |D_i u(y)| \, dy^i \quad \text{for } i = 1, 2, \ldots, d.$$

Multiplying these inequalities for $i = 1, \ldots, d$ yields

$$|u(y)|^{\frac{d}{d-1}} \leq \left(\prod_{i=1}^{d} \int_{-\infty}^{\infty} |D_i u(y)| \, dy^i \right)^{\frac{1}{d-1}}.$$

Using Hölder's inequality†, we compute

$$\int_{-\infty}^{\infty} |u(x)|^{\frac{d}{d-1}} \, dx^1$$

$$\leq \left(\int_{-\infty}^{\infty} |D_1 u(y)| \, dy^1 \right)^{\frac{1}{d-1}} \int_{-\infty}^{\infty} \left(\prod_{i=2}^{d} \int_{-\infty}^{\infty} |D_i u(y)| \, dy^i \right)^{\frac{1}{d-1}} dx^1$$

$$\leq \left(\int_{-\infty}^{\infty} |D_1 u(y)| \, dy^1 \right)^{\frac{1}{d-1}} \left(\prod_{i=2}^{d} \int_{-\infty}^{\infty} |D_i u(y)| \, dy^i dx^1 \right)^{\frac{1}{d-1}}.$$

† More precisely, one uses Exercise (2) below with $p_1 = \cdots = p_{d-1} = d - 1$.

Iteratively also integrating w.r.t $dx^2, \ldots, dx^d$ finally yields

$$\|u\|_{L^{\frac{d}{d-1}}(\Omega)} \leq \left(\prod_{i=1}^{d} \int_{\Omega} |D_i u(x)|\, dx\right)^{\frac{1}{d}}$$

$$\leq \frac{1}{d} \int_{\Omega} \sum_{i=1}^{d} |D_i u(x)|\, dx$$

$$= \frac{1}{d} \|Du\|_{L^1(\Omega)}. \qquad (3.4.14)$$

This is (3.4.12) for $p = 1$. The case of general p may now be obtained by applying (3.4.14) to $|u|^\mu$ for suitable $\mu > 1$ and using Hölder's inequality. Namely, from (3.4.14) for $|u|^\mu$ in place of u

$$\||u|^\mu\|_{L^{\frac{d}{d-1}}} \leq \frac{\mu}{d} \int_{\Omega} |u(x)|^{\mu-1} |Du(x)|\, dx$$

$$\leq \frac{\mu}{d} \||u|^{\mu-1}\|_{L^q} \|Du\|_{L^p} \quad \text{for } \frac{1}{p} + \frac{1}{q} = 1$$

by Hölder's inequality.

For $p < d$, we may take $\mu = \frac{(d-1)p}{d-p}$ and obtain

$$\|u\|_{L^{\frac{\mu d}{d-1}}}^{\mu} \leq \frac{\mu}{d} \|u\|_{L^{\frac{\mu d}{d-1}}}^{\mu-1} \|Du\|_{L^p},$$

which yields (3.4.12), since $\frac{\mu d}{d-1} = \frac{dp}{d-p}$.

q.e.d.

As a consequence, we obtain the theorem of Kondrachev:

Corollary 3.4.1. *Let $\Omega \in \mathbb{R}^d$ be open and bounded. Let $(u_n)_{n \in \mathbb{N}} \subset H_0^{1,p}(\Omega)$ be bounded for some $1 \leq p \leq d$. Then a subsequence converges in $L^q(\Omega)$ for any $1 \leq q < \frac{dp}{d-p}$.*

Proof. From Theorem 3.4.1 we know already that a subsequence converges in $L^p(\Omega)$. We may assume $q > p$ as otherwise the result is an easy consequence of Hölder's inequality since Ω is bounded. We denote this converging subsequence again by (u_n).
From Hölder's inequality, we obtain

$$\|u_n - u_m\|_{L^q(\Omega)} \leq \|u_n - u_m\|_{L^1(\Omega)}^{\mu} \|u_n - u_m\|_{L^{\frac{dp}{d-p}}(\Omega)}^{1-\mu}$$

$$\text{if } \mu \text{ satisfies } \frac{1}{q} = \mu + (1-\mu)\left(\frac{1}{p} - \frac{1}{d}\right)$$

$$\leq c \|u_n - u_m\|_{L^1(\Omega)}^{\mu} \|D(u_n - u_m)\|_{L^p(\Omega)}^{1-\mu} \qquad (3.4.15)$$

by Theorem 3.4.3 (i).

Since Du_n is bounded in $L^p(\Omega)$ by assumption, and (u_n) is a Cauchy sequence in $L^p(\Omega)$, hence also in $L^1(\Omega)$, (3.4.15) then implies the Cauchy property in $L^q(\Omega)$.

q.e.d.

Exercises

3.1 Let
$$A_1 := \{x \in \mathbb{R}^d \mid \|x\| > 1\}, \quad A_2 := \{x \in \mathbb{R}^d \mid \|x\| < 1\},$$
and consider
$$f(x) = \|x\|^\lambda \quad \text{for } \lambda \in \mathbb{R}.$$
For which values of d, p, λ is $f \in L^p(A_1)$, or $f \in L^p(A_2)$?

3.2 Let $A \subset \mathbb{R}^d$ be measurable. Let $p_1, \ldots, p_k \geq 1$, $\sum_{i=1}^k \frac{1}{p_i} = 1$, $f_i \in L^{p_i}(A)$ for $i = 1, \ldots, k$. Show $f_1 \cdot \ldots \cdot f_k \in L^1(A)$, with
$$\left\| \prod_{i=1}^k f_i \right\|_1 \leq \prod_{i=1}^k \|f_i\|_{p_i}.$$

3.3 Let $A \subset \mathbb{R}^d$ be measurable, meas $A < \infty$, $1 \leq p \leq q \leq \infty$. Then $L^q(A) \subset L^p(A)$, and for $f \in L^q(A)$
$$\frac{1}{(\text{meas } A)^{\frac{1}{p}}} \|f\|_{L^p(A)} \leq \frac{1}{(\text{meas } A)^{\frac{1}{q}}} \|f\|_{L^q(A)}.$$
(Hint: Apply Hölder's inequality with $f_1 = 1$, $f_2 = f$)

3.4 Let $A \subset \mathbb{R}^d$ be measurable, $1 \leq p \leq q \leq r$, $\frac{1}{q} = \frac{\alpha}{p} + \frac{(1-\alpha)}{r}$, $f \in L^p(A) \cap L^r(A)$. Then $f \in L^q(A)$, and
$$\|f\|_{L^q(A)} \leq \|f\|_{L^p(A)}^\alpha \|f\|_{L^r(A)}^{1-\alpha}.$$

3.5 Let $A \subset \mathbb{R}^d$ be measurable, meas $A < \infty$, $f : A \to \mathbb{R} \cup \{\pm\infty\}$ measurable. Then
$$\lim_{p \to \infty} \frac{1}{(\text{meas } A)^{\frac{1}{p}}} \|f\|_{L^p(A)} = \|f\|_{L^\infty(A)}$$
(where we allow these quantities to be infinite).

3.6 Let $A \subset \mathbb{R}^d$ be measurable, $(f_n)_{n\in\mathbb{N}} \subset L^p(A)$ with
$$\|f_n\|_p \leq \text{constant}.$$
Suppose f_n converges pointwise almost everywhere on A to some f. Is $f \in L^p(A)$, and do we necessarily get
$$\|f_n - f\|_p \to 0 \quad \text{as } n \to \infty?$$

3.7 Let A_1, A_2, f be as in exercise 1). For which d, k, p, λ is f in $W^{k,p}(A_1)$ or in $W^{k,p}(A_2)$?

3.8 Consider the sequence $(\sin(nx))_{n\in\mathbb{N}}$ in $L^2((0,1))$. Does it converge in the L^2-norm? Does it converge weakly? If so, what is the limit?

4
The direct methods in the calculus of variations

4.1 Description of the problem and its solution

The typical problem of the calculus of variations is to minimize an integral of the form

$$F(u) := \int_\Omega f(x, u(x), Du(x)) dx$$

where Ω is some open subset of $\mathbb{R}^d$ (in most cases, Ω is bounded), among functions

$$u : \Omega \to \mathbb{R}$$

belonging to some suitable class of functions and satisfying a boundary condition, for example a Dirichlet boundary condition

$$u(y) = g(y) \quad \text{for } y \in \partial\Omega$$

for some given $g : \partial\Omega \to \mathbb{R}$. Thus, the problem is

$$F(u) \to \min \quad \text{for } u \in \mathcal{C},$$

where $\mathcal{C}$ is some space of functions. The strategy of the direct method is very simple: Take a minimizing sequence $(u_n)_{n \in \mathbb{N}} \subset \mathcal{C}$, i.e.

$$\lim_{n \to \infty} F(u_n) = \inf_{u \in \mathcal{C}} F(u),$$

and show that some subsequence of (u_n) converges to a minimizer $u \in \mathcal{C}$. To make this strategy be successful, several conditions should be met:

(1) Some compactness condition has to hold so that a minimizing sequence contains a convergent subsequence. This requires the careful selection of a suitable topology on $\mathcal{C}$.

(2) The limit u of such a subsequence should be contained in $\mathcal{C}$. This is a closedness condition on $\mathcal{C}$.

In particular, for (1) and (2) to hold, $\mathcal{C}$ should not be too restrictive. In other words, one should not specify too many properties for a solution u in advance.

(3) Some lower semicontinuity condition of the form

$$F(u) \leq \liminf_{n \to \infty} F(u_n) \quad \text{if } u_n \text{ converges to } u$$

has to hold, in order to ensure that the limit of a minimizing sequence is indeed a minimizer for F.

The lower semicontinuity condition becomes easier if the topology of $\mathcal{C}$ is more restrictive, because the stronger the convergence of u_n to u is, the easier that condition is satisfied. That is at variance, however, with the requirement of (1) since for too strong a topology, sequences do not always contain convergent subsequences. Therefore, we expect that the topology for $\mathcal{C}$ has to be carefully chosen so as to balance these various requirements. In order to gain some insights into this aspect, it is useful to approach the problem from an abstract point of view. Thus, we shall return to the concrete integral variational problem raised in the beginning only later.

4.2 Lower semicontinuity

We say that a topological space X satisfies the first axiom of countability, if the neighbourhood system of each point $x \in X$ has a countable base, i.e. there exists a sequence $(U_\nu)_{\nu \in \mathbb{N}}$ of open subsets of X with $x \in U_\nu$ with the property that for every open set $U \subset X$ with $x \in U$ there exists $n \in \mathbb{N}$ with

$$U_n \subset U.$$

X satisfies the second axiom of countability if its topology has a countable base, i.e. there exists a family $(U_\nu)_{\nu \in \mathbb{N}}$ of open subsets of X with the property that for every open subset V of X, there exists $n \in \mathbb{N}$ with

$$U_n \subset V.$$

We note that separable metric spaces X satisfy the second axiom of countability. In fact, let $(x_\nu)_{\nu \in \mathbb{N}}$ be a dense subset of X, and let $(r_\mu)_{\mu \in \mathbb{N}}$ be dense in $\mathbb{R}^+$. Then

$$U(x_\nu, r_\mu) := \{x \in X : d(x, x_\nu) < r_\mu\}$$

4.2 Lower semicontinuity

($d(\cdot,\cdot)$ the distance function of X) forms a countable base for the topology.

If the first countability axiom is satisfied, topological notions usually admit sequential characterizations. For example, if $(x_n)_{n\in\mathbb{N}} \subset X$ is a sequence in a topological space X satisfying the first axiom of countability, then any accumulation point of (x_n) (i.e. any $x \in X$ with the property that for every neighbourhood U of x and any $m \in \mathbb{N}$, there exists $n \geq m$ with $x_n \in U$) can be obtained as the limit of some subsequence of (x_n). Although we shall often employ weak topologies which typically do not satisfy the first axiom of countability, for our purposes it will usually be sufficient to use sequential versions of topological properties. For that reason, we shall define our topological notions in sequential terms, without adding the word 'sequentially'.

Definition 4.2.1. *Let X be a topological space. A function $F : X \to \bar{\mathbb{R}} := \mathbb{R} \cup \{\pm\infty\}$ is called* lower semicontinuous *(lsc) at x if*

$$F(x) \leq \liminf_{n\to\infty} F(x_n)$$

for any sequence $(x_n)_{n\in\mathbb{N}} \subset X$ converging to x. F is called lower semicontinuous *if it is lsc at every $x \in X$.*

The following properties are immediate:

Lemma 4.2.1.

(i) *If $F : X \to \bar{\mathbb{R}}$ is lsc, $\lambda \geq 0$, then λF is lsc.*
(ii) *If $F, G : X \to \bar{\mathbb{R}}$ are lsc, and if their sum $F + G$ is well defined (i.e. there is no $x \in X$ for which one of the values $F(x), G(x)$ is $+\infty$ and the other one is $-\infty$), then $F + G$ is also lsc.*
(iii) *For $F, G : X \to \bar{\mathbb{R}}$ lsc, $\inf(F, G)$ is also lsc.*
(iv) *If $(F_i)_{i\in I}$ is a family of lsc functions, then $\sup_{i\in I} F_i$ is also lsc.*

Examples.

(1) Any continuous function is lower semicontinuous.
(2) If X satisfies the first axiom of countability, then $A \subset X$ is open if and only if its characteristic function χ_A is lsc.

Definition 4.2.2.

(i) *Let X be a normed space, with norm $\|\cdot\|$. $F : X \to \mathbb{R}$ is* weakly proper, *if for every sequence $(x_n)_{n\in\mathbb{N}} \subset X$ with $\|x_n\| \to \infty$ we have $F(x_n) \to \infty$ for $n \to \infty$.*

(ii) *Let X be a topological space. $F : X \to \mathbb{R}$ is coercive if every sequence $(x_n) \subset X$ with $F(x_n) \leq$ constant (independent of n) has an accumulation point.*

We now formulate the following general existence theorem for minimizers:

Theorem 4.2.1. *Let X be a separable reflexive Banach space, $F : X \to \mathbb{R}$ weakly proper and lower semicontinuous w.r.t. weak convergence. Then there exists a minimizer x_0 for F, i.e.*

$$F(x_0) = \inf_{x \in X} F(x) \quad (> -\infty).$$

Proof. Let $(x_n)_{n \in \mathbb{N}}$ be a minimizing sequence for F, i.e.

$$\lim_{n \to \infty} F(x_n) = \inf_{x \in X} F(x).$$

Since F is weakly proper, $\|x_n\|$ is bounded. Since X is reflexive, after selection of a subsequence, x_n converges weakly to some $x_0 \in X$ by Corollary 2.2.1. By lower semicontinuity of F,

$$F(x_0) \leq \lim_{n \to \infty} F(x_n) = \inf_{x \in X} F(x),$$

and since $x_0 \in X$, we must have in fact equality. Also, since F assumes only finite values by assumption, this implies that

$$\inf_{x \in X} F(x) > -\infty.$$

q.e.d.

Remark 4.2.1. The argument of the preceding proof also shows that in a separable reflexive Banach space, a weakly proper functional is coercive w.r.t. the weak topology.

Lower semicontinuity w.r.t. weak convergence is a rather strong property, in fact much stronger than lower semicontinuity w.r.t. to the Banach space topology of X. Fortunately, there exists a general class of functionals, namely the convex ones for which the latter property implies the former.

Definition 4.2.3. *Let V be a convex subset of a vector space; $F : V \to \bar{\mathbb{R}}$ is called convex if for any $x, y \in V$, $0 \leq t \leq 1$,*

$$F(tx + (1-t)y) \leq tF(x) + (1-t)F(y)$$

(convexity of V means that $tx + (1-t)y \in V$ whenever $x, y \in V$, $0 \leq t \leq 1$).

Lemma 4.2.2. *Let V be a convex subset of a separable reflexive Banach space, $F : V \to \bar{\mathbb{R}}$ convex and lower semicontinuous. Then F is also lower semicontinuous w.r.t. weak convergence.*

Proof. Let $(x_n)_{n \in \mathbb{N}} \subset V$ converge weakly to $x \in V$. We may assume that $F(x_n)$ converges to some $\kappa \in \bar{\mathbb{R}}$. By Theorem 2.2.4, for every $m \in \mathbb{N}$ and every $\varepsilon > 0$, we may find a convex combination

$$y_m := \sum_{n=m}^{N} \lambda_n x_n \quad (\lambda_n > 0, \sum_{n=m}^{N} \lambda_n = 1)$$

with

$$\|y_m - x\| \leq \varepsilon.$$

Since F is convex,

$$F(y_m) \leq \sum_{n=m}^{N} \lambda_n F(x_n). \qquad (4.2.1)$$

Given $\varepsilon > 0$, we choose $m = m(\varepsilon) \in \mathbb{N}$ so large that for all $n \geq m$,

$$F(x_n) < \kappa + \varepsilon.$$

Letting ε tend to 0, we get from (4.2.1)

$$\limsup_{m \to \infty} F(y_m) \leq \kappa.$$

Since F is lower semicontinuous

$$F(x) \leq \liminf_{m \to \infty} F(y_m) \leq \limsup_{m \to \infty} F(y_m) \leq \kappa = \lim F(x_n).$$

This shows weak lower semicontinuity of F.

q.e.d.

4.3 The existence of minimizers for convex variational problems

We return to the concrete variational problem discussed in Section 4.1 and begin with:

Lemma 4.3.1. *Let $\Omega \subset \mathbb{R}^d$ be open, $f : \Omega \times \mathbb{R}^d \to \mathbb{R}$, with $f(\cdot, v)$ measurable for all $v \in \mathbb{R}^d$, $f(x, \cdot)$ continuous for all $x \in \Omega$, and*

$$f(x, v) \geq -a(x) + b |v|^p$$

for almost all $x \in \Omega$, and all $v \in \mathbb{R}^d$, with $a \in L^1(\Omega)$, $b \in \mathbb{R}$, $p \geq 1$. Then

$$\Phi(v) := \int_\Omega f(x, v(x))dx$$

is a lower semicontinuous functional on $L^p(\Omega)$, $\Phi : L^p(\Omega) \to \mathbb{R} \cup \{\infty\}$.

Proof. Since f is continuous in v, $f(x, v(x))$ is a measurable function, and so Φ is well-defined on $L^p(\Omega)$, by Theorem 1.1.2. Suppose $(v_n)_{n \in \mathbb{N}}$ converges to v in $L^p(\Omega)$. Then a subsequence converges pointwise almost everywhere to v by Lemma 3.1.3. We shall denote this subsequence again by (v_n), noting that the subsequent arguments may also be applied to any remaining subsequence. Since f is continuous in v (actually, it would suffice to have f lower semicontinuous in v), we have

$$f(x, v(x)) - b|v(x)|^p \leq \liminf_{n \to \infty} \left(f(x, v_n(x)) - b|v_n(x)|^p \right).$$

Because of the lower bound

$$f(x, v_n(x)) - b|v_n(x)|^p \geq -a(x)$$

with $a \in L^1(\Omega)$, we may apply the Theorem 1.2.2 of Fatou to conclude

$$\int_\Omega \left(f(x, v(x)) - b|v(x)|^p \right) dx \leq \liminf_{n \to \infty} \int_\Omega \left(f(x, v_n(x)) - b|v_n(x)|^p \right) dx.$$

Since v_n converges to v in $L^p(\Omega)$,

$$\int_\Omega b|v(x)|^p \, dx = \lim_{n \to \infty} \int_\Omega b|v_n(x)|^p \, dx,$$

and we conclude lower semicontinuity, namely

$$\int_\Omega f(x, v(x))dx \leq \liminf_{n \to \infty} \int_\Omega f(x, v_n(x))dx.$$

q.e.d.

Lemma 4.3.2. *Under the assumptions of Lemma 4.3.1, assume that $f(x, \cdot)$ is a convex function on $\mathbb{R}^d$ for every $x \in \Omega$. Then $\Phi(v) := \int_\Omega f(x, v(x))dx$ defines a convex functional on $L^p(\Omega)$.*

4.3 Existence of minimizers

Proof. Let $v, w \in L^p(\Omega)$, $0 \le t \le 1$. Then

$$\Phi(tv + (1-t)w) = \int_\Omega f(x, tv(x) + (1-t)w(x))\,dx$$
$$\le \int_\Omega \{tf(x, v(x)) + (1-t)f(x, w(x))\}\,dx$$

by the convexity of f

$$= t\Phi(v) + (1-t)\Phi(w).$$

q.e.d.

We may now obtain a general existence result for the minimizer of a convex variational problem.

Theorem 4.3.1. *Let $\Omega \subset \mathbb{R}^d$ be open, and suppose $f : \Omega \times \mathbb{R}^d \to \mathbb{R}$ satisfies*

(i) $f(\cdot, v)$ *is measurable for all $v \in \mathbb{R}^d$.*
(ii) $f(x, \cdot)$ *is convex for all $x \in \Omega$.*
(iii) $f(x, v) \ge -a(x) + b|v|^p$ *for almost all $x \in \Omega$, all $v \in \mathbb{R}^d$, with $a \in L^1(\Omega)$, $b > 0$, $p > 1$.*

Let $g \in H^{1,p}(\Omega)$, and let $A := g + H_0^{1,p}(\Omega)$. Then

$$F(u) := \int_\Omega f(x, Du(x))\,dx$$

assumes its infimum on A, i.e. there exists $u_0 \in A$ with

$$F(u_0) = \inf_{u \in A} F(u).$$

Proof. By Lemma 4.3.1, F is lower semicontinuous w.r.t. $H^{1,p}(\Omega)$ convergence†, and by Lemma 4.2.2, F then is also lower semicontinuous w.r.t. weak $H^{1,p}(\Omega)$ convergence, since $H^{1,p}(\Omega)$ is separable and reflexive for $p > 1$ (see Theorems 3.3.1 and 3.3.3). Let $(u_n)_{n \in \mathbb{N}}$ be a minimizing sequence in A, i.e.

$$\lim_{n \to \infty} F(u_n) = \inf_{u \in A} F(u).$$

Since

$$\int_\Omega |Du_n|^p \le \frac{1}{b} F(u_n) + \frac{1}{b} \int_\Omega a(x)\,dx,$$

$(Du_n)_{n \in \mathbb{N}}$ is bounded in $L^p(\Omega)$, hence $(u_n)_{n \in \mathbb{N}} \subset g + H_0^{1,p}(\Omega)$ is bounded in $H^{1,p}(\Omega)$ by the Poincaré inequality (see Theorem 3.4.2). Since $H_0^{1,p}(\Omega)$

† Note that convex functions on $\mathbb{R}^d$ are continuous.

is a separable reflexive Banach space, by Theorem 3.3.5, after selection of a subsequence, $(u_n)_{n\in\mathbb{N}}$ converges weakly to some $u_0 \in A$ (A is closed under weak convergence, Theorem 3.3.4). Since F is convex by Lemma 4.3.2 and lower semicontinuous by Lemma 4.3.1, it is also lower semicontinuous w.r.t. weak $H^{1,p}(\Omega)$ convergence by Lemma 4.4.2. Therefore

$$F(u_0) \le \lim_{n\to\infty} F(u_n) = \inf_{u\in A} F(u),$$

and since $u_0 \in A$, we must have equality.

q.e.d.

Remark 4.3.1. The condition $u \in g + H_0^{1,p}(\Omega)$, i.e. $u - g \in H_0^{1,p}(\Omega)$, is a (generalized) Dirichlet boundary condition. It means that $u = g$ on $\partial\Omega$ in the sense of Sobolev spaces.

4.4 Convex functionals on Hilbert spaces and Moreau–Yosida approximation

In this section, we develop a more abstract method for showing the existence of minimizers of variational problems. It has the advantages that it does not need the concept of weak convergence and that it provides a constructive approach for finding the minimizer. In order to concentrate on the essential aspects, we shall only treat a special situation.

Definition 4.4.1. *Let X be a metric space with metric $d(\cdot,\cdot)$, and let $F : X \to \mathbb{R} \cup \{\infty\}$ be a functional. For $\lambda > 0$, we define the Moreau–Yosida approximation F^λ of F as*

$$F^\lambda(x) := \inf_{y\in X} (\lambda F(y) + d^2(x,y)) \tag{4.4.1}$$

for $x \in X$.

Remark 4.4.1. This is different from the definition in Section 5.1 where we shall take $d(x,y)$ instead of $d^2(x,y)$. Here, one might take $d^\alpha(x,y)$ for any exponent $\alpha > 1$. For our present purposes, it is most convenient to work with $\alpha = 2$.

We now let H be a Hilbert space with scalar product $\langle\cdot,\cdot\rangle$ and norm $\|\cdot\|$ and induced metric $d(x,y) = \|x-y\|$. Let $D(F) \subset H$, and let $F : D(F) \to \mathbb{R}$ be a functional. We say that F is densely defined if $D(F)$

4.4 Convex functionals

is dense in H. For $x \notin D(F)$, we put $F(x) = \infty$. We say that F is convex if whenever $\gamma : [0,1] \to H$ is a straight line segment, then for $0 \le t \le 1$

$$F(\gamma(t)) \le tF(\gamma(0)) + (1-t)F(\gamma(1)). \tag{4.4.2}$$

In particular, if $\gamma(0), \gamma(1) \in D(F)$, then also $\gamma(t) \in D(F)$ for $0 \le t \le 1$.

Lemma 4.4.1. *Let $F : H \to \mathbb{R} \cup \{\infty\}$ be convex, bounded from below, and lower semicontinuous. Then for every $x \in H$ and $\lambda > 0$, there exists a unique*

$$y^\lambda =: J^\lambda(x)$$

with

$$F^\lambda(x) = \lambda F(y^\lambda) + d^2(x, y^\lambda) \tag{4.4.3}$$

Proof. We have to show that the infimum in (4.4.1) is realized by a unique y^λ.

Uniqueness: Let y_1^λ, y_2^λ be solutions of (4.4.3), and let

$$y_0^\lambda = \frac{1}{2}(y_1^\lambda + y_2^\lambda)$$

be their mean value. By convexity of F

$$F(y_0^\lambda) \le \frac{1}{2}\left(F\left(y_1^\lambda\right) + F\left(y_2^\lambda\right)\right), \tag{4.4.4}$$

and by Euclidean geometry, if $y_1^\lambda \ne y_2^\lambda$, we have

$$\|x - y_0^\lambda\|^2 < \frac{1}{2}\left(\|x - y_1^\lambda\|^2 + \|x - y_2^\lambda\|^2\right), \tag{4.4.5}$$

hence

$$\lambda F(y_0^\lambda) + \|x - y_0^\lambda\|^2 < \lambda F(y_1^\lambda) + \|x - y_2^\lambda\|^2$$
$$= \lambda F(y_2^\lambda) + \|x - y_2^\lambda\|^2,$$

contradicting the minimizing property of y_1^λ and y_2^λ. Thus, we must have $y_1^\lambda = y_2^\lambda$, proving uniqueness.

Existence: (4.4.5) may be refined as follows: For $y_1, y_2 \in H$ and

$$y_0 := \frac{1}{2}(y_1 + y_2)$$

we have for any $x \in H$

$$\|x - y_0\|^2 = \frac{1}{2}\left(\|x - y_1\|^2 + \|x - y_2\|^2\right) - \frac{1}{4}\|y_1 - y_2\|^2. \tag{4.4.6}$$

We now let $(y_n)_{n\in\mathbb{N}}$ be a minimizing sequence, i.e.

$$\lambda F(y_n) + ||x - y_n||^2 \to \inf_{y \in H} \left(\lambda F(y) + ||x - y||^2\right) =: \kappa_\lambda. \quad (4.4.7)$$

We claim that (y_n) is a Cauchy sequence. For $l, k \in \mathbb{N}$, we put

$$y_{l,k} := \frac{1}{2}(y_k + y_l).$$

Using the convexity of F as in (4.4.4) and (4.4.6), we obtain

$$\lambda F(y_{k,l}) + ||x - y_{k,l}||^2$$

$$\leq \frac{1}{2}\left(\lambda F(y_k) + ||x - y_k||^2\right) + \frac{1}{2}\left(\lambda F(y_l) + ||x - y_l||^2\right) - \frac{1}{4}||y_k - y_l||^2. \quad (4.4.8)$$

By definition of κ_λ (see (4.4.7)), the left hand side of (4.4.8) cannot be smaller than κ_λ, and so we conclude that

$$||y_k - y_l||^2 \to 0$$

as $k, l \to \infty$, establishing the Cauchy property. Since the norm is continuous and F is assumed to be lower semicontinuous, the limit y^λ of $(y_n)_{n\in\mathbb{N}}$ then solves (4.4.3). q.e.d.

Lemma 4.4.2. *Let F and $y^\lambda = J^\lambda(x)$ be as in Lemma 4.4.1. Let x be in the closure of $D(F)$. Then*

$$x = \lim_{\lambda \to 0} J^\lambda(x). \quad (4.4.9)$$

Proof. Since x is in the closure of $D(F)$, for every $\delta > 0$, we may find

$$x_\delta \in B(x, \delta) := \{y \in H : ||x - y|| \leq \delta\}$$

with

$$F(x_\delta) < \infty.$$

Then

$$\lim_{\lambda \to 0} \left(\lambda F(x_\delta) + ||x - x_\delta||^2\right) \leq \delta^2$$

and therefore

$$\limsup_{\lambda \to 0} \kappa_\lambda \leq 0 \quad (4.4.10)$$

(see (4.4.7) for the definition of κ_λ).

Let us now assume that there exists a sequence $\lambda_n \to 0$ for $n \to \infty$ with

$$||x - y^{\lambda_n}||^2 \geq \alpha > 0 \quad \text{for all } n. \quad (4.4.11)$$

4.4 Convex functionals

Then from (4.4.10)

$$\limsup_{n \to \infty} \left(\lambda_n F(y^{\lambda_n}) + ||x - y^{\lambda_n}||^2 \right) \leq 0, \qquad (4.4.12)$$

hence

$$F(y^{\lambda_n}) \to -\infty \quad \text{as } n \to \infty. \qquad (4.4.13)$$

(4.4.12) and (4.4.13) imply

$$F(y^1) + ||x - y^1||^2 \leq F(y^{\lambda_n}) + ||x - y^{\lambda_n}||^2 \to -\infty \quad \text{as } n \to \infty$$

which is impossible. Thus, (4.4.11) cannot hold, and (4.4.9) follows.
q.e.d.

Theorem 4.4.1. *Let* $F : H \to \mathbb{R} \cup \{\infty\}$ *be convex, bounded from below, and lower semicontinuous, and* $F \not\equiv \infty$. *For* $x \in M$, *we let* $y^\lambda = J^\lambda(x)$ *as in Lemma 4.4.1. If* $(y^{\lambda_n})_{n \in \mathbb{N}}$ *is bounded for some sequence* $\lambda_n \to \infty$, *then* $(y^\lambda)_{\lambda > 0}$ *converges to a minimizer of* F *as* $\lambda \to \infty$.

Proof. Since $(y^{\lambda_n})_{n \in \mathbb{N}}$ is bounded and since y^{λ_n} minimizes

$$F(y) + \frac{1}{\lambda_n} ||x - y||^2,$$

we obtain

$$F(y^{\lambda_n}) \to \inf_{y \in H} F(y)$$

so that $(y^{\lambda_n})_{n \in \mathbb{N}}$ is a minimizing sequence for F. We now claim that

$$||x - y^\lambda||^2$$

is monotonically increasing in λ. Indeed, let $0 < \mu_1 < \mu_2$. Then by definition of y^{μ_1}

$$F(y^{\mu_2}) + \frac{1}{\mu_1} ||x - y^{\mu_2}||^2 \geq F(y^{\mu_1}) + \frac{1}{\mu_1} ||x - y^{\mu_1}||^2,$$

hence

$$F(y^{\mu_2}) + \frac{1}{\mu_2} ||x - y^{\mu_2}||^2 \geq F(y^{\mu_1}) + \frac{1}{\mu_2} ||x - y^{\mu_1}||^2$$
$$+ \left(\frac{1}{\mu_1} - \frac{1}{\mu_2} \right) \left(||x - y^{\mu_1}||^2 - ||x - y^{\mu_2}||^2 \right).$$

This is compatible with the minimizing property of y^{μ_2} only if

$$||x - y^{\mu_1}||^2 \geq ||x - y^{\mu_2}||^2$$

194   *Direct methods*

and monotonicity follows. This monotonicity then implies that
$$||x - y^\lambda||^2$$
is bounded independently of λ since it is assumed to be bounded for the sequence $\lambda_n \to \infty$. We next claim that
$$F(y^\lambda)$$
monotonically decreases towards
$$\inf_{y \in H} F(y).$$
Indeed, from the definition of y^λ,
$$F(y^\lambda) = \inf_{\{y:||x-y|| \leq ||x-y^\lambda||\}} F(y),$$
and therefore y^λ has to decrease since $||x - y^\lambda||$ increases. The limit has to be $\inf_{y \in H} F(y)$ since this is so for the subsequence $(y^{\lambda_n})_{n \in \mathbb{N}}$. We now claim that $(y^\lambda)_{\lambda > 0}$ satisfies the Cauchy property, i.e. for every $\epsilon > 0$, there exists $\lambda_0 > 0$ such that for all $\lambda, \mu > \lambda_0$
$$||y^\lambda - y^\mu||^2 < \epsilon.$$
For that purpose, we choose λ_0 so large that for $\lambda, \mu > \lambda_0$
$$\left| ||x - y^\lambda||^2 - ||x - y^\mu||^2 \right| < \frac{\epsilon}{2} \tag{4.4.14}$$
which is possible by the preceding monotonicity and boundedness results. We may also assume
$$F(y^\lambda) \geq F(y^\mu). \tag{4.4.15}$$
We let
$$y^{\lambda,\mu} := \frac{1}{2}(y^\lambda + y^\mu).$$
Then from the convexity of F, (4.4.15), and (4.4.6),
$$F(y^{\lambda,\mu}) + \frac{1}{\lambda}||x - y^{\lambda,\mu}||^2$$
$$\leq F(y^\lambda) + \frac{1}{\lambda}\left(\frac{1}{2}||x - y^\lambda||^2 + \frac{1}{2}||x - y^\mu||^2 - \frac{1}{4}||y^\lambda - y^\mu||^2 \right)$$
$$< F(y^\lambda) + \frac{1}{\lambda}\left(||x - y^\lambda||^2 + \frac{\epsilon}{4} - \frac{1}{4}||y^\lambda - y^\mu||^2 \right)$$
by (4.4.14).

This, however, is compatible with the minimizing property of y^λ only if
$$||y^\lambda - y^\mu||^2 < \epsilon.$$
Thus $(y^\lambda)_{\lambda>0}$ satisfies the Cauchy property for $\lambda \to \infty$, and it therefore converges to some $\bar{y} \in H$. $\bar{y}$ then minimizes F, because $F(y^\lambda)$ decreases towards $\inf_{y \in H} F(y)$ for $\lambda \to \infty$, and F is lower semicontinuous.

q.e.d.

The preceding reasoning is adapted from J. Jost, Convex functionals and generalized harmonic maps between metric spaces. *Comment. Math. Helv.* 70 (1995), 659–673.

For a more general construction, see J. Jost, *Nonpositive Curvature: Geometric and Analytic Aspects*, Birkhäuser, Basel, 1997, pp. 61–4. In particular, the method also works in uniformly convex Banach spaces.

General references for Moreau–Yosida approximation are the books of Attouch and dal Maso quoted in Chapter 6.

Theorem 4.4.1 yields an alternative proof of Theorem 4.3.1 in case $p = 2$. Namely, Lemma 4.3.1 implies the lower semicontinuity, Lemma 4.3.2 the convexity of the functional, and the Poincaré inequality the boundedness of any minimizing sequence, as described in the proof of Theorem 4.3.1. The present proof, however, does not need the concept of weak convergence. As mentioned, the method extends to uniformly convex Banach spaces, and thus can handle also arbitrary values of $p > 1$ (see Remark 3.1.2).

4.5 The Euler–Lagrange equations and regularity questions

In this section, we return to the variational problems considered in Sections 4.1 and 4.3; we consider variational integrals of the form
$$\Phi(u) := \int_\Omega f(x, u(x), Du(x))dx, \quad \text{for } u \in H^{1,p}(\Omega)$$
on a bounded, open subset Ω of $\mathbb{R}^d$, and we make the following assumptions on $f : \Omega \times \mathbb{R} \times \mathbb{R}^d \to \bar{\mathbb{R}} = \mathbb{R} \cup \{\pm\infty\}$:

(i) $f(\cdot, u, v)$ is measurable for all $u \in \mathbb{R}$, $v \in \mathbb{R}^d$.
(ii) $f(x, \cdot, \cdot)$ is differentiable for almost all $x \in \Omega$.
(iii) $|f(x, u, v)| \leq c_0 + c_1 |u|^p + c_2 |v|^p$, c_0, c_1, c_2 constants, for almost all $x \in \Omega$, and all $u \in \mathbb{R}$, $v \in \mathbb{R}^d$.

Condition (iii) implies that $\Phi(u)$ is finite for $u \in H^{1,p}(\Omega)$, since Ω is bounded. (If Ω is unbounded, this still holds provided $c_0 = 0$.) In the preceding section, we have obtained some results on the existence of a minimizer for Φ in the class $g + H_0^{1,p}(\Omega)$, for given $g \in H^{1,p}(\Omega)$. In the present section, we wish to characterize such minimizers by necessary conditions. These conditions will assume the form of differential equations. In fact, these differential equations will hold for arbitrary critical points of Φ (as specified in the assumptions of our subsequent results), and not only for minimizers.

Theorem 4.5.1. *Let f satisfy (in addition to (i)–(iii))*

(iv)
$$\left|\frac{\partial f}{\partial u}(x,u,v)\right| + \sum_{i=1}^{d}\left|\frac{\partial f}{\partial v^i}(x,u,v)\right| \leq c_3 + c_4\,|u|^p + c_5\,|v|^p,$$

c_3, c_4, c_5 *constants, for almost all $x \in \Omega$, and all $u \in \mathbb{R}$, $v \in \mathbb{R}^d$.*

Let u be a minimizer for Φ in the class $g + H_0^{1,p}(\Omega)$ ($g \in H^{1,p}(\Omega)$ given). We then have for all $\varphi \in C_0^\infty(\Omega)$

$$\int_\Omega \left\{ \frac{\partial f}{\partial u}(x, u(x), Du(x))\varphi(x) + \sum_{i=1}^{d} \frac{\partial f}{\partial v^i}(x, u(x), Du(x)) \frac{\partial}{\partial x^i}\varphi(x) \right\} dx$$
$$= 0. \quad (4.5.1)$$

Proof. Since u is a minimizer for Φ in $g + H_0^{1,p}(\Omega)$,

$$\Phi(u) \leq \Phi(u + t\varphi) \quad \text{for } t \in \mathbb{R}, \varphi \in C_0^\infty(\Omega). \quad (4.5.2)$$

We have

$$\Phi(u+t\varphi) = \int_\Omega f\big(x, u(x) + t\varphi(x), Du(x) + tD\varphi(x)\big)dx.$$

By (ii), (iii), (iv), we may apply Corollary 1.2.2 to conclude that $\Phi(u+t\varphi)$ is differentiable w.r.t. φ, and

$$\frac{d}{dt}\Phi(u+t\varphi)$$
$$= \int_\Omega \left\{ \frac{\partial f}{\partial u}(x, u(x) + t\varphi(x), Du(x) + tD\varphi(x))\varphi(x) \right.$$
$$\left. + \sum_{i=1}^d \frac{\partial f}{\partial v^i}(x, u(x) + t\varphi(x), Du(x) + tD\varphi(x)) \frac{\partial \varphi(x)}{\partial x^i} \right\} dx. \quad (4.5.3)$$

4.5 Euler–Lagrange equations

Furthermore, (4.5.2) implies

$$\frac{d}{dt}\Phi(u+t\varphi)|_{t=0}=0. \quad (4.5.4)$$

Equations (4.5.3) and (4.5.4) imply (4.5.1).

q.e.d.

Remark 4.5.1. From the preceding proof, it is clear that we do not need to assume that u is a minimizer for Φ. If suffices that u is a critical point for Φ in the sense that

$$\frac{d}{dt}\Phi(u+t\varphi)|_{t=0} \quad \text{for all } \varphi \in C_0^\infty(\Omega). \quad (4.5.5)$$

Corollary 4.5.1. *Suppose that f satisfies (i)–(iv), and in addition, $f \in C^2$. If $u \in C^2(\Omega)$ minimizes Φ in the class $g + H_0^{1,p}(\Omega)$ (or, more generally, satisfies (4.5.5)), then*

$$\sum_{i,j=1}^d \frac{\partial^2 f}{\partial v^i \partial x^j}(x,u(x),Du(x))\frac{\partial^2 u}{\partial x^i \partial x^j} + \sum_{i=1}^d \frac{\partial^2 f}{\partial v^i \partial u}(x,u(x),Du(x))\frac{\partial u}{\partial x^i}$$

$$+ \sum_{i=1}^d \frac{\partial^2 f}{\partial v^i \partial x^i}(x,u(x),Du(x)) - \frac{\partial f}{\partial u}(x,u(x),Du(x)) = 0.$$

$$(4.5.6)$$

Definition 4.5.1. *Equation (4.5.6) is called the* Euler–Lagrange *equation for Φ.*

Proof (Corollary 4.5.1). By the differentiability assumptions made, we may integrate (4.5.1) by parts to obtain

$$\int_\Omega \left\{ \frac{\partial f}{\partial u}(x,u(x),Du(x)) - \sum_{i,j=1}^d \frac{\partial^2 f}{\partial v^i \partial v^j}(x,u(x),Du(x))\frac{\partial}{\partial x^i}\left(\frac{\partial u(x)}{\partial x^j}\right) \right.$$

$$- \sum_{i=1}^d \frac{\partial^2 f}{\partial v^i \partial u}(x,u(x),Du(x))\frac{\partial u}{\partial x^i}(x)$$

$$\left. - \sum_{i=1}^d \frac{\partial^2 f}{\partial v^i \partial x^i}(x,u(x),Du(x)) \right\} \varphi(x) dx. \quad (4.5.7)$$

From Lemma 3.2.3 (applied to $\operatorname{supp}\varphi \subset\subset \Omega$ so that the term in $\{\cdots\}$ is in L^2), we then obtain (4.5.6).

q.e.d.

Equations (4.5.6) constitutes a quasilinear partial differential equation of second order for u. Many such partial differential equations arise as Euler–Lagrange equations of variational problems. Therefore, if one wants to solve such an equation, one might try to find a minimizer of the associated variational problem. However, the existence theory for minimizers as described in Section 4.3 naturally yields an element u of the Sobolev space $H_0^{1,p}(\Omega)$, whereas in Corollary 4.5.1 it is required that u be of class $C^2(\Omega)$. Thus, there exists a gap, since in general elements of $H_0^{1,p}(\Omega)$ are not of class C^2. It is the task of regularity theory to bridge this gap, i.e. to show that under suitable assumptions on f, any minimizer of Φ is smooth, and specifically here of class C^2. The theory of partial differential equations indicates that such a result does not hold without additional assumptions on f, like an ellipticity assumption, meaning that the matrix $(a^{ij}(x))_{i,j=1,\ldots,d}$ with coefficients $a^{ij}(x) = \frac{\partial^2 f}{\partial v^i \partial v^j}(x, u(x), Du(x))$ is positive definite. Indeed, examples show that without such an assumption, in general one does not get smoothness of minimizers. On the positive side, however, we do have de Giorgi's and Nash's:

Theorem 4.5.2. Let Ω be open and bounded in $\mathbb{R}^d$, $f : \Omega \times \mathbb{R}^d \to \mathbb{R}$ be of class C^∞, with

(i)
$$\lambda |v|^2 \leq f(x,v) \leq \Lambda(1+|v|^2)$$

and

$$\lambda |\xi|^2 \leq \sum_{i,j=1}^d \frac{\partial^2 f}{\partial v^i \partial v^j}(x,v) \xi^i \xi^j \leq \Lambda |\xi|^2$$

for all $x \in \Omega$, $u \in \mathbb{R}$, $v, \xi \in \mathbb{R}^d$, with constants $\lambda > 0$, $\Lambda < \infty$,

(ii)
$$\left|\frac{\partial f}{\partial v^i}(x,v)\right| \leq M(1+|v|) \quad \text{for a constant } M < \infty.$$

Let $u \in g + H_0^{1,2}(\Omega)$ be a bounded minimizer of $F(u) := \int_\Omega f(x, Du(x))dx$ ($g \in H^{1,p}(\Omega)$ given). Then u is smooth in Ω ($u \in C^\infty(\Omega)$).

The proof of the theorem of de Giorgi and Nash is too long to be presented here. We refer to M. Giaquinta, *Introduction to Regularity Theory for Nonlinear Elliptic Systems*, Birkhäuser, Basel, 1993, pp. 76–99 and

4.5 Euler–Lagrange equations

J. Jost, *Partielle Differentialgleichungen*, Springer, Berlin, 1998 where a detailed proof is given. Of course, there also exist extensions of this result to more general integrands of the form $f(x, u, v)$. We refer the interested reader to O. Ladyzhenskaya, N. Ural'tseva, *Linear and Quasilinear Elliptic Equations*, Academic Press, New York, 1968 (translated from the Russian), Chapters IV–VI.

One *remark* is in order here: Since Sobolev functions are only equivalence classes of functions (in the sense specified at the beginning of Section 3.1), a more precise version of Theorem 4.5.2 is: Under the stated assumptions, the equivalence class of u contains a function of class C^∞. This point, however, usually is assumed to be implicitly understood in statements of regularity theorems.

In order to display at least one regularity result, however, we consider a particular example:

For a bounded, open $\Omega \subset \mathbb{R}^d$, $g \in H^{1,2}(\Omega)$, we wish to minimize Dirichlet's integral

$$D(u) := \int_\Omega |Du(x)|^2 \, dx \tag{4.5.8}$$

in the class $g + H_0^{1,p}(\Omega)$. By Theorem 4.3.1, a minimizer u exists, and by Theorem 4.5.1, it satisfies

$$\int_\Omega Du(x) \cdot D\varphi(x) dx = 0 \quad \text{for all } \varphi \in C_0^\infty(\Omega) \tag{4.5.9}$$

(here $Du(x) \cdot D\varphi(x) := \sum_{i=1}^d D_i u(x) D_i \varphi(x)$). If u can be shown to be of class C^2, it would satisfy

$$\Delta u(x) = 0 \quad \text{in } \Omega \quad \left(\Delta := \sum_{i=1}^d \frac{\partial^2}{(\partial x^i)^2}\right)$$

(Δ is called Laplace operator.) by Corollary 4.5.1, i.e. it is harmonic. This is the famous **Dirichlet principle**: obtain a harmonic function u in Ω with boundary values g by minimizing the Dirichlet integral among all functions with those boundary values.

In order to justify Dirichlet's principle it thus remains to show that any solution of (4.5.9) is of class C^2. Actually, one can show more, namely, $u \in C^\infty$ (in fact, u is even real analytic in Ω but this will not be demonstrated here), and at the same time weaken the assumption. Namely, we have:

Theorem 4.5.3 (Weyl's lemma). Let $u \in L^1(\Omega)$ satisfy

$$\int_\Omega u(x)\Delta\varphi(x) = 0 \quad \text{for all } v \in C_0^\infty(\Omega). \tag{4.5.10}$$

Then $u \in C^\infty(\Omega)$.

Remark 4.5.2.

(1) Clearly, (4.5.9) implies (4.5.10) by definition of Du.
(2) The remark made after Theorem 4.5.2 again applies.

Proof (Theorem 4.5.3). We consider the mollifications with a rotationally symmetric ρ (and we express this by writing ρ as a function of $|x|$)

$$u_h(x) = \frac{1}{h^d}\int_\Omega \varrho\left(\frac{|x-y|}{h}\right) u(y) dy$$

as in Section 3.2. Given $\varphi \in C_0^\infty(\Omega)$, we restrict h to be smaller than $\text{dist}(\text{supp}\,\varphi, \partial\Omega)$. We obtain

$$\int_\Omega u_h(x)\Delta\varphi(x) dx = \int_\Omega \frac{1}{h^d}\int_\Omega \varrho\left(\frac{|x-y|}{h}\right) u(y) dy \Delta\varphi(x) dx$$

$$= \int_\Omega u(x)\Delta\varphi_h(x) dx, \tag{4.5.11}$$

using Fubini's theorem.

q.e.d.

Remark 4.5.3. We have also used the fact that Δ commutes with mollification, i.e.

$$(\Delta\varphi)_h = \Delta(\varphi_h). \tag{4.5.12}$$

For this, one needs that ϱ is a function of $|x|$ only, i.e. rotationally symmetric. Also, this point needs the rotational invariance of the Laplace operator Δ. Therefore, the present proof does not generalize to other variational problems.

After this interruption, we return to (4.5.11) and conclude that

$$\int_\Omega u_h(x)\Delta\varphi(x) dx = 0 \tag{4.5.13}$$

by applying (4.5.10) to $\varphi_h \in C_0^\infty(\Omega)$ (by our choice of h). Since u_h is smooth, we obtain e.g. from Corollary 4.5.1

$$\Delta u_h = 0$$

4.5 Euler–Lagrange equations

in $\Omega_h := \{x \in \Omega \mid \text{dist}(x, \partial\Omega) > h\}$. Also

$$\int_{\Omega_h} |u_h(y)| \, dy$$

$$\leq \int_{\Omega_h} \frac{1}{h^d} \int_{\Omega} \varrho\left(\frac{|x-y|}{h}\right) |u(x)| \, dx dy$$

$$\leq \int_{\Omega} |u(x)| \, dx \tag{4.5.14}$$

by Fubini's theorem, using $\frac{1}{h^d} \int \varrho\left(\frac{|x-y|}{h}\right) dy = 1$ by (3.2.3)

$< \infty$ since $u \in L^1(\Omega)$.

Therefore, the functions u_h are uniformly bounded in L^1. We now need

Lemma 4.5.1. Let $f \in C^2(\Omega)$ be harmonic, i.e.

$$\Delta f(x) = 0 \quad \text{in } \Omega.$$

Then f satisfies the mean value property, i.e. for every ball $B(x_0, r) \subset \Omega$,

$$f(x_0) = \frac{1}{\omega_d r^d} \int_{B(x_0,r)} f(x) dx = \frac{1}{d\omega_d r^{d-1}} \int_{\partial B(x_0,r)} f(x) d\sigma(x) \tag{4.5.15}$$

where ω_d is the volume of the unit ball in $\mathbb{R}^d$.

Proof. For $0 < \varrho \leq r$

$$0 = \int_{B(x_0,\varrho)} \Delta f(x) dx$$

$$= \int_{\partial B(x_0,\varrho)} \frac{\partial f}{\partial \nu}(x) d\sigma(x),$$

where ν denotes the exterior normal of $B(x_0, \varrho)$

$$= \int_{\partial B(0,1)} \frac{\partial f}{\partial \varrho}(y + \varrho\omega) \varrho^{d-1} d\omega$$

in polar coordinates $\omega = \frac{x-y}{\varrho}$

$$= \varrho^{d-1} \frac{\partial}{\partial \varrho} \int_{\partial B(0,1)} f(y + \varrho\omega) d\omega$$

$$= \varrho^{d-1} \frac{\partial}{\partial \varrho} \left(\varrho^{1-d} \int_{\partial B(0,1)} f(x) d\sigma(x) \right)$$

$$= d\omega_d \varrho^{d-1} \frac{\partial}{\partial \varrho} \left(\frac{1}{d\omega_d \varrho^{d-1}} \int_{\partial B(x_0,\varrho)} f(x) d\sigma(x) \right).$$

Thus,
$$\frac{1}{d\omega_d \varrho^{d-1}} \int_{B(x_0,\varrho)} f(x) d\sigma(x)$$
is constant in ϱ, and since its limit for $\varrho \to 0$ is $f(x_0)$ as f is continuous, it has to coincide with $f(x_0)$ for all $0 < \varrho \le r$. Since
$$\frac{1}{\omega_d r^d} \int_{B(x_0,\varrho)} f(x) dx = \frac{d}{r^d} \int_0^r \left(\frac{1}{d\omega_d \varrho^{d-1}} \int_{\partial B(x_0,\varrho)} f(x) d\sigma(x) \right) \varrho^{d-1} d\varrho,$$
the first inequality in (4.5.15) also follows.

q.e.d.

We return to the proof of Theorem 4.5.3: Since u_h is harmonic, it satisfies the mean value properties of Lemma 4.5.1. Since the family u_h is bounded in L^1,
$$u_h(x_0) = \frac{1}{\omega_d r^d} \int_{B(x_0,r)} u_h(x) dx$$
is bounded for fixed r with $B(x_0,r) \subset \Omega_h$. Therefore, the u_h are uniformly bounded in Ω_{h_0} for $0 < h \le \frac{h_0}{2}$. Furthermore, from (4.5.15)
$$|u_h(x_1) - u_h(x_2)| \le \frac{1}{d\omega_d} \left(\frac{1}{r}\right)^d \int_{\substack{B(x_1,r) \setminus B(x_2,r) \\ \cup B(x_2,r) \setminus B(x_1,r)}} |u_h(x)| dx$$
$$\le c(r) |x_1 - x_2| \qquad (4.5.16)$$
for some constant depending on r, if $B(x_1,r), B(x_2,r) \subset \Omega_{h_0}$. Therefore, the gradient of u_h is also uniformly bounded on Ω_{h_0}. Likewise, derivatives of u_h of all orders can be uniformly bounded on Ω_{h_0} ($0 < h \le \frac{h_0}{2}$), either by repeating the same procedure, or by observing that together with u_h, also all derivatives of u_h are harmonic so that (4.5.16) can be iteratively applied to all derivatives in order to convert a bound on some derivative into a bound for a higher one. Therefore, a subsequence of u_h converges towards some smooth function v, together with all its derivatives, as $h \to 0$. Since all the u_h satisfy $\Delta u_h = 0$ so then does v:
$$\Delta v = 0 \quad \text{in } \Omega.$$
Since on the other hand u_h converges to u in $L^1(\Omega)$ by Theorem 3.2.1, the two limits have to coincide (e.g. by Lemma 3.1.3). Therefore $u = v$, and consequently u is smooth and harmonic.

q.e.d.

As an application, we consider the following

Example 4.5.1. Let $a : \mathbb{R} \to \mathbb{R}$ be Lipschitz continuous with
$$0 < \lambda \leq a(y) \leq \Lambda < \infty \quad \text{for all } y \in \mathbb{R}.$$
Let $\Omega \subset \mathbb{R}^d$ be open. We want to minimize
$$F(u) := \int_\Omega \sum_{i=1}^d a(u(x)) D_i u(x) D_i u(x) dx \qquad (4.5.17)$$
in the class $A := g + H_0^{1,p}(\Omega)$, with given $g \in H^{1,p}(\Omega)$. By the Picard–Lindelöf theorem, the ordinary differential equation
$$\frac{du}{dv} = \frac{1}{\sqrt{a(u)}} \qquad (4.5.18)$$
admits a solution $u(v)$ of class $C^{1,1}$. We then have
$$a(u) \frac{du}{dv} \frac{du}{dv} = 1. \qquad (4.5.19)$$
Since $\frac{du}{dv} \geq \Lambda^{-\frac{1}{2}} > 0$, the inverse function $v(u)$ exists and is of class $C^{1,1}$ as well, and we have by (4.5.19) and a chain rule for Sobolev functions that easily follows from the chain rule for differentiable functions by an approximation argument that
$$\sum_{i=1}^d a(u) D_i u D_i u = \sum_{i=1}^d D_i v D_i v.$$
Therefore, (4.5.17) is transformed into Dirichlet's integral
$$F(u) = D(v).$$
Since the latter admits a smooth minimizer, the original problem (4.5.17) then admits a minimizer that is of class $C^{1,1}$ in Ω.

Exercises

4.1 Weaken the growth assumption required for $\left|\frac{\partial f}{\partial u}\right|$ in (iv) of Theorem 4.5.1. Hint: Use the Sobolev Embedding Theorem.

4.2 Compute the Euler–Lagrange equations for the variational integral
$$A(u) := \int_\Omega \sqrt{1 + |Du(x)|^2} dx.$$
($A(u)$ represents the volume of the graph of u over Ω. Critical points are minimal hypersurfaces that can be represented as graphs over Ω.)

4.3 Compute the Euler–Lagrange equations for
$$E(u) := \int_\Omega g^{ij}(x) D_i u(x) D_j u(x) \left(\det g_{ij}(x)\right)^{\frac{1}{2}} dx,$$
where $(g^{ij}(x))_{i,j=1,\ldots,d}$ is the inverse matrix of $(g_{ij}(x))_{i,j=1,\ldots,d}$. Assume that $(g_{ij}(x))_{i,j=1,\ldots,d}$ is positive definite for all $x \in \Omega$. Show that for given $g \in H^{1,2}(\Omega)$, there exists a unique minimizer of E among all $u \in H^{1,2}(\Omega)$ with $u - g \in H^{1,2}(\Omega)$. (Minimizers for E are harmonic functions w.r.t. the metric $g_{ij}(x)$.)

5
Nonconvex functionals. Relaxation

5.1 Nonlower semicontinuous functionals and relaxation

From Section 4.3, we recall the following

Theorem 5.1.1. *Let $\Omega \subset \mathbb{R}^d$ be open, $1 < p < \infty$, $f : \Omega \times \mathbb{R}^d \to \bar{\mathbb{R}}$ measurable and suppose:*

(i) *For almost all $x \in \Omega$, $f(x, \cdot)$ is convex on $\mathbb{R}^d$*
(ii) *There exist $a \in L^1(\Omega)$, $b \in \mathbb{R}$ with*

$$f(x,v) \geq -a(x) + b|v|^p$$

for almost all $x \in \Omega$ and all $v \in \mathbb{R}^d$.

Then

$$F(u) := \int_\Omega f(x, Du(x))dx$$

is lsc and convex on $H^{1,p}(\Omega)$ equipped with its weak topology and assumes its infimum in the class of all $f \in H^{1,p}(\Omega)$ with $f - g \in H_0^{1,p}(\Omega)$ for some given $g \in H^{1,p}(\Omega)$.

Here, (ii) is just a coercivity condition ensuring that a minimizing sequence stays bounded w.r.t. the $H^{1,p}$ -norm (w.l.o.g. $F \not\equiv \infty$) (i) implies that F is lsc, w.r.t. the norm topology of $H^{1,p}$, and the convexity then implies that F is also lsc w.r.t. the weak $H^{1,p}$ topology. Since bounded sequences in $H^{1,p}$ have weakly convergent subsequences, any minimizing sequence has a convergent subsequence, and a limit of such a subsequence then minimizes F by lower semicontinuity.

Not all functionals that one wishes to consider in the calculus of variations are convex, however. As a motivation for what follows, we consider the following example of Bolza:

$$\Omega = (0,1) \subset \mathbb{R}, \quad u: (0,1) \to \mathbb{R}, \quad u(0) = 0 = u(1)$$

$$F(u) = \int_0^1 \left(u^2(x) + \left(u'(x)^2 - 1\right)^2 \right) dx.$$

We claim that

$$\inf\{F(u) : u \in H_0^{1,4}((0,1))\} = 0. \tag{5.1.1}$$

For the proof, we consider 'sawtooth'-functions: Let $n \in \mathbb{N}$,

$$u_n(x) := \begin{cases} x - \dfrac{i}{n} & \text{for } \dfrac{2i}{2n} \leq x \leq \dfrac{2i+1}{2n} \\ -x + \dfrac{i+1}{n} & \text{for } \dfrac{2i+1}{2n} \leq x \leq \dfrac{2i+2}{2n} \end{cases}$$

$(i = 0, 1, ..., n-1)$.
u_n is contained in $H^{1,\infty}((0,1)) \subset H^{1,4}((0,1))$ and satisfies:

$$\text{For all } x \in (0,1) \quad 0 \leq u_n(x) \leq \frac{1}{2n}, \tag{5.1.2}$$

$$u_n(0) = 0 = u_n(1), \tag{5.1.3}$$

$$\text{for almost all } x \in (0,1) \quad |u_n'(x)| = 1. \tag{5.1.4}$$

Consequently

$$\lim_{n \to \infty} F(u_n) = 0.$$

Since $F(u)$ is nonnegative for every u, (5.1.1) follows. The infimum of F therefore cannot be realized by any $H_0^{1,4}$ function, because if we had

$$F(u) = 0,$$

then $u(x) = 0$ for almost all $x \in (0,1)$ and $|u'(x)| = 1$ for almost all $x \in (0,1)$, and these two conditions are not compatible. (In fact, since $d = 1$ here, any $u \in H_0^{1,4}((0,1))$ is absolutely continuous, and so $u \equiv 0$ if $u(x) = 0$ a.e., hence u is differentiable and $u' \equiv 0$. (More generally, any Sobolev function that is constant on some set A has a representative u whose derivative Du vanishes on A.)

We have thus shown that the problem

$$F(u) \to \min \text{ in } H_0^{1,4}(\Omega)$$

does not have a solution.

5.1 Nonlower semicontinuous functionals and relaxation

We observe that our minimizing sequence (u_n) converges to zero weakly in $H_0^{1,4}$, by (5.1.2) and

$$\int_0^1 u_n'(x)\varphi(x)dx = -\int_0^1 u_n(x)\varphi'(x)dx \to 0 \quad \text{for all } \varphi \in C_0^\infty((0,1)).$$

However,

$$F(0) = 1 > 0 = \lim_{n\to\infty} F(u_n).$$

Therefore, F is not lsc w.r.t. weak $H^{1,4}$-convergence although the integrand is continuous in u'. As we shall see this results from the lack of convexity of the integrand. We also observe that any sequence of sawtooth functions u_n, i.e. satisfying

$$|u_n'| = 1 \text{ a.e.}$$

that converges to 0 in L^2 is a minimizing sequence for F.

Remark 5.1.1. Functionals of the type of our example often arise in optimal control theory as described in Section 5.2 of Part I. For example, one considers problems of the following type

$$\int_0^T f(t, u(t), \sigma(t))dt \to \min \tag{5.1.5}$$

under the side conditions

$$u(0) = u_0, \quad u(T) = u_T \tag{5.1.6}$$

$$u'(t) = g(t, u(t), \sigma(t)) \tag{5.1.7}$$

with given functions f and g. u is called a state variable, σ a control variable. This means that one assumes that u describes the state of some system evolving in time t whose derivative or rate of change can be controlled through a parameter σ. The aim then is to choose σ in such a manner that the functional, often considered as 'cost function', is minimized.

Thus, one needs to find some equation

$$\bar{\sigma}(t) = \varphi(t, u(t))$$

for an optimal control $\bar{\sigma}$ at time t assuming a given state $u(t)$ of the system. If one knows the optimal control, one can reconstruct the evolution $u(t)$ of the state of the system from (5.1.6) and (5.1.7) under appropriate assumptions. The simplest control equation (5.1.7) is

$$u'(t) = \sigma(t),$$

Nonconvex functionals. Relaxation

and this leads to minimizing functionals of the type

$$\int_0^T f(t, u(t), u'(t))dt.$$

Expressions of the type $(u'(t)^2 - 1)^2$ can occur in many technical examples, like boats sailing against the wind.

Faced with a problem that one cannot solve, one may contemplate several options:

One could try to modify the problem, or one might generalize the concept of a solution, or both.

We shall discuss several such strategies. We first modify the problem via relaxation. This is an important method in the calculus of variations, and we therefore discuss it in some generality.

Definition 5.1.1. Let X be a topological space, $F : X \to \bar{\mathbb{R}}$. We define the lower semicontinuous envelope or relaxed function $sc^- F$ of F as follows:

$$(sc^- F)(x) := \sup \ \{\Phi(x) : \Phi : X \to \bar{\mathbb{R}} \text{ is lower semicontinuous} \\ \text{with } \Phi(y) \leq F(y) \quad \text{for all } y \in X\}$$

Lemma 5.1.1. $sc^- F$ is the largest lsc function on X that is $\leq F$ everywhere.

In particular, F is lower semicontinuous if and only if $F = sc^- F$.

Proof. $sc^- F$ is lsc as a supremum of lsc functions, see Lemma 4.2.1 (iv). Obviously, $sc^- F \leq F$, and for all lsc Φ with $\Phi \leq F$, we have

$$\Phi \leq sc^- F$$

by definition of $sc^- F$.

q.e.d.

Theorem 5.1.2. Let X be a topological space, $F : X \to \bar{\mathbb{R}}$ a function. Then every accumulation point of a minimizing sequence for F is a minimum point for $sc^- F$. Consequently, if F is coercive, then $sc^- F$ assumes its minimum, and

$$\min_X sc^- F = \inf_X F.$$

5.1 Nonlower semicontinuous functionals and relaxation

Proof. Let $(x_n)_{n \in \mathbb{N}} \subset X$ be a minimizing sequence for F with accumulation point x_0. Then

$$(sc^- F)(x_0) \leq \liminf_{n \to \infty} (sc^- F)(x_n) \quad \text{by lower semicontinuity of } sc^- F \text{ (see Lemma 5.1.1)}$$

$$\leq \liminf_{n \to \infty} F(x_n) \quad \text{since } sc^- F \leq F$$

$$= \inf_{y \in X} F(y) \quad \text{since } (x_n) \text{ is a minimizing sequence for } F. \tag{5.1.8}$$

On the other hand, the constant function

$$\Phi(x) \equiv \inf_{y \in X} F(y)$$

is lsc and $\leq F$, hence by Lemma 5.1.1 for every $x \in X$

$$\inf_{y \in X} F(y) \leq (sc^- F)(x). \tag{5.1.9}$$

From (5.1.8) and (5.1.9) we conclude

$$(sc^- F)(x_0) = \inf_{y \in X} F(y) = \min_{x \in X} (sc^- F)(x). \tag{5.1.10}$$

This implies the first claim. If F is coercive, then every minimizing sequence has an accumulation point, and the second claim also follows.

q.e.d.

What does Theorem 5.1.2 tell us for our example? It simply says that if we cannot minimize our original functional F due to its lack of lower semicontinuity, we then minimize another functional instead, one that is lower semicontinuous and as close as possible to F. Theorem 5.1.2 then says that limits (or more generally, accumulation points) of minimizing sequences for F do not minimize F, but the relaxed functional $sc^- F$. Since $sc^- F$ is the largest lsc functional $\leq F$ by Lemma 5.1.1 that is the best one can hope for.

It then remains the task to determine the relaxed functional of some given F. Before proceeding to do so for our example, let us relax ourselves a little and derive some easy consequences of the definition of the relaxed functional and consider some easier examples first.

Lemma 5.1.2. *Let X satisfy the first axiom of countability. Then $sc^- F$ is the relaxed function for $F : X \to \bar{\mathbb{R}}$ iff the following two conditions are satisfied:*

(i) whenever $x_n \to x$
$$(sc^- F)(x) \leq \liminf_{n \to \infty} F(x_n)$$

(ii) *for every $x \in X$, there exists a sequence $x_n \to x$ with*
$$(sc^- F)(x) \geq \lim_{n \to \infty} F(x_n)$$

Proof. We claim that, since X satisfies the first axiom of countability,
$$(sc^- F)(x) = \inf\{\liminf F(x_n) : x_n \to x \text{ in } X\}. \tag{5.1.11}$$

We denote the right hand side of (5.1.11) by $F^-(x)$. Then F^- is lsc. In order to verify this, we have to check
$$\liminf_{\nu \to \infty}(\inf\{\liminf F(y_{\nu,n}) : y_{\nu,n} \to y_\nu\}) \geq \inf\{\liminf F(x_n) : x_n \to x\} \tag{5.1.12}$$

whenever $y_\nu \to x$. Indeed, otherwise, for some $\delta > 0$, we would find some diagonal sequence $y_{\nu,n_\nu} \to x$ as $\nu \to \infty$ with
$$F(y_{\nu,n_\nu}) < \inf\{\liminf F(x_n) : x_n \to x\} - \delta$$

which is impossible. Thus, F^- is sequentially lsc, hence lsc, because X is assumed to satisfy the first axiom of countability. Also, $F^- \leq F$, and for every lsc $\Phi \leq F$, we have for $x_n \to x$
$$\Phi(x) \leq \liminf_{n \to \infty} \Phi(x_n) \leq \liminf_{n \to \infty} F(x_n),$$

and hence
$$\Phi(x) \leq F^-(x).$$

Thus, F^- is the largest lsc functional $\leq F$, and (5.1.11) follows from Lemma 5.1.1. It is then easy to see (and left as an exercise) that $F^-(x)$ satisfies and is characterized by the properties (i) and (ii).

q.e.d.

Example 5.1.1. Let X be a topological space, $A \subset X$ a subset. The indicator function ι_A is defined by
$$\iota_A(x) := \begin{cases} 0 & \text{if } x \in A \\ \infty & \text{if } x \notin A. \end{cases}$$

We then have
$$sc^- \iota_A = \iota_{\bar{A}},$$

where $\bar{A}$ is the closure of A in X.

5.1 Nonlower semicontinuous functionals and relaxation

The characteristic function χ_A is defined by

$$\chi_A(x) := \begin{cases} 1 & \text{if } x \in A \\ 0 & \text{if } x \notin A. \end{cases}$$

Then

$$sc^- \chi_A = \chi_{\mathring{A}}$$

where $\mathring{A}$ is the complement of $\overline{X \setminus A}$.

Example 5.1.2. Let $\Omega \subset \mathbb{R}^d$ be open, $1 < p < \infty$,

$$I : L^p(\Omega) \to \bar{\mathbb{R}}$$

defined by

$$I(u) := \begin{cases} \int_\Omega |Du|^p \, dx + \int_\Omega |u|^p \, dx & \text{if } u \in C^1(\Omega) \\ \infty & \text{otherwise.} \end{cases}$$

(Note that $I(u)$ may also be infinite for some $u \in C^1(\Omega)$.) We claim

$$(sc^- I)(u) = \begin{cases} \int_\Omega |Du|^p \, dx + \int_\Omega |u|^p \, dx & \text{if } u \in H^{1,p}(\Omega) \\ \infty & \text{otherwise.} \end{cases}$$

In order to show this, we shall verify the conditions of Lemma 5.1.2:

(i) $(sc^- I)$ is lower semicontinuous on L^p which yields condition (i). The lower semicontinuity is seen as follows:
Suppose $u_n \to u$ in $L^p(\Omega)$. For the purpose of lower semicontinuity, we may select a subsequence $(w_\nu)_{\nu \in \mathbb{N}} \subset (u_n)_{n \in \mathbb{N}}$ with

$$\lim_{\nu \to \infty} (sc^- I)(w_\nu) = \liminf_{n \to \infty} (sc^- I)(u_n),$$

and we may also assume that this limit is finite. $(w_\nu)_{\nu \in \mathbb{N}}$ then is bounded in $H^{1,p}(\Omega)$. A subsequence of (w_ν) then converges weakly in $H^{1,p}(\Omega)$ (Theorem 3.3.5), and by the Rellich–Kondrachev compactness Theorem 3.4.1, it also converges strongly in $L^p(\Omega)$. The limit has to be u, because the original sequence (u_n) was assumed to converge to this limit. Since the $H^{1,p}$-norm is lsc w.r.t. weak $H^{1,p}$ convergence (Lemma 2.2.7), we have

$$(sc^- I)(u) \leq \lim_{\nu \to \infty} (sc^- I)(w_\nu)$$
$$= \liminf_{n \to \infty} (sc^- I)(u_n).$$

(ii) Let $u \in H^{1,p}(\Omega)$. Since $C^1(\Omega) \cap H^{1,p}(\Omega)$ is dense in $H^{1,p}(\Omega)$, we may find a sequence $(u_n)_{n \in \mathbb{N}} \subset C^1(\Omega) \cap H^{1,p}(\Omega)$ with

$$\lim_{n \to \infty} \left(\int_\Omega |Du_n|^p + \int_\Omega |u_n|^p \right) = \int_\Omega |Du|^p + |u|^p,$$

i.e.

$$\lim I(u_n) = (sc^- I)(u).$$

If $u \notin H^{1,p}(\Omega)$, then

$$I(u) = (sc^- I)(u) = \infty.$$

This verifies condition (ii).

Example 5.1.3. Similarly, for

$$I_0(u) := \begin{cases} \int |Du|^p & \text{if } u \in C_0^1(\Omega) \\ \infty & \text{if } u \in L^p(\Omega) \setminus C_0^1(\Omega), \end{cases}$$

the relaxed functional is

$$\left(sc^- I_0 \right)(u) = \begin{cases} \int |Du|^p & \text{if } u \in H_0^{1,p}(\Omega) \\ \infty & \text{otherwise.} \end{cases}$$

Remark 5.1.2. We may also define the above functionals I, I_0 on $L_{loc}^p(\Omega)$ instead of $L^p(\Omega)$. The relaxed functionals will be given by the same formulae.

Remark 5.1.3. For $p = 1$, the relaxations of I and I_0 are not given anymore by the $H^{1,1}$-norm, but by the BV-norm which is defined in Chapter 7.

In metric spaces, there is an alternative useful characterization of the relaxation of a given functional which we now want to describe.

Definition 5.1.2. Let X be a metric space with distance function $d(\cdot, \cdot)$, $F : X \to \mathbb{R} \cup \{\infty\}$ be bounded from below, $F \neq \infty$. For $\lambda > 0$, we define the Moreau-Yosida transform of F as

$$F_\lambda(x) := \inf_{y \in X} \left(F(y) + \lambda d(x, y) \right). \tag{5.1.13}$$

Theorem 5.1.3. The functionals F_λ satisfy

$$|F_\lambda(x_1) - F_\lambda(x_2)| \leq \lambda d(x_1, x_2) \tag{5.1.14}$$

for every $\lambda > 0$, $x_1, x_2 \in X$. In particular, they are Lipschitz continuous. For any $x \in X$

$$(sc^- F)(x) = \lim_{\lambda \to \infty} F_\lambda(x). \tag{5.1.15}$$

5.2 Representation of relaxed functionals via convex envelopes

Proof. For $x_1, x_2, y \in X$, $\lambda > 0$, we obtain from the triangle inequality

$$F(y) + \lambda d(x_1, y) \leq F(y) + \lambda d(x_2, y) + \lambda d(x_1, x_2).$$

The definition of $F_\lambda(x_2)$ implies then

$$\inf_{y \in X} (F(y) + \lambda d(x_1, y)) \leq F_\lambda(x_2) + \lambda d(x_1, x_2),$$

hence

$$F_\lambda(x_1) \leq F_\lambda(x_2) + \lambda d(x_1, x_2).$$

Interchanging the rôles of x_1 and x_2, we conclude

$$|F_\lambda(x_1) - F_\lambda(x_2)| \leq \lambda d(x_1, x_2).$$

Since we have now shown that F_λ is Lipschitz continuous, and since $F_\lambda \leq F$, we obtain

$$F_\lambda \leq sc^- F,$$

hence for all $x \in X$

$$\sup_{\lambda > 0} F_\lambda(x) \leq (sc^- F)(x). \tag{5.1.16}$$

For any $\lambda > 0$, we find $x_\lambda \in X$ with

$$F(x_\lambda) + \lambda d(x, x_\lambda) \leq F_\lambda(x) + \frac{1}{\lambda}.$$

Therefore

$$\lim_{\lambda \to \infty} x_\lambda = x$$

and

$$(sc^- F)(x) \leq \liminf_{\lambda \to \infty} F(x_\lambda) \leq \liminf_{\lambda \to \infty} F_\lambda(x). \tag{5.1.17}$$

Equations (5.1.16) and (5.1.17) imply (5.1.15).

q.e.d.

5.2 Representation of relaxed functionals via convex envelopes

Theorem 5.2.1. *Let $\Omega \subset \mathbb{R}^d$ be open, $1 < p < \infty$, $f : \mathbb{R}^d \to \mathbb{R}$ continuous with*

$$c_0 |v|^p \leq f(v) \leq c_1 |v|^p + c_2 \quad \text{for some constants } c_0, c_1, c_2.$$

Let $F : \{u \in H^{1,p}(\Omega) : u - u_0 \in H_0^{1,p}(\Omega)\} \to \mathbb{R}$ be given by

$$F(u) := \int_\Omega f(Du(x))\, dx, \quad (u_0 \in H^{1,p}(\Omega) \text{ given}).$$

Then the relaxed function of F w.r.t. the weak $H^{1,p}$ topology is given by

$$(sc^- F)(u) = \int_\Omega (cvx^- f)(Du(x))\, dx$$

where

$$(cvx^- f)(v) := \sup\{g(v) : g \leq f, g \text{ convex}\}$$

is the largest convex function $\leq f$.

For the proof, we shall need the following:

Lemma 5.2.1. Let $W = \prod_{i=1}^d (\alpha_i, \beta_i) \subset \mathbb{R}^d$ be an open rectangle, $1 < p < \infty$. We let $f \in L^p(W)$ and extend f periodically to $\mathbb{R}^d$, i.e.

$$f(x^1 + m_1(\beta_1 - \alpha_1), \ldots, x^d + m_d(\beta_d - \alpha_d)) = f(x^1, \ldots, x^d)$$

for $m_1, \ldots, m_d \in \mathbb{Z}$, $x = (x^1, \ldots, x^d) \in W$, and put

$$f_n(x) := f(nx) \quad \text{for } n \in \mathbb{N}.$$

Then we get the weak convergence

$$f_n \rightharpoonup \bar{f} \equiv \frac{1}{\operatorname{meas} W} \int_W f(x)\, dx \quad \text{in } L^p(W) \quad \text{for } n \to \infty. \tag{5.2.1}$$

Proof. First

$$\int_W |f_n(x)|^p\, dx = \int_W |f(nx)|^p\, dx = \frac{1}{n^d}\int_{nW} |f(y)|^p\, dy = \int_W |f(x)|^p\, dx$$

by the periodicity of f. Thus

$$\|f_n\|_{L^p(W)} = \|f\|_{L^p(W)}. \tag{5.2.2}$$

In the same manner,

$$\int_W f_n(x)\, dx = \int_W f(x)\, dx = \int_W \bar{f}\, dx. \tag{5.2.3}$$

Let now W_0 be a subrectangle of W, written in the form

$$W_0 = \prod_{i=1}^d (a_i + b_i \alpha_i, a_i + b_i \beta_i),$$

or more compactly

$$W_0 = a + bW \quad (a = (a_1, \ldots, a_d), b = (b_1, \ldots, b_d)).$$

5.2 Representation of relaxed functionals via convex envelopes

Then
$$\int_{W_0} (f_n(x) - \bar{f}) \, dx = \int_{a+bW} (f(nx) - \bar{f}) \, dx$$
$$= \frac{1}{n^d} \int_{na+nbW} (f(y) - \bar{f}) \, dy$$
$$= \frac{1}{n^d} \int_{na+[nb]W} (f(y) - \bar{f}) \, dy$$
$$+ \frac{1}{n^d} \int_{na+(nb-[nb])W} (f(y) - \bar{f}) \, dy$$
$$= \left(\frac{[nb]}{n}\right)^d \int_W (f(y) - \bar{f}) \, dy$$
$$+ \frac{1}{n^d} \int_{na+(nb-[nb])W} (f(y) - \bar{f}) \, dy$$

by periodicity of f.

The first term in the right-hand side vanishes by (5.2.3), and thus, again using the periodicity of f,
$$\left| \int_{W_0} (f_n(x) - \bar{f}) \, dx \right| \leq \frac{1}{n^d} \int_W |f(y) - \bar{f}| \, dy.$$

Letting $n \to \infty$, we obtain for every subrectangle W_0 of W
$$\lim_{n \to \infty} \int_{W_0} (f_n(x) - \bar{f}) \, dx = 0. \tag{5.2.4}$$

Let now $g \in L^q(W)$, with $\frac{1}{p} + \frac{1}{q} = 1$. We have to show
$$\lim_{n \to \infty} \int_W f_n(x) g(x) \, dx = \int_W \bar{f} g(x) \, dx. \tag{5.2.5}$$

Given $\epsilon > 0$, we then find subrectangles $W_1, \ldots, W_k$ ($k = k(\epsilon)$) and $\lambda_i \in \mathbb{R}$ ($i = 1, \ldots, k$) with
$$\left\| g - \sum_i \lambda_i \chi_{W_i} \right\|_{L^q(W)} < \epsilon \tag{5.2.6}$$

(The possibility of approximating $L^q(\Omega)$ functions g (Ω open in $\mathbb{R}^d$) in such a manner by step functions can easily be seen as follows: Since $C_0^\infty(\Omega)$ is dense in $L^q(\Omega)$, there exist $\varphi_\epsilon \in C_0^\infty(\Omega)$ with $\|g - \varphi_\epsilon\|_{L^q(\Omega)} < \frac{\epsilon}{2}$. It is then easy to construct a step function $\sum \lambda_i \chi_{W_i}$ ($\lambda_i \in \mathbb{R}$, W_i disjoint rectangles contained in $\operatorname{supp} \varphi_\epsilon$) with
$$\sup_{\operatorname{supp} \varphi_\epsilon} \left| \varphi_\epsilon(x) - \sum \lambda_i \chi_{W_i}(x) \right| \leq \frac{\epsilon}{2 \operatorname{meas} \operatorname{supp} \varphi_\epsilon}.$$

Then indeed
$$\left\|g - \sum \lambda_i \chi_{W_i}\right\|_{L^p(\Omega)} < \epsilon.$$

Then

$$\left|\int_W (f_n(x) - \bar{f}) g(x) dx\right|$$

$$\leq \left|\int_W (f_n(x) - \bar{f}) \sum_i \lambda_i \chi_{W_i}(x)\right|$$

$$+ \left|\int_W (f_n(x) - \bar{f}) \left(g(x) - \sum_i \lambda_i \chi_{W_i}(x)\right)\right|$$

$$\leq \sum_{i=1}^k |\lambda_i| \left|\int_{W_i} (f_n(x) - \bar{f}) dx\right| + \epsilon \|f_n - \bar{f}\|_{L^p(W)}$$

by (5.2.6) and Hölder's inequality (Lemma 3.1.1).

The first term tends to zero as $n \to \infty$ by (5.2.4), whereas the second one is bounded by $2\epsilon \|f\|_{L^p(W)}$ by (5.2.2) and can hence be made arbitrarily small. Therefore, (5.2.5) holds.

q.e.d.

The *proof of Theorem 5.2.1* will be broken up into several steps:

(1) We put

$$(q^- f)(v) := \inf \left\{ \tfrac{1}{\operatorname{meas} U} \int_U f(v + D\varphi(x)) dx : \varphi \in H_0^{1,p}(U), \ U \text{ bounded domain in } \mathbb{R}^d \right\}, \tag{5.2.7}$$

and we claim:

Lemma 5.2.2.

$$(sc^- F)(u) \leq \int_\Omega (q^- f)(Du(x)) dx.$$

Proof. Replacing $F(u)$ by $G(v) := F(v + u_0)$ for $v = u - u_0$, we may assume $u_0 = 0$, i.e. $u \in H_0^{1,p}(\Omega)$. Since the piecewise affine functions, i.e. those u for which Du is constant on disjoint rectangles $W_i \subset \Omega$, with $\Omega \setminus \bigcup W_i$ arbitrarily small, are dense in $H^{1,p}$ (for the same reason that the functions that are piecewise constant on disjoint rectangles W_i are dense in L^p), and since F

5.2 Representation of relaxed functionals via convex envelopes

is continuous under strong $H^{1,p}$-convergence, it suffices to treat the case where

$$Du \equiv v_0 = \text{constant}$$

on some rectangle W. We next observe that for a given constant vector v, $(q^- f)(v)$ is independent of the choice of U in (5.2.7). First, the value of the inf on the right hand side of (5.2.7) does not change under translations or homotheties of U. The general case of U_1 and U_2 then is handled by approximating U_1 by disjoint homothetical translations of U_2 and vice versa. We may therefore take $U = W$ in (5.2.7). We now choose a sequence $(\varphi_n)_{n \in \mathbb{N}} \subset H_0^{1,p}(W)$ with

$$(q^- f)(v_0) + \frac{1}{n} \geq \frac{1}{\text{meas } W} \int_W f(v_0 + D\varphi_n(x))dx \geq (q^- f)(v_0). \tag{5.2.8}$$

We extend φ_n periodically from W to $\mathbb{R}^d$ and put

$$\bar{u}(x) := v_0 x \quad (\text{then } D\bar{u} = v_0)$$

and

$$u_n(x) := \bar{u}(x) + \frac{1}{n}\varphi_n(nx).$$

By Lemma 5.2.1, u_n converges to $\bar{u}$ weakly in $H^{1,p}$. Then $u_n = \bar{u}$ on ∂W by periodicity of φ_n and $\varphi_n|_{\partial W} = 0$. We have

$$\int_W f(Du_n(x))dx = \int_W f(v_0 + \frac{1}{n}D\varphi_n(nx))dx$$
$$= \frac{1}{n^d} \int_{nW} f(v_0 + D\varphi_n(y))dy$$
$$= \int_W f(v_0 + D\varphi_n(y))dy \tag{5.2.9}$$

since φ_n is periodic.

Equations (5.2.8) and (5.2.9) imply

$$\lim_{n \to \infty} F(u_n) = \lim_{n \to \infty} \int_W f(Du_n(x))dx$$
$$= \int_W (q^- f)(v_0) = (q^- f)(v_0) \text{ meas } W.$$

The claim then follows from the characterization of $(sc^- F)$, see e.g. Lemma 5.1.2(i).

q.e.d.

(2) We observe
$$(q^-f)(v) \le f(v) \quad \text{(put } \varphi = 0 \text{ in } (5.2.7)). \tag{5.2.10}$$

With
$$(q^-F)(u) := \int_\Omega (q^-f)(Du(x))dx,$$

we obtain from Lemma 5.2.2 and (5.2.10)
$$sc^-F = sc^-(q^-F), \tag{5.2.11}$$

and upon iteration
$$sc^-F = sc^-((q^-)^nF), \tag{5.2.12}$$

where $(q^-)^n$ means performing the construction q^- iteratively n times. From the growth conditions on f assumed in Theorem 5.2.1, we conclude that
$$(q^{-n}f)(v)$$
is monotonically decreasing and bounded from below in n, hence converges to some limit
$$(Qf)(v).$$

From B. Levi's Theorem 1.2.1, we conclude
$$\lim_{n \to \infty} (q^{-n}F)(u) = \lim_{n \to \infty} \int (q^{-n}f)(Du(x))dx$$
$$= \int (Qf)(Du(x))dx =: (QF)(u). \tag{5.2.13}$$

Since by (5.2.12)
$$(sc^-F)(u) \le (q^{-n}F)(u) \quad \text{for all } n,$$
we conclude from (5.2.13)
$$(sc^-F)(u) \le (QF)(u).$$

From the definition of Qf, we also conclude
$$Qf(v) = \inf \left\{ \frac{1}{\text{meas } U} \int_U (Qf)(v + D\varphi(x))dx, \right.$$
$$\left. \varphi \in H_0^{1,p}(U), U \subset \mathbb{R}^d \text{ open, bounded} \right\}. \tag{5.2.14}$$

As before, this expression is independent of the choice of U.

5.2 Representation of relaxed functionals via convex envelopes 219

Definition 5.2.1. $g : \mathbb{R}^d \to \mathbb{R}$ is called quasiconvex if for all $v \in \mathbb{R}^d$, $\varphi \subset H_0^{1,p}(U)$, $U \subset \mathbb{R}^d$ bounded and open

$$g(v) \leq \frac{1}{\operatorname{meas} U} \int_U g(v + D\varphi(x)) dx. \tag{5.2.15}$$

Equation (5.2.14) then implies that Qf is quasiconvex.

(3) **Lemma 5.2.3.** $f : \mathbb{R}^d \to \mathbb{R}$ is convex if and only if it is quasiconvex.

Proof. '$\Longrightarrow$':
Jensen's inequality says that if f is convex, for every $\psi \in L^1(\mathbb{R}^d, \mathbb{R}^d)$

$$f\left(\fint \psi(x) dx\right) \leq \fint f(\psi(x)) \, dx \tag{5.2.16}$$

(see Theorem 1.1.6). Since, as observed above, in Definition 5.2.1 it suffices to consider one fixed domain U, we may assume

$$\operatorname{meas} U = 1$$

and put

$$\psi(x) = v + D\varphi(x).$$

Since $\varphi \in H_0^{1,p}$, $\int \psi(x) = v \operatorname{meas} U = v$, and (5.2.16) therefore implies that f is quasiconvex.
'$\Longleftarrow$'
We assume that f is quasiconvex, i.e.

$$\begin{aligned} f(v_0) &= \frac{1}{\operatorname{meas} U} \int_U f(v_0) dx \\ &\leq \frac{1}{\operatorname{meas} U} \int_U f(v_0 + D\varphi(y)) dy \end{aligned} \tag{5.2.17}$$

for all $\varphi \in H_0^{1,p}(U)$.

We have to show that for all $v_1, v_2 \in \mathbb{R}^d$, $0 \leq t \leq 1$

$$f(tv_1 + (1-t)v_2) \leq tf(v_1) + (1-t)f(v_2). \tag{5.2.18}$$

Equation (5.2.17) implies

$$f(tv_1 + (1-t)v_2) \leq \frac{1}{\operatorname{meas} U} \int_U f(tv_1 + (1-t)v_2 + D\varphi(y)) \, dy \tag{5.2.19}$$

for all U and all $\varphi \in H_0^{1,p}(U)$. After a rotation, we may assume that $v_1 - v_2$ is a positive multiple of the first basis vector of our standard basis of $\mathbb{R}^d$, i.e. $v_1 - v_2$ points in the x^1-direction. We shall take a cube $W := (a,b)^d \subset \mathbb{R}^d$ as our set U and construct a family of functions

$$(\varphi_n)_{n \in \mathbb{N}} \subset H_0^{1,p}(W)$$

with

$$\nabla \varphi_n(x) = \begin{cases} (1-t)(v_1 - v_2) & \text{on a set } W_1^n \subset W \text{ with meas} \\ & W_1^n = t(b-a)(b-a-\frac{2}{n})^{d-1} \\ -t(v_1 - v_2) & \text{on a set } W_2^n \subset W \text{ with meas} \\ & W_2^n = (1-t)(b-a)(b-a-\frac{2}{n})^{d-1} \end{cases}$$

and

$$\|\nabla \varphi_n\|_{L^\infty(W)} \leq c_0 \text{ for some fixed constant } c_0 \text{ that does not depend on } n.$$

Using these φ_n in (5.2.19) yields

$$f(tv_1 + (1-t)v_2) \leq tf(v_1) + (1-t)f(v_2) + \rho_n$$

with $\rho_n \to 0$ as $n \to \infty$, hence (5.2.18).

It remains to construct φ_n. We divide the interval (a,b) into 2^{n+1} subintervals as follows:

$$I_1 = (a, a + \frac{t}{2^n}(b-a))$$

$$I_2 = (a + \frac{t}{2^n}(b-a), a + \frac{1}{2^n}(b-a))$$

$$I_3 = (a + \frac{1}{2^n}(b-a), a + \frac{1}{2^n}(b-a) + \frac{t}{2^n}(b-a))$$

.
.
.

i.e. the intervals $I_{2\nu-1}$ have length $\frac{t}{2^n}(b-a)$, and they alternate with the intervals $I_{2\nu}$ of length $\frac{(1-t)}{2^n}(b-a)$. We then put

$$W_1^n := (\bigcup_{\nu=1}^{n} I_{2\nu-1}) \times (a + \frac{1}{n}, b - \frac{1}{n})^{d-1}$$

$$W_2^n := (\bigcup_{\nu=1}^{n} I_{2\nu}) \times (a + \frac{1}{n}, b - \frac{1}{n})^{d-1}.$$

5.2 Representation of relaxed functionals via convex envelopes

We then put $\varphi_n(a, x^2, ..., x^d) = 0$,

$$\frac{\partial \varphi_n}{\partial x^1}(x) = \begin{cases} (1-t)|v_1 - v_2| & \text{for } x \in W_1^n \\ -t|v_1 - v_2| & \text{for } x \in W_2^n, \end{cases}$$

$$\frac{\partial \varphi_n(x)}{\partial x^i} = 0 \text{ for } i = 2, \ldots, d.$$

(Remember that we assume that $v_1 - v_2$ points in the positive x^1-direction.)

We then have $\varphi_n(b, x^2, ..., x^d) = 0$. We also put

$$\varphi_n = 0 \text{ on } \partial W,$$

and on $W \setminus (W_1^n \cup W_2^n)$ we choose an interpolation that is affine linear in $x^2, ..., x^d$. Since

$$\sup_{x \in W_1^n \cup W_2^n} |\varphi_n(x)| \leq \frac{t(1-t)}{2^n}(b-a)|v_1 - v_2| =: \frac{c_1}{2^n}$$

we get

$$\sup_{x \in W \setminus (W_1^n \cup W_2^n)} \left| \frac{\partial \varphi_n(x)}{\partial x^i} \right| \leq n \frac{c_1}{2^n}.$$
$$i=2,\ldots,d$$

Thus, for large enough n,

$$\sup_{x \in W} |\nabla \varphi_n(x)| = \sup_{x \in W_1^n \cup W_2^n} \left| \frac{\partial \varphi_n(x)}{\partial x^1} \right| \leq |v_1 - v_2| =: c_0.$$

This completes the construction of φ_n and the proof of Lemma 5.2.3.

q.e.d.

(4) We may now complete the proof of Theorem 5.2.1 From (2), we know

$$(sc^- F)(u) \leq QF(u) = \int Qf(Du(x))dx.$$

By Lemma 5.2.3, Qf is convex. By Lemma 4.3.1, $Qf(u)$ therefore is lsc w.r.t. weak $H^{1,p}$ convergence. Since $QF \leq F$ (see (5.2.10) and the definition of QF), we must also have from the definition of $sc^- F$ that

$$QF(u) \leq (sc^- F)(u).$$

Hence equality. Thus

$$(sc^- F)(u) = \int (Qf)(Du(x))dx.$$

Moreover, for every convex function $g \leq f$,

$$G(u) := \int g(Du(x))dx$$

is a weakly $H^{1,p}$ lsc functional $\leq F$. Therefore, from the definition of $sc^- F$, the convex function Qf must in fact be the largest convex function $\leq f$. This completes the proof.

q.e.d.

Corollary 5.2.1. *F as in Theorem 5.2.1 is weakly lower semicontinuous in $H^{1,p}$ if and only if f is convex.*

Proof. Lemma 4.3.1 says that convex functionals are weakly lower semicontinuous. If f is not convex, then by Theorem 5.2.1 $sc^- F \neq F$, hence F is not weakly lsc by Lemma 5.1.1.

q.e.d.

Remark 5.2.1. One may also consider variational problems for vector valued functions $u : \Omega \subset \mathbb{R}^d \to \mathbb{R}^n$,

$$F(u) := \int_\Omega f(Du(x))dx.$$

Again, f is called quasiconvex if for all open and bounded $U \subset \mathbb{R}^d$ and all $\varphi \in H_0^{1,p}(U; \mathbb{R}^n)$, $v \in \mathbb{R}^{nd}$

$$f(v) \leq \frac{1}{\text{meas } U} \int f(v + D\varphi(x))dx.$$

In this case, however, while convex functions are still quasiconvex, the converse is no longer true. Theorem 5.2.1 continues to hold but with convexity replaced by quasiconvexity. Also, one may consider more general problems of the form

$$F(u) = \int f(x, u(x), Du(x))dx$$

with similar results and conceptually similar, but technically more involved proofs.

Remark 5.2.2. The notation of quasiconvexity and many of the basic corresponding lower semicontinuity results are due to C. Morrey. In fact, the quasiconvex functionals are precisely the weakly lower semicon-

5.2 Representation of relaxed functionals via convex envelopes

tinuous ones. For detailed references to the work of Morrey and other researchers, see the book of Dacorogna quoted at the end of this chapter.

Remark 5.2.3. Theorem 5.2.1 can be considered as a representation theorem for relaxed functionals. In particular, it says that a functional on $H^{1,p}$ obtained by integrating an integrand $f(Du(x))$ (with certain technical assumptions on f) has a relaxed functional of the same type, i.e. again representable by integration w.r.t. to some integrand $g(Du(x))$ of the same type. Furthermore, g may be computed explicitly from f.

We now return to our initial example

$$F(u) = \int_0^1 \left\{ u^2(x) + \left(u'(x)^2 - 1\right)^2 \right\} dx$$

for $u \in H_0^{1,4}((0,1))$. $F(u)$ is the sum of a functional which is continuous w.r.t. strong L^2-convergence, hence also w.r.t. weak $H^{1,4}$ convergence, and another one to which Theorem 5.2.1 applies. We conclude that

$$(sc^- F)(u) = \int_0^1 \left\{ u^2(x) + Q(u'(x)) \right\} dx,$$

with

$$Q(v) = \begin{cases} (v^2 - 1)^2 & \text{if } |v| \geq 1 \\ 0 & \text{otherwise,} \end{cases}$$

the largest convex function $\leq (v^2 - 1)^2$.

References

For the definition of relaxation and its general properties:
G. dal Maso, *An Introduction to Γ-Convergence*, Birkhäuser, Boston 1993, pp. 28–37.
G. Buttazzo, *Semicontinuity, Relaxation and Integral Representation in the Calculus of Variations*, Pitman Research Notes in Math. 207, Longman Scientific, Harlow, Essex, 1989, pp. 7–28.
For Theorem 5.2.1 and generalizations thereof:
B. Dacorogna, *Direct Methods in the Calculus of Variations*, Springer, Berlin, 1989, pp. 197–249.

Exercises

5.1 Determine sc^-F and discuss the relaxation for

$$F(u) = \int_{-1}^{1} (1 - u'(x))^2 u(x)^2 \, dx \quad \text{for } u \in H^{1,4}$$

with $u(-1) = 0, u(1) = 1$,

$$F(u) = \int_{-1}^{1} (2x - u'(x))^2 u(x)^2 \, dx \quad \text{for } u \in H^{1,4}$$

with $u(-1) = 0, u(1) = 1$,

and

$$F(u) = \int_{-1}^{1} \left((u(x)^2 - \alpha) + (u'(x)^2 - 1) \right) dx \quad \text{for } u \in H^{1,4}$$

with $\alpha \in \mathbb{R}$.

5.2 Determine sc^-I for $I : L^p(\Omega) \to \bar{\mathbb{R}}$ ($\Omega \in \mathbb{R}^d$ open and bounded),

$$I(u) := \begin{cases} \int_\Omega |Du|^p \, dx + \int_\Omega |u|^{\frac{dp}{d-p}} \, dx & \text{if } u \in C^1(\Omega) \\ \infty & \text{otherwise.} \end{cases}$$

5.3 Why does the proof of Lemma 5.3.3 not work for vector-valued mappings $\mathbb{R}^d \to \mathbb{R}^n$ with $n > 1$, i.e. $g : \mathbb{R}^{dn} \to \mathbb{R}, v \in \mathbb{R}^{dn}, \varphi \in H_0^{1,p}(\mathbb{R}, \mathbb{R}^n)$ as in Remark 5.2.1?

6
Γ-convergence

6.1 The definition of Γ-convergence

In this chapter, we treat the important concept of Γ-convergence, introduced and developed by de Giorgi and his school.

Definition 6.1.1. *Let X be a topological space satisfying the first axiom of countability, $F_n : X \to \bar{\mathbb{R}}$ functions ($n \in \mathbb{N}$). We say that F_n Γ-converges to F,*

$$F = \Gamma\text{-}\lim_{n \to \infty} F_n$$

if

(i) *for every sequence $(x_n)_{n \in \mathbb{N}}$ converging to some $x \in X$,*

$$F(x) \leq \liminf_{n \to \infty} F_n(x_n)$$

and

(ii) *for every $x \in X$, there exists a sequence x_n converging to x with*

$$F(x) = \lim_{n \to \infty} F_n(x_n).$$

Example 6.1.1. $F_n : \mathbb{R} \to \mathbb{R}$

$$F_n(x) := \begin{cases} 1 & \text{for } x \geq \dfrac{1}{n} \\ nx & \text{for } -\dfrac{1}{n} \leq x \leq \dfrac{1}{n} \\ -1 & \text{for } x \leq -\dfrac{1}{n}. \end{cases}$$

Then

$$(\Gamma\text{-}\lim F_n)(x) = \begin{cases} 1 & \text{for } x > 0 \\ -1 & \text{for } x \leq 0 \end{cases}$$

while the pointwise limit is 0 for $x = 0$, $1(-1)$ for $x > 0(< 0)$.

Example 6.1.2. $F_n : \mathbb{R} \to \mathbb{R}$

$$F_n(x) := \begin{cases} nx & \text{for } 0 \leq x \leq \dfrac{1}{n} \\ 2 - nx & \text{for } \dfrac{1}{n} \leq x \leq \dfrac{2}{n} \\ 0 & \text{otherwise.} \end{cases}$$

Then

$$(\Gamma\text{-}\lim F_n)(x) = 0$$

which is the same as the pointwise limit.

Example 6.1.3. $F_n : \mathbb{R} \to \mathbb{R}$

$$F_n(x) := \begin{cases} -nx & \text{for } 0 \leq x \leq \dfrac{1}{n} \\ nx - 2 & \text{for } \dfrac{1}{n} \leq x \leq \dfrac{2}{n} \\ 0 & \text{otherwise.} \end{cases}$$

Then

$$(\Gamma\text{-}\lim F_n)(x) = \begin{cases} -1 & \text{for } x = 0 \\ 0 & \text{otherwise.} \end{cases}$$

whereas the pointwise limit is again identically 0. Note that the F_n of 6.1.3 is the negative of the F_n of 6.1.2. Thus, in general

$$-(\Gamma\text{-}\lim F_n) \neq \Gamma\text{-}\lim(-F_n).$$

Example 6.1.4. $F_n : \mathbb{R} \to \mathbb{R}$

$$F_n(x) := \begin{cases} -nx & \text{for } 0 \leq x \leq \dfrac{1}{n} \\ nx - 2 & \text{for } \dfrac{1}{n} \leq x \leq \dfrac{2}{n} & \text{for odd } n \\ 0 & \text{otherwise} \\ 0 & & \text{for even } n. \end{cases}$$

F_n then converges pointwise to 0, but does not Γ-converge at $x = 0$.

Example 6.1.5. $F_n : \mathbb{R} \to \mathbb{R}$

$$F_n(x) = \sin nx.$$

Then

$$(\Gamma\text{-}\lim F_n)(x) = -1,$$

whereas F_n does not converge pointwise.

6.1 The definition of Γ-convergence

From Examples 6.1.4 and 6.1.5, we see that among the two notions of pointwise convergence and Γ-convergence, neither one implies the other.

Example 6.1.6. $F_n : X \to \bar{\mathbb{R}}$ converges continuously to $F : X \to \bar{\mathbb{R}}$ if for every $x \in X$ and every neighbourhood V of $F(x)$ in $\bar{\mathbb{R}}$ (i.e. $V = \{y \in \mathbb{R} : |F(x) - y| < \epsilon\}$ for some $\epsilon > 0$ in case $F(x) \in \mathbb{R}$, $V = \{y \in \mathbb{R} : y > K\} \cup \{\infty\}$ for some $K \in \mathbb{R}$ in case $F(x) = \infty$, and analogously for $F(x) = -\infty$), there exist $n_0 \in \mathbb{N}$ and a neighbourhood U of x with

$$F_n(y) \in V$$

for all $n \geq n_0$, $y \in U$.

F_n converges continuously if and only if both F_n and $-F_n$ converge to F and $-F$, respectively. Continuous convergence implies pointwise convergence, and we conclude from Examples 6.1.2 and 6.1.3 that Γ-convergence is weaker than continuous convergence.

Example 6.1.7. Let X satisfy the first axiom of countability,

$$F_n \equiv F : X \to \mathbb{R}$$

a constant sequence. Then

$$\Gamma\text{-}\lim F_n = (sc^- F)$$

is the relaxed function of F. Thus, we have the remarkable phenomenon that a constant sequence may converge to a limit different from the constant sequence element.

Remark 6.1.1. Without changing the content of the definition of Γ-convergence, condition (ii) may be replaced by the following condition which is weaker and therefore easier to verify:

(ii') for every $x \in X$, there exists a sequence x_n converging to x with

$$\limsup_{n \to \infty} F_n(x_n) \leq F(x).$$

The following result is useful in approximation arguments:

Lemma 6.1.1. Let X satisfy the first axiom of countability. Suppose $(x_m)_{m \in \mathbb{N}}$ converges to x in X, and

$$\limsup_{m \to \infty} F(x_m) \leq F(x).$$

Suppose that (ii') is satisfied for every x_m (i.e. for every m, there exists a sequence $(x_{m,n})_{n \in \mathbb{N}}$ converging to x_m with

$$\limsup_{n \to \infty} F_n(x_{m,n}) \leq F(x_m)).$$

Then (ii') also holds for x.

Proof. Since X satisfies the first axiom of countability, we may take a neighbourhood system $(U_\nu)_{\nu \in \mathbb{N}}$ of x and renumber it and take intersections so that
$$x_m \in U_m \quad \text{for all } m \in \mathbb{N},$$
and that every sequence $(y_\nu)_{\nu \in \mathbb{N}}$ with $y_\nu \in U_{\mu(\nu)}$ for all ν and some sequence $\mu(\nu) \to \infty$ as $\nu \to \infty$ converges to x. For $n \in \mathbb{N}$, we let
$$m_n := \max\left\{m \in \mathbb{N} : x_{m,n} \in U_m, \; F_n(x_{m,n}) \leq \frac{1}{m} + F(x_m)\right\}.$$
Then
$$\lim_{n \to \infty} m_n = \infty.$$
Namely, otherwise, we would find $k_0 \in \mathbb{N}$ with
$$F_{n_\nu}(x_{k,n_\nu}) > \frac{1}{k} + F(x_k)$$
or
$$x_{k,n_\nu} \notin U_k \quad \text{for all } k \geq k_0 \text{ and some}$$
$$\text{sequence } n_\nu \to \infty \text{ for}$$
$$\nu \to \infty.$$
To see that this is impossible we simply observe that since $x_{k_0} \in U_{k_0}$ and since $x_{k_0,n}$ converges to x_{k_0} as $n \to \infty$ we have
$$x_{k_0,n} \in U_{k_0} \quad \text{for all sufficiently large } n,$$
and likewise since we assume
$$\limsup_{n \to \infty} F_n(x_{k_0,n}) \leq F(x_{k_0}),$$
we have
$$F_n(x_{k_0,n}) \leq F(x_{k_0}) + \frac{1}{k_0} \quad \text{for all sufficiently large } n.$$
We then have
$$x_{m_n,n} \in U_{m_n}$$
$$F_n(x_{m_n,n}) \leq F(x_{m_n}) + \frac{1}{m_n}.$$
Therefore $y_n := x_{m_n,n}$ converges to x as $n \to \infty$, and
$$\limsup_{n \to \infty} F_n(y_n) \leq \limsup_{n \to \infty} \left(F(x_{m_n}) + \frac{1}{m_n}\right) \leq F(x)$$

6.1 The definition of Γ-convergence

by assumption and since $m_n \to \infty$ as $n \to \infty$. Thus, $(y_n)_{n \in \mathbb{N}}$ is the desired sequence.

q.e.d.

Let $F : X \to \mathbb{R} \cup \{\infty\}$ satisfy

$$\inf_{y \in X} F(y) > -\infty.$$

Given $\epsilon > 0$, we say that $x \in X$ is an ϵ-minimizer of X if

$$F(x) < \inf_{y \in X} F(y) + \epsilon.$$

Note that x is a minimizer of F if it is an ϵ-minimizer for every $\epsilon > 0$. In contrast to minimizers, ϵ-minimizers always exist for any $\epsilon > 0$.

The following result is a trivial consequence of the definition of Γ-convergence, but quite important.

Theorem 6.1.1. (*Let X satisfy the first axiom of countability*). *Let the sequence of functions $F_n : X \to \bar{\mathbb{R}}$ Γ-converge to $F : X \to \bar{\mathbb{R}}$. Let $\inf_{y \in X} F_n(y) > -\infty$ for every $n \in \mathbb{N}$. Let x_n be an ϵ_n-minimizer for F_n. Assume $\epsilon_n \to 0$ and $x_n \to x$ for some $x \in X$. Then x is a minimizer for F, and*

$$F(x) = \lim_{n \to \infty} F_n(x_n). \tag{6.1.1}$$

Proof. If x were not a minimizer for F, there would exist $x' \in X$ with

$$F(x') < F(x). \tag{6.1.2}$$

Since F_n Γ-converges to F, there exists a sequence $(x'_n) \subset X$ with

$$\lim x'_n = x'$$

$$\lim F_n(x'_n) = F(x').$$

We put $\delta := \frac{1}{4}(F(x) - F(x'))$. We may choose n so large that

$$\epsilon_n < \delta \tag{6.1.3}$$

$$F_n(x'_n) < F(x') + \delta \tag{6.1.4}$$

$$F_n(x_n) > F(x) - \delta \quad \text{(by property (i) of Definition 6.1.1)}. \tag{6.1.5}$$

Since x_n is an ϵ_n-minimizer of F_n,

$$F_n(x'_n) > F_n(x_n) - \epsilon_n \qquad (6.1.6)$$
$$> F_n(x_n) - \delta \quad \text{by (6.1.3)}$$
$$> F(x) - 2\delta \quad \text{by (6.1.5).}$$

From (6.1.4) and (6.1.6), we get

$$F(x) < F(x') + 3\delta$$

contradicting (6.1.2) by definition of δ. Thus, x is a minimizer for F. If (6.1.1) did not hold, then after selection of a subsequence,

$$F(x) < \lim F_n(x_n)$$

whereas by property (ii) of Definition 6.1.1, there would exist a sequence (x'_n) converging to x with

$$F(x) = \lim F_n(x'_n),$$

and we would again contradict the ϵ_n-minimizing property of x_n.

q.e.d.

Corollary 6.1.1. (*Let X satisfy the first axiom of countability.*) *Let $F_n : X \to \bar{\mathbb{R}}$ Γ-converge to $F : X \to \bar{\mathbb{R}}$. Let x_n be a minimizer for F_n. If $x_n \to x$, then x minimizes F, and*

$$F(x) = \liminf F_n(x_n).$$

The following result is similarly both trivial and important.

Theorem 6.1.2. (*Let X satisfy the first axiom of countability.*) *Let F_n Γ-converge to F. Then F is lower semicontinuous.*

Proof. Otherwise, there exist some $x \in X$ and some sequence $(x_m)_{m \in \mathbb{N}}$ with

$$\lim_{m \to \infty} x_m = x$$

$$\lim_{m \to \infty} F(x_m) < F(x). \qquad (6.1.7)$$

By Γ-convergence, for every m, there exists a sequence $(x_{m,n})_{n \in \mathbb{N}} \subset X$ with

$$\lim_{n \to \infty} x_{m,n} = x_m$$

$$\lim_{n \to \infty} F_n(x_{m,n}) = F(x_m). \qquad (6.1.8)$$

We assume $-\infty < \lim F(x_m)$, $F(x) < \infty$ simply to avoid case distinctions. We let

$$\delta := \frac{1}{4}\left(F(x) - \lim_{m\to\infty} F(x_m)\right) > 0 \quad \text{by (6.1.7)}.$$

For every $m \in \mathbb{N}$, we may find $n_m \in \mathbb{N}$ with

$$F_{n_m}(x_{m,n_m}) - F(x_m) < \delta \tag{6.1.9}$$

$$\lim_{m\to\infty} x_{m,n_m} = x, \quad \lim_{m\to\infty} n_m = \infty.$$

Then by Γ-convergence

$$F(x) \leq \lim_{m\to\infty} \inf F_{n_m}(x_{m,n_m}). \tag{6.1.10}$$

We may then choose m so large that

$$F(x_m) < F(x) - 3\delta \tag{6.1.11}$$

and

$$F_{n_m}(x_{m,n_m}) > F(x) - \delta. \tag{6.1.12}$$

Equations (6.1.9), (6.1.11) and (6.1.12) are not compatible, and the resulting contradiction proves the lower semicontinuity.

q.e.d.

Remark. As a consequence of Corollary 3.2.2 and Theorems 3.1.3, 3.3.1, and 3.3.3, in combination with Lemma 2.2.4, the weak topology of $L^p(\Omega)$ and $W^{k,p}(\Omega)$ for $1 < p < \infty$ satisfies the first axiom of countability so that the preceding notions are applicable.

The reference for this section is
G. dal Maso, *An Introduction to Γ-Convergence*, Birkhäuser, Boston, 1993

6.2 Homogenization

In this section and the next one, we describe two important examples of Γ-convergence. They are taken from H. Attouch, *Variational Convergence for Functions and Operators*, Pitman, Boston, 1984.

In the discussion of these two examples, we shall be more sketchy about some technical details than in the rest of the book, because the main point of these examples is to show how the concept of Γ-convergence can be usefully applied to concrete problems that arise in various applications of the calculus of variations.

Let M be a smooth subset of the open unit cube $(0,1)^d$ of $\mathbb{R}^d$. M is considered as a hole. Let
$$M_\epsilon := \bigcup_{m \in \mathbb{Z}^d} \epsilon(M+m)$$
$(\epsilon(M+m) := \{x = y + \epsilon m \text{ with } \frac{y}{\epsilon} \in M\})$ be a periodic lattice of 'holes' of scale ϵ.

Let $\Omega \subset \mathbb{R}^d$, $\Omega_\epsilon := \Omega \setminus (M_\epsilon \cap \Omega)$, i.e. a domain with many small holes. Such domains occur in many physical problems like crushed ice, porous media etc. Often, the physical value of ϵ is so small that it is useful to perform the mathematical analysis for $\epsilon \to 0$. This is called homogenization. Let
$$a(x) := \imath_{\mathbb{R}^d \setminus M_1}(x) := \begin{cases} 0 & \text{for } x \in \mathbb{R}^d \setminus M_1 \\ \infty & \text{for } x \in M_1 \end{cases}$$
be the indicator function of $\mathbb{R}^d \setminus M_1$. $a(\frac{x}{\epsilon})$ then is the indicator function of $\mathbb{R}^d \setminus M_\epsilon$. We consider the functional
$$F_\epsilon(u) := \frac{1}{2}\epsilon^2 \int_\Omega |Du(x)|^2\, dx + \int_\Omega a\left(\frac{x}{\epsilon}\right) u^2(x)\, dx \tag{6.2.1}$$
for $u \in H_0^{1,2}(\Omega)$. A minimizer of the functional
$$F_\epsilon(u) - \int_\Omega f(x)u(x)\, dx$$
(for given $f \in L^2(\Omega)$) satisfies
$$\Delta u = -\frac{f}{\epsilon^2} \text{ in } \Omega_\epsilon \text{ and } u = 0 \text{ on } \partial\Omega_\epsilon. \tag{6.2.2}$$
Here $\partial\Omega_\epsilon = \partial\Omega \cup (\partial M_\epsilon \cap \Omega)$. The boundary condition on $\partial\Omega$ comes from the requirement that $u \in H_0^{1,2}(\Omega)$, while the boundary condition on ∂M_ϵ is forced by the functional.

Theorem 6.2.1. *With respect to weak $L^2(\Omega)$ convergence*
$$\Gamma\text{-}\lim_{\epsilon \to 0} F_\epsilon = F, \tag{6.2.3}$$
with
$$F(u) = \frac{1}{2\mu(M)} \int_\Omega u^2(x)\, dx,$$
where
$$\mu(M) := \int_{(0,1)^d \setminus M} |D\eta(x)|^2\, dx = \int_{(0,1)^d} \eta(x)\, dx,$$

6.2 Homogenization

and η is the solution of

$$\Delta \eta = -1 \quad \text{in } (0,1)^d \setminus M$$
$$\eta = 0 \quad \text{in } M \qquad (6.2.4)$$
$$\eta \text{ is } \mathbb{Z}^d\text{-periodic (i.e. } \eta(x+m) = \eta(x) \text{ for } x \in (0,1)^d,\ m \in \mathbb{Z}^d).$$

Proof. We put $\eta_\epsilon(x) := \eta(\frac{x}{\epsilon})$. By Lemma 5.2.1, η_ϵ converges weakly in $L^2(\Omega)$ to $\mu(M)$ as $\epsilon \to 0$. Let now $u \in L^2(\Omega)$. By approximation, we may assume that u is smooth, e.g. contained in $W^{1,2}(\Omega) \cap C^0(\Omega)$. We put

$$u_\epsilon := \frac{1}{\mu(M)} \eta_\epsilon u.$$

Then u_ϵ converges weakly in $L^2(\Omega)$ to u, and

$$u_\epsilon = 0 \quad \text{on } M_\epsilon.$$

Moreover

$$F_\epsilon(u_\epsilon) = \frac{\epsilon^2}{2} \int_{\Omega_\epsilon} |Du_\epsilon|^2 \qquad (6.2.5)$$
$$= \frac{\epsilon^2}{2} \frac{1}{\mu(M)^2} \int_{\Omega_\epsilon} \left(u^2 |D\eta_\epsilon|^2 + 2u\eta_\epsilon Du \cdot D\eta_\epsilon + \eta_\epsilon^2 |Du|^2 \right).$$

If $U \subset \Omega$ is open, because of the periodicity, $\int_U |D\eta_\epsilon|^2$ asymptotically behaves like

$$\frac{\text{meas } U}{\epsilon^d} \int_{(0,\epsilon)^d \setminus M_\epsilon} |D\eta_\epsilon|^2 = \frac{\text{meas } U}{\epsilon^2} \int_{(0,1)^d \setminus M} |D\eta|^2.$$

This means that

$$\lim_{\epsilon \to 0} \epsilon^2 \int_U |D\eta_\epsilon|^2 = \text{meas } U \int_{(0,1)^d \setminus M} |D\eta|^2 = \text{meas } U \cdot \mu(M) \qquad (6.2.6)$$

hence, approximating u by step functions, we also get

$$\lim_{\epsilon \to 0} \epsilon^2 \int_{\Omega_\epsilon} u^2 |D\eta_\epsilon|^2 = \mu(M) \int_\Omega u^2 \qquad (6.2.7)$$

(note that we assume u to be continuous). Moreover, since η_ϵ is bounded independently of ϵ,

$$\lim_{\epsilon \to 0} \epsilon^2 \int_{\Omega_\epsilon} \eta_\epsilon^2 |Du|^2 = 0, \qquad (6.2.8)$$

and from (6.2.6), (6.2.7) and the Schwarz inequality, also

$$\lim_{\epsilon \to 0} \epsilon^2 \int_{\Omega_\epsilon} u\eta_\epsilon Du \cdot D\eta_\epsilon = 0. \qquad (6.2.9)$$

Equations (6.2.5)–(6.2.9) imply
$$\lim_{\epsilon \to 0} F_\epsilon(u_\epsilon) = F(u). \tag{6.2.10}$$

In order to complete the proof of Γ-convergence, we need to verify that whenever functions v_ϵ that vanish on M_ϵ converge weakly in $L^2(\Omega)$ to u, then
$$\liminf_{\epsilon \to 0} F_\epsilon(v_\epsilon) \geq F(u). \tag{6.2.11}$$

By an approximation argument, we may assume $u \in C_0^\infty(\Omega)$. We put
$$u_\epsilon = \frac{1}{\mu(M)} \eta_\epsilon u$$

as before. We have
$$F_\epsilon(v_\epsilon) + F_\epsilon(u_\epsilon) \geq \epsilon^2 \int_{\Omega_\epsilon} Dv_\epsilon \cdot Du_\epsilon$$
$$= \frac{\epsilon^2}{\mu(M)} \int_{\Omega_\epsilon} (\eta_\epsilon Dv_\epsilon \cdot Du + u Dv_\epsilon \cdot D\eta_\epsilon). \tag{6.2.12}$$

Using (6.2.10), we obtain from (6.2.12) in the limit $\epsilon \to 0$
$$\liminf_{\epsilon \to 0} F_\epsilon(v_\epsilon) + \frac{1}{2\mu(M)} \int_\Omega u^2 \geq \liminf_{\epsilon \to 0} \frac{\epsilon^2}{\mu(M)} \int_{\Omega_\epsilon} u Dv_\epsilon D\eta_\epsilon \tag{6.2.13}$$

since the other term on the right hand side of (6.2.12) goes to 0 by a similar reasoning as above. Equation (6.2.4) implies
$$\epsilon^2 \Delta \eta_\epsilon = -1 \quad \text{in } \Omega_\epsilon. \tag{6.2.14}$$

Moreover
$$\epsilon^2 \int_{\Omega_\epsilon} Duv_\epsilon D\eta_\epsilon \leq \epsilon^2 \left(\int |Du|^2 v_\epsilon^2 \right)^{\frac{1}{2}} \left(\int |D\eta_\epsilon|^2 \right)^{\frac{1}{2}} \to 0, \tag{6.2.15}$$

since v_ϵ as a weakly converging sequence is bounded in L^2, $|Du|$ is bounded by our approximation assumption that u is smooth enough, and since we may use (6.2.6).

Integrating the right-hand side of (6.2.13) by parts, and using (6.2.14) and (6.2.15), we obtain
$$\liminf_{\epsilon \to 0} F_\epsilon(v_\epsilon) + \frac{1}{2\mu(M)} \int_\Omega u^2 \geq \frac{1}{\mu(M)} \liminf_{\epsilon \to 0} \int_{\Omega_\epsilon} v_\epsilon u$$
$$= \frac{1}{\mu(M)} \int_\Omega u^2,$$

since v_ϵ converges weakly in L^2 to u. This implies (6.2.11) and concludes the proof.

q.e.d.

6.3 Thin insulating layers

We consider an insulating layer of width 2ϵ and conductivity λ, and we want to analyse the limit where ϵ and λ tend to 0.

Let $\Omega \subset \mathbb{R}^3$ be bounded and open, S a smooth complete surface in $\mathbb{R}^3$, e.g. a plane, $\Sigma := \Omega \cap S$,

$$S_\epsilon := \{x \in \mathbb{R}^3 : \operatorname{dist}(x, S) < \epsilon\}$$

$$\Sigma_\epsilon := \Omega \cap S_\epsilon$$

$$\Omega_\epsilon := \Omega \setminus \Sigma_\epsilon.$$

Conductivity coefficient

$$a_{\epsilon,\lambda} := \begin{cases} 1 & \text{on } \Omega_\epsilon \\ \lambda & \text{on } \Sigma_\epsilon \end{cases} \quad (\lambda > 0).$$

Variational problem:

$$I^{\epsilon,\lambda} : H_0^{1,2}(\Omega) \to \mathbb{R}$$

$$I^{\epsilon,\lambda}(u) := \frac{1}{2} \int_{\Omega_\epsilon} |Du|^2 \, dx + \frac{\lambda}{2} \int_{\Sigma_\epsilon} |Du|^2 \, dx \tag{6.3.1}$$

$$I^{\epsilon,\lambda}(u) - \int_\Omega fu \to \min \quad (f \in L^2(\Omega) \text{ given}).$$

The Euler–Lagrange equations are

$$\Delta u_{\epsilon,\lambda} + f = 0 \quad \text{on } \Omega_\epsilon \tag{6.3.2}$$

$$\lambda \Delta u_{\epsilon,\lambda} + f = 0 \quad \text{on } \Sigma_\epsilon \tag{6.3.3}$$

$$u_{\epsilon,\lambda}|_{\Omega_\epsilon} = u_{\epsilon,\lambda}|_{\Sigma_\epsilon} \quad \text{on } \partial\Omega_\epsilon \cap \partial\Sigma_\epsilon \tag{6.3.4}$$

$$\lambda \left.\frac{\partial u_{\epsilon,\lambda}}{\partial \vec{n}_{\Sigma_\epsilon}}\right|_{\partial\Sigma_\epsilon} = - \left.\frac{\partial u_{\epsilon,\lambda}}{\partial \vec{n}_{\Omega_\epsilon}}\right|_{\partial\Omega_\epsilon} \quad \text{on } \partial\Omega_\epsilon \cap \partial\Sigma_\epsilon \tag{6.3.5}$$

(where $\vec{n}_U$ denotes the exterior normal of a set U)

$$u_{\epsilon,\lambda} = 0 \quad \text{on } \partial\Omega. \tag{6.3.6}$$

Theorem 6.3.1. *We let $\epsilon \to 0$, $\lambda \to 0$. If $\frac{\lambda}{\epsilon} \to \alpha$ with $0 \leq \alpha \leq \infty$, then $u_{\epsilon,\lambda} \rightharpoonup u$ weakly in $L^2(\Omega)$ $u_{\epsilon,\lambda} \rightrightarrows u$ uniformly on every $\Omega_0 \subset\subset \Omega \setminus \Sigma$, where u solves*

$$\Delta u + f = 0 \quad \text{on } \Omega \setminus \Sigma$$

$$u|_{\partial\Omega} = 0$$

$$\frac{\partial u}{\partial \vec{n}}\bigg|_1 = \frac{\partial u}{\partial \vec{n}}\bigg|_2 = \alpha [u]_\Sigma$$

where $[u]_\Sigma$ is the jump of u across Σ, and $\frac{\partial u}{\partial \vec{n}}|_1$ and $\frac{\partial u}{\partial \vec{n}}|_2$ are the exterior normal derivatives for the two components of $\Omega \setminus \Sigma$. (In case $\alpha = \infty$, u is continuous across Σ, and $\Delta u = f$ in Ω.) Furthermore

$$\frac{1}{2}\int_{\Omega_\epsilon}|Du_{\epsilon,\lambda}|^2 + \frac{\lambda}{2}\int_{\Sigma_\epsilon}|Du_{\epsilon,\lambda}|^2$$

$$\to \frac{1}{2}\int_{\Omega\setminus\Sigma}|Du|^2 + \frac{\alpha}{4}\int_\Sigma [u]_\Sigma^2 d\Sigma.$$

$I^{\epsilon,\lambda}$ Γ-*converges w.r.t. the weak L^2-topology to $I(u)$:*

$$0 < \alpha < \infty : I(u) = \begin{cases} \frac{1}{2}\int_{\Omega\setminus\Sigma}|Du|^2 + \frac{\alpha}{4}\int [u]_\Sigma^2 d\Sigma & \text{if } u \in H_0^{1,2}(\Omega \setminus \Sigma) \\ \infty & \text{otherwise} \end{cases}$$

$$\alpha = \infty : I(u) = \begin{cases} \frac{1}{2}\int_\Omega |Du|^2 & \text{if } u \in H^{1,2}(\Omega) \\ \infty & \text{otherwise} \end{cases}$$

$$\alpha = 0 : I(u) = \begin{cases} \frac{1}{2}\int_{\Omega\setminus\Sigma}|Du|^2 & \text{if } u \in H^{1,2}(\Omega \setminus \Sigma) \\ \infty & \text{otherwise} \end{cases}$$

(*in this case, the result holds for the strong L^2-topology in place of the weak one*)

Thus, in case $\alpha = 0$, we obtain a perfect insulation in the limit, whereas for $\alpha = \infty$, the limiting layer does not insulate at all.

We assume for simplicity $S = \{x^3 = 0\}$.

Lemma 6.3.1. *There exists a constant c_1 (depending on f, Ω, S, but not on ϵ, λ) such that for all sufficiently small ϵ, λ*

$$\int_\Omega u_{\epsilon,\lambda}^2 \leq c_1 \left(1 + (\frac{\epsilon}{\lambda})^2\right)$$

$$\int_{\Omega_\epsilon}|Du_{\epsilon,\lambda}|^2 + \lambda \int_{\Sigma_\epsilon}|Du_{\epsilon,\lambda}|^2 \leq c_1 \left(1 + \frac{\epsilon}{\lambda}\right).$$

6.3 Thin insulating layers

Proof.

$$\int_{\Omega_\epsilon} |Du_{\epsilon,\lambda}|^2 + \lambda \int_{\Sigma_\epsilon} |Du_{\epsilon,\lambda}|^2$$

$$= -\int_{\Omega_\epsilon} \Delta u_{\epsilon,\lambda} \cdot u_{\epsilon,\lambda} + \int_{\partial\Omega_\epsilon} u_{\epsilon,\lambda} \frac{\partial u_{\epsilon,\lambda}}{\partial \vec{n}_{\Omega_\epsilon}} - \lambda \int_{\Sigma_\epsilon} \Delta u_{\epsilon,\lambda} \cdot u_{\epsilon,\lambda}$$

$$+ \lambda \int_{\partial\Sigma_\epsilon} u_{\epsilon,\lambda} \frac{\partial u_{\epsilon,\lambda}}{\partial \vec{n}_{\Sigma_\epsilon}}$$

$$= \int_\Omega f \cdot u_{\epsilon,\lambda} \quad \text{because of the Euler–Lagrange equations}$$

$$\leq |f|_{L^2(\Omega)} \cdot |u_{\epsilon,\lambda}|_{L^2(\Omega)}.$$

By the Poincaré inequality (Theorem 3.4.2),

$$\int_{\Omega_\epsilon} u_{\epsilon,\lambda}^2 \leq c_2 \int_{\Omega_\epsilon} |Du_{\epsilon,\lambda}|^2$$

$$\int_{\Sigma_1} u_{\epsilon,\lambda}^2 \leq c_3 \int_{\Sigma_1} |Du_{\epsilon,\lambda}|^2.$$

By a change of scale

$$y^3 = \epsilon x^3 \quad (y^1 = x^1, y^2 = x^2),$$

$$\int_{\Sigma_\epsilon} u_{\epsilon,\lambda}^2 \leq c_3 \epsilon \int_{\Sigma_\epsilon} |Du_{\epsilon,\lambda}|^2$$

(we only get ϵ instead of ϵ^2, because the area of the portion of $\partial \Sigma_\epsilon$ on which $u_{\epsilon,\lambda}$ vanishes, namely $\partial\Omega \cap S_\epsilon$, is proportional to ϵ). Altogether

$$\int_\Omega u_{\epsilon,\lambda}^2 \leq c_4 \left(1 + \frac{\epsilon}{\lambda}\right) \left(\int_{\Omega_\epsilon} |Du_{\epsilon,\lambda}|^2 + \lambda \int_{\Sigma_\epsilon} |Du_{\epsilon,\lambda}|^2 \right)$$

$$\leq c_4 \left(1 + \frac{\epsilon}{\lambda}\right) \left(\int_\Omega f^2 \right)^{\frac{1}{2}} \left(\int_\Omega u_{\epsilon,\lambda}^2 \right)^{\frac{1}{2}},$$

and the estimates follow.

q.e.d.

Proof (Theorem 6.3.1). We only consider the case $0 < \alpha < \infty$ (the other cases follow from a limiting argument). We first observe

$$\Gamma\text{-}\lim I^{\epsilon,\lambda}(u) = \infty \quad \text{if } u \in L^2(\Omega) \setminus H^{1,2}(\Omega \setminus \Sigma).$$

We assume for simplicity

$$\lambda = \lambda(\epsilon) = \epsilon\alpha, \quad I^\epsilon := I^{\epsilon,\lambda(\epsilon)}.$$

Let $u \in H^{1,2}(\Omega \setminus \Sigma)$. We first check property (ii) of Γ-convergence:

We need to find $u_\epsilon \rightharpoonup u$ weakly in $L^2(\Omega)$ with $\lim I^\epsilon(u_\epsilon) = I(u)$. We define

$$u_\epsilon(x^1, x^2, x^3) := \begin{cases} u(x^1, x^2, x^3) & \text{if } |x^3| \geq \epsilon \\ \frac{1}{2}\{u(x^1, x^2, \epsilon) + u(x^1, x^2, -\epsilon)\} \\ \quad + \frac{x^3}{2\epsilon}\{u(x^1, x^2, \epsilon) - u(x^1, x^2, -\epsilon)\} & \text{if } |x^3| \leq \epsilon. \end{cases}$$

Then $u_\epsilon \in H_0^{1,2}(\Omega \setminus \Sigma)$, $u_\epsilon \rightharpoonup u$ weakly in $L^2(\Omega)$ for $\epsilon \to 0$,

$$I^\epsilon(u_\epsilon)$$
$$= \frac{1}{2}\int_{\Omega_\epsilon} |Du|^2 + \frac{\epsilon\alpha}{2}\int_{|x^3|<\epsilon} \left|\frac{1}{2}D\left(u(x^1,x^2,\epsilon) + u(x^1,x^2,-\epsilon)\right)\right.$$
$$\left. + D\left(\frac{x^3}{2\epsilon}\left(u(x^1,x^2,\epsilon) - u(x^1,x^2,-\epsilon)\right)\right)\right|^2$$
$$\sim \frac{1}{2}\int_{\Omega_\epsilon} |Du|^2 + \frac{\alpha}{8\epsilon}\int_{|x^3|<\epsilon} \left|D\left(x^3\left(u(x^1,x^2,\epsilon) - u(x^1,x^2,-\epsilon)\right)\right)\right|^2$$
$$\sim \frac{1}{2}\int_{\Omega_\epsilon} |Du|^2 + \frac{\alpha}{8\epsilon}\int_{|x^3|<\epsilon} \left|u(x^1,x^2,\epsilon) - u(x^1,x^2,-\epsilon)\right|^2$$
$$+ \text{ terms that contain } x^3 \text{ and go to zero as } \epsilon \to 0 \ (|x^3| < \epsilon).$$

If u is smooth (which we may assume by an approximation argument), therefore for $\epsilon \to 0$

$$I^\epsilon(u_\epsilon) \to \frac{1}{2}\int_{\Omega \setminus \Sigma} |Du|^2 + \frac{\alpha}{4}\int_\Sigma [u]_\Sigma^2.$$

We now check property (i) of Γ-convergence:
Let $v_\epsilon \rightharpoonup u$ weakly in $L^2(\Omega)$. We need to show

$$\liminf_{\epsilon \to 0} I^\epsilon(v_\epsilon) \geq I(u).$$

For u_ϵ as above,

$$\frac{\alpha\epsilon}{2}\int_{\Sigma_\epsilon} |Dv_\epsilon|^2 + \frac{\alpha\epsilon}{2}\int_{\Sigma_\epsilon} |Du_\epsilon|^2 \geq \alpha\epsilon \int_{\Sigma_\epsilon} Du_\epsilon \cdot Dv_\epsilon$$
$$\geq \frac{\alpha\epsilon}{2}\int_{\Sigma_\epsilon} \left(D\{u(\cdot,\epsilon) + u(\cdot,-\epsilon)\}\right.$$
$$+ \frac{x^3}{\epsilon}D\{u(\cdot,\epsilon) - u(\cdot,-\epsilon)\}$$
$$\left. + \frac{\vec{e}_3}{\epsilon}\left(u(\cdot,\epsilon) - u(\cdot,-\epsilon)\right)\right) \cdot Dv_\epsilon,$$

where $\vec{e}_3$ of course is the unit vector in the x^3 direction.

6.3 Thin insulating layers

We may assume u smooth (otherwise, we use an approximation argument). Then as above

$$\liminf_{\epsilon \to 0} \frac{\alpha \epsilon}{2} \int_{\Sigma_\epsilon} |Dv_\epsilon|^2 + \frac{\alpha}{4} \int_\Sigma [u]_\Sigma^2$$

$$\geq \liminf_{\epsilon \to 0} \frac{\alpha}{2} \int_{\Sigma_\epsilon} \left(D\left(u(\cdot,\epsilon) - u(\cdot,-\epsilon)\right) \right) \cdot Dv_\epsilon \, x^3$$

$$+ \liminf_{\epsilon \to 0} \frac{\alpha}{2} \int_{\Sigma_\epsilon} \vec{e}_3 \cdot Dv_\epsilon \left(u(\cdot,\epsilon) - u(\cdot,-\epsilon)\right).$$

Without loss of generality $\liminf_{\epsilon \to 0} I^\epsilon(v_\epsilon) < \infty$. Then

$$\sup_{\epsilon > 0} \epsilon \int_{\Sigma_\epsilon} |Dv_\epsilon|^2 < \infty. \tag{6.3.7}$$

Consequently,

$$\lim_{\epsilon \to 0} \alpha \epsilon \int_{\Sigma_\epsilon} \left(\frac{1}{2} Du(\cdot,\epsilon) + \frac{1}{2} Du(\cdot,-\epsilon) \right) \cdot Dv_\epsilon \leq c\alpha\epsilon \left(\int_{\Sigma_\epsilon} |Dv_\epsilon|^2 \right)^{\frac{1}{2}}$$

$$\leq \frac{c\alpha\epsilon}{\sqrt{\epsilon}}$$

$$\to 0 \quad \text{for } \epsilon \to 0.$$

Similarly,

$$\limsup_{\epsilon \to 0} \alpha \int_{\Sigma_\epsilon} \left(D\left(u(\cdot,\epsilon) - u(\cdot,-\epsilon)\right) \right) \cdot Dv_\epsilon \, x^3 \to 0$$

since $|x^3| < \epsilon$. Thus

$$\liminf_{\epsilon \to 0} \frac{\alpha \epsilon}{2} \int_{\Sigma_\epsilon} |Dv_\epsilon|^2 + \frac{\alpha}{4} \int_\Sigma [u]_\Sigma^2$$

$$\geq \liminf_{\epsilon \to 0} \frac{\alpha}{2} \int_{\Sigma_\epsilon} \vec{e}_3 \cdot Dv_\epsilon \left(u(\cdot,\epsilon) - u(\cdot,-\epsilon)\right).$$

Since $u(\cdot,\epsilon)$ and $u(\cdot,-\epsilon)$ do not depend on x^3, we obtain by integration

$$\frac{\alpha}{2} \int_{\Sigma_\epsilon} \vec{e}_3 \cdot Dv_\epsilon \left(u(\cdot,\epsilon) - u(\cdot,-\epsilon)\right)$$

$$= \frac{\alpha}{2} \int_{(\partial \Sigma_\epsilon)^+} v_\epsilon \left(u(\cdot,\epsilon) - u(\cdot,-\epsilon)\right) - \frac{\alpha}{2} \int_{(\partial \Sigma_\epsilon)^-} v_\epsilon \left(u(\cdot,\epsilon) - u(\cdot,-\epsilon)\right),$$

where here of course $\partial \Sigma_\epsilon^\pm = \Omega \cap \{x^3 = \pm\epsilon\}$. Since we may assume $\liminf_{\epsilon \to 0} I^\epsilon(v_\epsilon) < \infty$, v_ϵ is bounded in $H^{1,2}(\Omega_\epsilon)$. Therefore, we may assume that the traces of v_ϵ on $\partial \Sigma_\epsilon$ converge†. Since u is assumed smooth and v_ϵ converges to u weakly in L^2, we may assume

$$v_{\epsilon,(\partial \Sigma_\epsilon)^\pm} \rightharpoonup u(\cdot, 0^\pm) \quad \text{weakly in } L^2(\partial \Sigma^\epsilon).$$

We then get

$$\lim_{\epsilon \to 0} \frac{\alpha}{2} \int_{\Sigma_\epsilon} \vec{e}_3 \cdot Dv_\epsilon \left(u(\cdot, \epsilon) - u(\cdot, -\epsilon) \right) = \frac{\alpha}{2} \int_\Sigma [u]_\Sigma^2.$$

Altogether

$$\liminf_\epsilon \frac{\alpha \epsilon}{2} \int_{\Sigma_\epsilon} |Dv_\epsilon|^2 + \frac{\alpha}{4} \int_\Sigma [u]_\Sigma^2 \geq \frac{\alpha}{2} \int_\Sigma [u]_\Sigma^2.$$

Therefore

$$\liminf_{\epsilon \to 0} I^\epsilon(v_\epsilon) \geq \frac{1}{2} \int_{\Omega \setminus \Sigma} |Du|^2 + \frac{\alpha}{4} \int_\Sigma [u]_\Sigma^2.$$

q.e.d.

Exercises

6.1 Determine the Γ-limits of the following sequences of functions $F_n : \mathbb{R} \to \mathbb{R}$:

$$F_n(x) := n(\sin nx + 1)$$

$$F_n(x) := \begin{cases} 0 & \text{for } x \neq 0 \\ 1 & \text{for } x = 0 \end{cases}$$

$$F_n(x) := \begin{cases} n^2 x & \text{for } 0 \leq x \leq \frac{1}{n} \\ 2n - n^2 x & \text{for } \frac{1}{n} \leq x \leq \frac{2}{n} \end{cases}$$

$$F_n(x) := \sin nx + \cos nx.$$

6.2 Show the following result: Let X be a topological space satisfying the first axiom of countability, $F_n, G_n : X \to \bar{\mathbb{R}}$. Suppose that F_n Γ-converges to F, G_n Γ-converges to G, $F_n + G_n$ Γ-converges to H (assume that the sums $F_n + G_n$, $F + G$ are always well defined; for example, there must not exist $x \in X$ with $F(x) = \infty$, $G(x) = -\infty$ or vice versa). Then

$$F + G \leq H.$$

Does one get equality '=' instead of '≤' here? (Hint: Consider $F_n(x) = \sin nx$, $G_n(x) = -\sin nx$.)

---

† For this technical point, see e.g. W. Ziemer, *Weakly Differentiable Functions*, Springer, GTM 120, New York, 1989, pp. 189ff.

7
BV-functionals and Γ-convergence: the example of Modica and Mortola

7.1 The space $BV(\Omega)$

Let $C_0^0(\mathbb{R}^d)$ be the space of continuous functions on $\mathbb{R}^d$ with compact support. For each Radon measure μ and each μ-measurable function $\nu : \mathbb{R}^d \to \mathbb{R}$ with $|\nu| = 1$ μ-almost everywhere, we can form a linear functional

$$L : C_0^0(\mathbb{R}^d) \to \mathbb{R}$$

$$L(f) = \int_{\mathbb{R}^d} f\nu d\mu.$$

Conversely, we have the Riesz representation theorem, given here without proof (see e.g. N. Dunford, J. Schwartz, *Linear Operators*, Vol. I, Interscience, New York, 1958, p. 265).

Theorem 7.1.1. *Let $L : C_0^0(\mathbb{R}^d) \to \mathbb{R}$ be a linear functional with*

$$\|L\|_K := \sup\{L(f) : f \in C_0^0(\mathbb{R}^d), |f| \leq 1, \operatorname{supp} f \subset K\} < \infty \quad (7.1.1)$$

for each compact $K \subset \mathbb{R}^d$. Then there exist a Radon measure μ on $\mathbb{R}^d$ and a μ-measurable function $\nu : \mathbb{R}^d \to \mathbb{R}$ with $|\nu| = 1$ μ-almost everywhere with

$$L(f) = \int_{\mathbb{R}^d} f\nu d\mu \quad \text{for all } f \in C_0^0(\mathbb{R}^d). \quad (7.1.2)$$

If L is nonnegative, i.e. $L(f) \geq 0$ whenever $f \geq 0$ everywhere, then $\nu = 1$, i.e.

$$L(f) = \int_{\mathbb{R}^d} f d\mu. \quad (7.1.3)$$

Thus, the Radon measures on $\mathbb{R}^d$ are precisely the nonnegative linear functionals on $C_0^0(\mathbb{R}^d)$. (Note that (7.1.1) automatically holds if L is nonnegative; namely

$$||L||_K = L(\chi_K)$$

in that case where χ_K is the characteristic function of K.)

The same result more generally holds for $C_0^0(\mathbb{R}^d, H)$ where H is a finite dimensional Hilbert space with scalar product $\langle \cdot, \cdot \rangle$. Then linear functionals $L: C_0^0(\mathbb{R}^d, H) \to \mathbb{R}$ satisfying (7.1.1) are represented as

$$L(f) = \int_{\mathbb{R}^d} \langle f, \nu \rangle \, d\mu, \qquad (7.1.4)$$

where μ again is a Radon measure and $\nu : \mathbb{R}^d \to H$ is μ-measurable with $|\nu| = 1$ μ-almost everywhere. Also, in the situation of Theorem 7.1.1, one has

$$\mu(\Omega) = \sup\{L(f) : f \in C_0^0(\Omega), |f| \leq 1\}$$

for any open $\Omega \subset \mathbb{R}^d$.

The expression $\nu d\mu$ in (7.1.4) ($|\nu| = 1$ μ-almost everywhere) is called a vector-valued signed measure. (μ is supposed to be a Radon measure and ν a μ-measurable function with values in H.)

Definition 7.1.1. Let $\Omega \in \mathbb{R}^d$ be open. The space $BV(\Omega)$ consists of all functions $u \in L^1(\Omega)$ for which there exists a vector-valued signed measure $\nu\mu$ with $\mu(\Omega) < \infty$ and

$$\int_\Omega u \operatorname{div} g = -\int g\nu d\mu \qquad (7.1.5)$$

for all $g \in C_0^\infty(\Omega, \mathbb{R}^d)$. In this case, we write $Du = \nu\mu$, $D_i u = \nu_i \mu$ ($\nu = (\nu_1, \ldots, \nu_d)$, $i = 1, \ldots, d$). For $u \in BV(\Omega)$, we put

$$||Du||(\Omega) := \mu =$$
$$\sup\left\{\int_\Omega u \operatorname{div} g\, dx : g = (g^1, \ldots, g^d) \in C_0^\infty(\Omega, \mathbb{R}^d), |g(x)| \leq 1\right.$$
$$\left. \text{for all } x \in \Omega\right\}$$
$$< \infty$$

and

$$||u||_{BV(\Omega)} := ||u||_{L^1(\Omega)} + ||Du||(\Omega).$$

7.1 The space $BV(\Omega)$

For $u \in BV(\Omega)$, $\|Du\|$ is a Radon measure on Ω:

$$\|Du\|(\Omega_0) = \sup\left\{\int_\Omega u \operatorname{div} g\, dx : g \in C_0^\infty(\Omega_0, \mathbb{R}^d), |g| \leq 1\right\}$$

for Ω_0 open in Ω. We write

$$\|Du\|(\Omega_0) =: \int_{\Omega_0} \|Du\|,$$

and also

$$\|Du\|(f) =: \int_\Omega f\, \|Du\| \quad \text{for a nonnegative Borel measurable function } f \text{ on } \Omega.$$

We have for $f \in C_0^0(\Omega)$, $f \geq 0$:

$$\|Du\|(f)$$

$$= \sup\left\{\int_\Omega u \operatorname{div} g : g \in C_0^\infty(\Omega, \mathbb{R}^d), |g(x)| \leq f(x) \ \forall\, x \in \Omega\right\}. \quad (7.1.6)$$

Lemma 7.1.1. *If $u \in W^{1,1}(\Omega)$, then $u \in BV(\Omega)$, and*

$$d\mu = |Du|\, dx \quad \text{where } Du \text{ is the weak derivative of } u \text{ and } dx \text{ is } d\text{-dimensional Lebesgue measure,}$$

and

$$\nu(x) = \begin{cases} \dfrac{Du(x)}{|Du(x)|} & \text{if } Du(x) \neq 0 \\ 0 & \text{otherwise.} \end{cases}$$

The *proof* is obvious. q.e.d.

On a compact hypersurface $S \subset \mathbb{R}^d$ of class C^∞, we have an induced metric and in particular a volume form dS. The $(d-1)$-dimensional volume of S then is

$$|S|_{d-1} = \int_S dS.$$

Lemma 7.1.2. *Let E be a bounded open set in $\mathbb{R}^d$ with a boundary ∂E of class C^∞. Then*

$$|\partial E|_{d-1} = \|D\chi_E\|(\mathbb{R}^d), \quad (7.1.7)$$

where χ_E is the characteristic function of E.

Proof. We have to show

$$|\partial E|_{d-1} = \sup\left\{\int_E \operatorname{div} g : g \in C_0^\infty(\mathbb{R}^d, \mathbb{R}^d), |g| \le 1\right\}.$$

By the Gauss theorem

$$\int_E \operatorname{div} g = \int_{\partial E} g(x)n(x)d(\partial E)$$

where $n(x)$ is the exterior normal. Therefore

$$|\partial E|_{d-1} \ge \sup\left\{\int_E \operatorname{div} g : g \in C_0^\infty(\mathbb{R}^d, \mathbb{R}^d), |g| \le 1\right\}.$$

For the converse inequality, we use a partition of unity to extend n to a C^∞-vector field V on $\mathbb{R}^d$ with $|V(x)| \le 1$ for all $x \in \mathbb{R}^d$. For $\varphi \in C_0^\infty(\mathbb{R}^d)$ with $|\varphi| \le 1$, we put $g = \varphi V$ and get

$$\int_E \operatorname{div} g = \int_{\partial E} \varphi d(\partial E).$$

Consequently

$$\sup\left\{\int_E \operatorname{div} g : g \in C_0^\infty(\mathbb{R}^d, \mathbb{R}^d), |g| \le 1\right\}$$
$$\ge \sup\left\{\int_{\partial E} \varphi d(\partial E) : \varphi \in C_0^\infty(\mathbb{R}^d), |\varphi| \le 1\right\}$$
$$= |\partial E|_{d-1}.$$

This completes the proof.

q.e.d.

The same conclusion holds if $E \subset\subset \Omega$ for some bounded open set; namely

$$|\partial E|_{d-1} = \|D\chi_E\|(\Omega) = \sup\left\{\int_E \operatorname{div} g : g \in C_0^\infty(\Omega, \mathbb{R}^d), |g| \le 1\right\}$$

in that case.

Definition 7.1.2. *A Borel set $E \subset \mathbb{R}^d$ has finite perimeter in an open set Ω if $\chi_{E_{|\Omega}} \in BV(\Omega)$. The perimeter of E in Ω in that case is*

$$P(E, \Omega) := \|D\chi_E\|(\Omega)$$
$$\left(= \sup\left\{\int_E \operatorname{div} g : g \in C_0^\infty(\Omega, \mathbb{R}^d), |g| \le 1\right\}\right). \quad (7.1.8)$$

E is a set of finite perimeter if $\chi_E \in BV(\mathbb{R}^d)$.

The following lower semicontinuity result is easy to prove and very useful.

Theorem 7.1.2. *Let $\Omega \subset \mathbb{R}^d$ be open, $(u_n)_{n \in \mathbb{N}} \subset BV(\Omega)$, and suppose*

$$u_n \to u \quad \text{in } L^1(\Omega).$$

Then for every open $U \subset \Omega$

$$\|Du\|(U) \leq \liminf_{n \to \infty} \|Du_n\|(U). \tag{7.1.9}$$

If in addition

$$\sup\{\|Du_n\|(\Omega) : n \in \mathbb{N}\} < \infty, \tag{7.1.10}$$

then

$$u \in BV(\Omega).$$

Proof. Let $g \in C_0^\infty(U, \mathbb{R}^d)$ with $|g| \leq 1$. Then

$$\int_U u \operatorname{div} g = \lim_{n \to \infty} \int_U u_n \operatorname{div} g \leq \liminf_{n \to \infty} \|Du_n\|(U).$$

Taking the supremum over all such g, we obtain (7.1.9). If $\varphi \in C_0^\infty(\Omega)$, then for $i = 1, \ldots d$

$$\lim_{n \to \infty} \int_\Omega \varphi D_i u_n = -\lim_{n \to \infty} \int_\Omega u_n D_i \varphi = -\int_\Omega u D_i \varphi$$

and hence

$$\left| \int_\Omega u D_i \varphi \right| \leq \sup |\varphi| \liminf_{n \to \infty} \|Du_n\|(\Omega) < \infty$$

in case (7.1.10) holds. Since $C_0^\infty(\Omega)$ is dense in $C_0^0(\Omega)$, for $i = 1, \ldots, d$

$$D_i u(\varphi) := -\int_\Omega u D_i \varphi,$$

then is a bounded linear functional on $C_0^0(\Omega)$, and thus $u \in BV(\Omega)$.
$$\text{q.e.d.}$$

We next discuss the approximation of BV-functionals by smooth ones through mollification. As usually, we let $\rho \in C_0^\infty(\mathbb{R}^d)$ by a mollifier with $\rho \geq 0$, $\operatorname{supp} \rho \subset B(0,1)$, $\int_{\mathbb{R}^d} \rho(x) dx = 1$, and we also impose the symmetry condition

$$\rho(x) = \rho(-x). \tag{7.1.11}$$

We then put as in Section 3.2

$$\rho_h(x) := h^{-d}\rho\left(\frac{x}{h}\right)$$

and for $u \in L^1(\Omega)$, we extend u to $L^1(\mathbb{R}^d)$ by defining $u(x) = 0$ for $x \in \mathbb{R}^d \setminus \Omega$ and put

$$u_h(x) := \rho_h * u(x) := \int_{\mathbb{R}^d} \rho_h(x-y)u(y)dy \in C^\infty(\Omega).$$

Theorem 7.1.3. *If $u \in BV(\Omega)$, then $u_h \to u$ in $L^1(\Omega)$ and $\|Du_h\| \to \|Du\|$ in the sense of Radon measures as $h \to 0$, i.e. for every $f \in C_0^0(\Omega)$*

$$\lim_{h \to 0} \int_\Omega f \|Du_h\| \to \int_\Omega f \|Du\|. \qquad (7.1.12)$$

In particular,

$$\lim_{h \to 0} \|Du_h\|(\Omega) = \|Du\|(\Omega). \qquad (7.1.13)$$

Proof. $u_h \to u$ in $L^1(\Omega)$ by Theorem 3.2.1. It suffices to consider the case $f \geq 0$. From (7.1.3) it follows as in the proof of Theorem (7.1.2) that for every $f \in C_0^0(\Omega)$ with $f \geq 0$

$$\int_\Omega f \|Du\| \leq \liminf_{h \to 0} \int_\Omega f \|Du_h\|. \qquad (7.1.14)$$

It thus remains to prove that for such f

$$\limsup_{h \to 0} \int_\Omega f \|Du_h\| \leq \int_\Omega f \|Du\|. \qquad (7.1.15)$$

For that purpose, we first obtain from (7.1.6)

$$\int_\Omega f \|Du_h\| =$$

$$\sup\left\{\int_\Omega g(x) Du_h(x) dx : g \in C_0^\infty(\Omega, \mathbb{R}), |g(x)| \leq f(x) \ \forall \, x \in \Omega\right\}. \qquad (7.1.16)$$

Here, $Du_h = \left(\frac{\partial}{\partial x^1} u_h, \ldots, \frac{\partial}{\partial x^d} u_h\right)$ is the gradient of u_h, since u_h is smooth.

7.1 The space $BV(\Omega)$

For any such g as in (7.1.16)

$$\int_\Omega g(x) Du_h(x)dx = -\int u_h(x) \operatorname{div} g(x) dx$$
$$= -\int\int \rho_h(x-y) u(y) dy \operatorname{div} g(x) dx$$
$$= -\int\int \rho_h(y-x) \operatorname{div} g(x) dx\, u(y) dy \quad \text{by (7.1.11)}$$
$$= -\int u(y) \operatorname{div}(g_h)(y) dy. \tag{7.1.17}$$

Since we assume $|g| \leq f$, we have

$$|g_h| \leq |g|_h \leq f_h,$$

and since f is continuous, $f_h \rightrightarrows f$ uniformly as $h \to 0$ (see Lemma 3.2.2), i.e. $|f_h(x) - f(x)| \leq \eta_h$ for all $x \in \Omega$, with $\lim_{h \to 0} \eta_h = 0$. By definition of $\|Du\|$, the right hand side of (7.1.17) therefore is bounded by $\int_\Omega (f + \eta_h) \|Du\|$.

Thus, for every such g

$$\lim_{h \to 0} \int_\Omega g(x) Du_h(x) dx \leq \int_\Omega f \|Du\|,$$

and (7.1.15) follows (cf. (7.1.16)).

q.e.d.

Corollary 7.1.1. *Let Ω be a bounded, open subset of $\mathbb{R}^d$. Then any sequence $(u_n)_{n \in \mathbb{N}} \subset BV(\Omega)$ with*

$$\|u_n\|_{BV} \leq K \quad \text{for some } K$$

contains a subsequence that converges in $L^1(\Omega)$ to some $u \in BV(\Omega)$ with $\|u\|_{BV} \leq K$.

Proof. By Theorem 7.1.3, there exist functions $v_n \in C^\infty(\Omega)$ with

$$\|u_n - v_n\|_{L^1(\Omega)} < \tfrac{1}{n}$$
$$\|Dv_n\|(\Omega) \leq K + 1.$$

Therefore $(v_n)_{n \in \mathbb{N}}$ is bounded in $W^{1,1}(\Omega)$. By the Rellich–Kondrachev compactness theorem 3.4.1, after selection of a subsequence, $(v_n)_{n \in \mathbb{N}}$ converges in $L^1(\Omega)$ to some $u \in L^1(\Omega)$. (u_n) has to converge to u as well (in $L^1(\Omega)$). By Theorem 7.1.1, $u \in BV(\Omega)$, and

$$\|u\|_{BV} \leq K.$$

q.e.d.

A reference for the BV theory is W. Ziemer, *Weakly Differentiable Functions*, Springer, GTM 120, New York, 1989, Chapter 5.

7.2 The example of Modica–Mortola

We now come to the theorem of Modica–Mortola:

Theorem 7.2.1. Let

$$F_n(u) := \begin{cases} \int_{\mathbb{R}^d} \left\{ \dfrac{|Du|^2}{n} + n \sin^2(\pi n u) \right\} & \text{for } u \in H^{1,2} \cap L^1(\mathbb{R}^d) \\ \infty & \text{otherwise,} \end{cases}$$

$$F(u) := \begin{cases} \dfrac{4}{\pi} \int_{\mathbb{R}^d} |Du| = \dfrac{4}{\pi} ||Du|| & \text{for } u \in BV(\mathbb{R}^d) \\ \infty & \text{otherwise.} \end{cases}$$

Then w.r.t. to $L^1(\mathbb{R}^d)$ convergence

$$F = \Gamma\text{-}\lim_{n \to \infty} F_n. \tag{7.2.1}$$

Proof.

(i) We first want to show

$$F(u) \leq \liminf_{n \to \infty} F_n(u_n) \tag{7.2.2}$$

whenever

$$u_n \to u \quad \text{in } L^1(\mathbb{R}^d).$$

For that purpose, we put

$$h_n(t) := \frac{1}{n} \int_0^{nt} |\sin(\pi \tau)| \, d\tau.$$

We note that

$$|h_n(s) - h_n(t)| \leq |s - t| \quad \text{for all } n \in \mathbb{N}, s, t \in \mathbb{R}.$$

Therefore

$$||h_n \circ u_n - h_n \circ u||_{L^1} \leq ||u_n - u||_{L^1} \to 0 \quad \text{as } n \to \infty. \tag{7.2.3}$$

Also

$$\lim_{n \to \infty} h_n(t) = \frac{2}{\pi} t. \tag{7.2.4}$$

7.2 The example of Modica–Mortola

We now obtain
$$\left\|h_n \circ u_n - \frac{2}{\pi}u\right\|_{L^1} \leq \|h_n \circ u_n - h_n \circ u\|_{L^1} + \left\|h_n \circ u - \frac{2}{\pi}u\right\|_{L^1}$$
$$\to 0 \quad \text{as } n \to \infty \quad (7.2.5)$$

by (7.2.3), (7.2.4), and Lebesgue's Theorem 1.2.3 on dominated convergence. We may assume
$$u_n \in H^{1,2}(\mathbb{R}^d) \quad \text{for every } n \in \mathbb{N}, \quad (7.2.6)$$
because otherwise $F_n(u_n) = \infty$, and (7.2.2) is trivial. Then

$$\liminf_{n\to\infty} F_n(u_n) \geq 2\liminf_{n\to\infty} \int_{\mathbb{R}^d} |Du_n| \, |\sin(\pi n u_n)|$$
$$= 2\liminf_{n\to\infty} \int |D(h_n \circ u_n)|$$
$$\geq \frac{4}{\pi} \int_{\mathbb{R}^d} |Du| \quad \text{by (7.2.5) and Theorem 7.1.2}$$
$$= F(u).$$

This shows (7.2.2).

(ii) We want to show that for every $u \in L^1(\mathbb{R}^d)$, there exists a sequence $(u_n)_{n\in\mathbb{N}} \subset L^1(\mathbb{R}^d)$ converging to u in $L^1(\mathbb{R}^d)$ with
$$\limsup_{n\to\infty} F_n(u_n) \leq F(u), \quad (7.2.7)$$
thereby completing the proof of Γ-convergence. This inequality will be much harder to show than (7.2.2), however. We shall proceed in several steps:

(1) We may assume $u \in C_0^\infty(\mathbb{R}^d)$. By a slight extension of the reasoning of Theorem 7.1.3, we may find $u_h \in C_0^\infty(\mathbb{R}^d)$ (take a smooth φ_h with $\varphi_h \equiv 1$ on $B(0, \frac{1}{h})$, $\varphi(h) \equiv 0$ on $\mathbb{R}^d \setminus B(0, \frac{1}{h}+1)$, $|D\varphi_h| \leq 2$ and multiply the mollification of u with parameter h by φ_h) with
$$\lim_{h\to 0} \int_{\mathbb{R}^d} |u_h(x) - u(x)| \, dx = 0$$
$$\lim_{h\to 0} F(u_h) = F(u).$$

Applying Lemma 6.1.1, we may indeed assume $u \in C_0^\infty(\mathbb{R}^d)$.

(2) We now want to show that it suffices to verify the claim for certain step functions.

By (1), we assume $u \in C_0^\infty(\mathbb{R}^d)$. By Sard's theorem, for almost all $t \in \mathbb{R}$,

$$u^{-1}(t) = \{x : u(x) = t\}$$

is a hypersurface of class C^∞. For every $\nu \in \mathbb{Z}$, $n \in \mathbb{N}$, we may then choose $t_{\nu,n}$ with this property, with

$$\frac{\nu}{n} < t_{\nu,n} < \frac{\nu+1}{n},$$

and satisfying

$$\int_{\frac{\nu}{n}}^{\frac{\nu+1}{n}} |u^{-1}(t)|_{d-1}\, dt \geq \frac{1}{n} |u^{-1}(t_{\nu,n})|_{d-1}.$$

The coarea formula (Theorem A.1) then implies

$$\int_{\mathbb{R}^d} |Du(x)|\, dx = \int_{\mathbb{R}} |u^{-1}(t)|_{d-1}\, dt$$

$$\geq \sum_{\nu=-\infty}^{\infty} \int_{\frac{\nu}{n}}^{\frac{\nu+1}{n}} |u^{-1}(t)|_{d-1}\, dt$$

$$\geq \sum_{\nu=-\infty}^{\infty} \frac{1}{n} |u^{-1}(t_{\nu,n})|_{d-1}$$

$$= \sum_{\nu=-\infty}^{\infty} \frac{1}{n} \|D\chi_{\{u > t_{\nu,n}\}}\| \quad \text{by Lemma 7.1.2.}$$

We choose $N(n) \in \mathbb{N}$ with $N(n) > (n \max |u| + 1)$ and put

$$u_n := -\frac{N(n)}{n} + \frac{1}{n} \sum_{\nu=-N(n)}^{N(n)} \chi_{\{u > t_{\nu,n}\}}.$$

The preceding inequality implies

$$\limsup_{n \to \infty} F(u_n) \leq F(u).$$

If $t_{\nu,n} < u(x) < t_{\nu,n+1}$, then $u_n(x) = \dfrac{\nu}{n}$. Therefore

$$\operatorname{supp} u_n \subset \operatorname{supp} u,$$

and for all x

$$|u(x) - u_n(x)| \leq \frac{2}{n}.$$

7.2 The example of Modica–Mortola

Since u is assumed to have compact support, therefore

$$\lim_{n \to \infty} \int_{\mathbb{R}^d} |u_n(x) - u(x)|\, dx = 0.$$

Lemma 6.1.1 then implies that it suffices to prove the claim for the functions u_n.

(3) In (2), we have reduced the claim to step functions

$$u = \sum_{i=1}^{N} \alpha_i \chi_{\Omega_i},$$

where the Ω_i are disjoint bounded open sets with boundary $\partial \Omega_i$ of class C^∞. Since the general case is completely analogous, for simplicity, we only consider the case $N=1$, i.e.

$$u = \alpha \chi_\Omega \qquad (7.2.8)$$

with Ω bounded and $\partial\Omega$ of class C^∞. Thus

$$F(u) = \frac{4}{\pi} \alpha |\partial \Omega|_{d-1} \quad \text{(cf. Lemma 7.1.1).} \qquad (7.2.9)$$

We let $0 < \rho < \epsilon_0$, where ϵ_0 is given in Lemma B.1. Thus, the signed distance function $d(x)$ as defined in Appendix B is smooth on $\{x \in \mathbb{R}^d : \text{dist}(x, \partial\Omega) \leq \rho\}$. We need the following auxiliary result:

Lemma 7.2.1. *Let $n \in \mathbb{N}$, let $\alpha_n \in \mathbb{R}$, with $\lim_{n \to \infty} \alpha_n = \alpha \in \mathbb{R}$, $n\alpha_n \in \mathbb{Z}$,*

$$\phi_n(\chi) := \int_\mathbb{R} \left\{ \frac{\chi'(t)^2}{n} + n \sin^2(\pi n \chi(t)) \right\} dt$$

be the one-dimensional analogue of F_n. Then there exist Lipschitz functions $\chi_n : \mathbb{R} \to \mathbb{R}$ with

$$\chi_n(t) = 0 \quad \text{for } t \leq 0$$
$$\chi_n(t) = \alpha_n \quad \text{for } t \geq \frac{\rho}{\sqrt{n}}$$
$$0 \leq \chi_n(t) \leq \alpha_n \quad \text{for } 0 \leq t \leq \frac{\rho}{\sqrt{n}},$$

and

$$\limsup_{n \to \infty} \phi_n(\chi_n) \leq \frac{4}{\pi}\alpha. \qquad (7.2.10)$$

We postpone the proof of Lemma 7.2.1 and proceed with the proof of the theorem. We choose a sequence

$$(\alpha_n)_{n \in \mathbb{R}} \subset \mathbb{R}$$

with

$$\lim_{n \to \infty} \alpha_n = \alpha,$$

and $n\alpha_n \in \mathbb{Z}$ as in Lemma 7.2.1. We put

$$\Omega_n := \left\{ x \in \Omega : d(x) < \frac{\rho}{\sqrt{n}} \right\}$$

and

$$u_n(x) := \chi_n(d(x)) \quad \text{with } \chi_n \text{ as in Lemma 7.2.1.} \tag{7.2.11}$$

Then

$$u_n(x) = 0 \quad \text{for } x \in \mathbb{R}^d \setminus \bar{\Omega}$$
$$u_n(x) = \alpha_n \quad \text{for } x \in \Omega \setminus \bar{\Omega}_n$$
$$0 \leq u_n(x) \leq \alpha_n \quad \text{for } x \in \Omega_n.$$

We also note

$$\lim_{n \to \infty} |\Omega_n|_d = 0. \tag{7.2.12}$$

Thus (cf. (7.2.8))

$$\lim_{n \to \infty} \int_{\mathbb{R}^d} |u(x) - u_n(x)| \, dx = 0, \tag{7.2.13}$$

and u_n converges to u in L^1. We also let (as in Appendix B)

$$\Sigma_t := \{ x \in \mathbb{R}^d : d(x) = t \}.$$

We note

$$Du_n(x) = 0, \quad \sin(n\pi u_n(x)) = 0 \quad \text{for } x \in \mathbb{R}^d \setminus \Omega_n, \tag{7.2.14}$$

and

$$|Dd(x)| = 1 \quad \text{for } x \in \Omega_n \quad \text{by Lemma B.1.} \tag{7.2.15}$$

7.2 The example of Modica–Mortola

Then

$$\limsup_{n\to\infty} F_n(u_n)$$

$$= \limsup_{n\to\infty} \int_{\Omega_n} \left(\frac{|Du_n(x)|^2}{n} + n\sin^2(n\pi u_n(x))\right)|Dd(x)|\,dx$$

$$= \limsup_{n\to\infty} \int_0^{\frac{\rho}{\sqrt{n}}} \left(\frac{|D\chi_n(t)|^2}{n} + n\sin^2(n\pi\chi_n(t))\right)|\Sigma_t|_{d-1}\,dt$$

by Corollary B1 (coarea formula)

$$\leq \limsup_{n\to\infty} \left(\sup_{0\leq t\leq \frac{\rho}{\sqrt{n}}} \phi_n(\chi_n) * |\Sigma_t|_{d-1}\right)$$

$$\leq \frac{4}{\pi}\alpha|\partial\Omega|_{d-1} \quad \text{by Lemma 7.2.1 and Lemma B.1}$$

$$= F(u) \quad (\text{cf. } (7.2.9)).$$

This is (7.2.7).

(4) It only remains to prove Lemma 7.2.1:
The idea is of course to minimize $\phi_n(\chi)$ under the given side conditions on χ. The Euler–Lagrange equations for ϕ_n are

$$\frac{1}{n^2}\chi'' = \pi n \sin(\pi n\chi)\cos(\pi n\chi),$$

and these are implied by

$$\frac{1}{n^2}\chi'^2 = \sin^2(\pi n\chi) + c_1. \qquad (7.2.16)$$

We now construct a solution of (7.2.16) with the desired properties: w.l.o.g. $\alpha > 0$ (the case $\alpha < 0$ is analogous). We choose $c_1 = \frac{1}{n}$ in (7.2.16). We put

$$\Psi_n(t) := \int_0^t \frac{1}{n}\left(\frac{1}{\frac{1}{n} + \sin^2(n\pi s)}\right)^{\frac{1}{2}} ds$$

$$\eta_n := \psi_n(\alpha_n).$$

Then

$$0 < \eta_n \leq \frac{1}{\sqrt{n}}\alpha_n.$$

We let

$$\chi_n : [0, \eta_n] \to [0, \alpha_n]$$

be the inverse of ψ_n. Then χ_n is of class C^1 and

$$\frac{1}{n}\chi'_n(t) = \left(\frac{1}{n} + \sin^2(n\pi\chi_n(t))\right)^{\frac{1}{2}}. \qquad (7.2.17)$$

We extend χ_n to $\mathbb{R}$ as a Lipschitz function by putting

$$\chi_n(t) = 0 \quad \text{for } t < 0$$

$$\chi_n(t) = \alpha_n \quad \text{for } t > \eta_n.$$

Then

$$\phi_n(\chi_n)$$

$$= \int_0^{\eta_n} \left(\frac{\chi'_n(t)^2}{n} + n\sin^2(\pi n \chi_n(t))\right) dt$$

$$\leq \int_0^{\eta_n} \frac{\chi'_n(t)^2}{n} + n\left(\frac{1}{n} + \sin^2(\pi n \chi_n(t))\right) dt$$

$$= \int_0^{\eta_n} 2\left(\frac{1}{n} + \sin^2(\pi n \chi_n(t))\right)^{\frac{1}{2}} \chi'_n(t) dt \quad \text{by (7.2.17)}$$

$$= 2 \int_0^{\alpha_n} \left(\frac{1}{n} + \sin^2(\pi n s)\right)^{\frac{1}{2}} ds$$

and Lemma 7.2.1 follows.

q.e.d.

References

L. Modica and St. Mortola, Un esempio di Γ^--convergenza, Boll. U.M.I. (5), **14-B** (1977), 285–99.

L. Modica, The gradient theory of phase transitions and the minimal interface criterion, Arch. Rat. Mech. Anal. **98** (1987), 123–42

Let us also quote without proof the following result of L. Modica, loc. cit., which plays an important rôle in the theory of phase transitions:

Let $\Omega \subset \mathbb{R}^d$ be open and bounded with Lipschitz boundary, $W : \mathbb{R} \to \mathbb{R}^+$ be continuous with precisely two zeroes α, β (which then are absolute minima, because W is nonnegative)

$$F_n(u) := \begin{cases} \int_\Omega \left(\frac{1}{n}\|Du(x)\|^2 + nW(u(x))\right) dx & \text{for } u \in H^{1,2}(\Omega) \\ \infty & \text{otherwise} \end{cases}$$

and

$$F_0(u) = \begin{cases} 2c_0 \int_\Omega \|Du\| & \text{for } u \in BV(\Omega) \text{ and for almost all } x \in \mathbb{R} \\ \infty & \text{otherwise} \end{cases}$$

7.2 The example of Modica–Mortola

with
$$c_0 = \int_\alpha^\beta W^{\frac{1}{2}}(s)\,ds.$$

Then F_0 is the Γ-limit of F_n w.r.t. L^1-convergence.

The proof is similar to the one of Theorem 7.2.1, except that we cannot apply Sard's lemma anymore, because even for a smooth function u, α and β need not be regular values. Thus, one has to consider nonsmooth level sets as well and appeal to some general results about BV-functions and sets of finite perimeter.

The interpretation of Modica's theorem is the following:
Consider first the problem
$$\int_\Omega W(u(x))\,dx \to \min$$
under the constraint
$$\frac{1}{\operatorname{meas}\Omega}\int_\Omega u(x) = \gamma,$$
with $\alpha < \gamma < \beta$ (w.l.o.g. assume $\alpha < \beta$). A minimizer then is of the form
$$u_\gamma = \begin{cases} \alpha & \text{for } A_1 \subset \Omega \\ \beta & \text{for } A_2 \subset \Omega \end{cases} \tag{7.2.18}$$
such that $A_1 \cup A_2 = \Omega$,
$$\alpha \operatorname{meas} A_1 + \beta \operatorname{meas} A_2 = \gamma \operatorname{meas} \Omega. \tag{7.2.19}$$

u_γ thus jumps from the value α to the value β along $\partial A_1 \cap \Omega = \partial A_2 \cap \Omega =: \Gamma$. However, apart from the preceding relations (7.2.19), A_1 and A_2 and hence also Γ are completely arbitrary. In particular, Γ may be very irregular. In order to gain some control over the transition hypersurface Γ, one adds the the regularizing term $\int_\Omega \|Du(x)\|^2$ to the functional, albeit with an arbitrarily small weight, and in fact one passes to the limit where this weight vanishes so that one preserves (7.2.18), (7.2.19). Although this regularizing term disappears in the limit it still has the effect of regularizing the hypersurface Γ along which the transition from α to β occurs. Namely, the hypersurface of discontinuity of the minimizer u now is constrained by the requirement that the BV norm of u, $\int_\Omega \|Du\|$, be minimized. This means that Γ is a so-called minimal hypersurface. The existence and regularity theory for such minimal hypersurfaces may be found for example in E. Giusti, *Minimal Surfaces and Functions of Bounded Variation*, Birkhäuser, Boston 1984, pp. 3–134.

Exercises

7.1 Try to construct bounded sets in $\mathbb{R}^d$ that do not have a finite perimeter.

7.2 Prove the preceding theorem of L. Modica for $d = 1$.

Appendix A
The coarea formula

Theorem A.1 (coarea formula for smooth functions).
Let $u \in C_0^d(\mathbb{R}^d)$. Then by Sard's theorem,

$$C_u := \{t \in \mathbb{R} : \exists x \in \mathbb{R}^d : Du(x) = 0, u(x) = t\}$$

has one-dimensional Lebesgue measure zero, and thus, for almost all $t \in \mathbb{R}$, $u^{-1}(t)$ is a smooth hypersurface by the implicit function theorem. We then have for every open $\Omega \subset \mathbb{R}^d$

$$\int_\Omega |Du(x)|\, dx = \int_{-\infty}^{\infty} |u^{-1}(t) \cap \Omega|_{d-1}\, dt. \tag{A.1}$$

Proof.

(1) We first show the result for a linear map

$$l : \mathbb{R}^d \to \mathbb{R}$$

(w.l.o.g. $l \neq 0$). Let $\pi : \mathbb{R}^d \to \mathbb{R}$ be the projection onto the first coordinate. We may find $A \in Gl(1,\mathbb{R})$, $R \in O(d,\mathbb{R})$† with

$$l = A \circ \pi \circ R.$$

For every measurable subset E of $\mathbb{R}^d$, we have by Fubini's theorem

$$|E|_d = \int_{-\infty}^{\infty} |E \cap \pi^{-1}(t)|_{d-1}\, dt,$$

where

$$|E|_d = \int_{\mathbb{R}^d} \chi_E$$

† $Gl(d,\mathbb{R}) := \{d \times d\text{-matrices } A \text{ with real entries and } \det A \neq 0\}$, $O(d,\mathbb{R}) := \{A \in Gl(d,\mathbb{R})\mid A^t = A^{-1}\}$ (orthogonal group).

is the Lebesgue measure of E. Since R is orthogonal, we likewise have

$$|R(E)|_d = \int_{-\infty}^{\infty} |E \cap R^{-1} \circ \pi^{-1}(t)|_{d-1} \, dt.$$

We then change variables via $s = At$ and obtain

$$|A|\,|E|_d = \int_{-\infty}^{\infty} |E \cap R^{-1} \circ \pi^{-1} \circ A^{-1}(s)|_{d-1} \, ds$$
$$= \int_{-\infty}^{\infty} |E \cap l^{-1}(s)|_{d-1} \, ds. \qquad (A.2)$$

Since $|A| = |dl|$, and l is linear, this is the coarea formula for linear maps.

(2) Let

$$S_u = \{x \in \mathbb{R}^d : Du(x) = 0\}$$

$$U_t := \{x \in \mathbb{R}^d : u(x) > t\} \quad \text{for } t \in \mathbb{R}.$$

We put

$$u_t := \begin{cases} \chi_{U_t} & \text{if } t \geq 0 \\ -\chi_{\mathbb{R}^d \setminus U_t} & \text{if } t < 0. \end{cases}$$

Then

$$u(x) = \int_{\mathbb{R}} u_t(x) \, dt.$$

Let $\varphi \in C_0^\infty(\mathbb{R}^d \setminus S_u)$, $|\varphi| \leq 1$. Then

$$\int_{\mathbb{R}^d} u(x) \operatorname{div} \varphi(x) \, dx = \int_{\mathbb{R}^d} \int_{\mathbb{R}} u_t(x) \operatorname{div} \varphi(x) \, dt \, dx$$
$$= \int_{\mathbb{R}} \int_{\mathbb{R}^d} u_t(x) \operatorname{div} \varphi(x) \, dx \, dt$$
$$\text{by Fubini's theorem.} \qquad (A.3)$$

By definition of S_u and the implicit function theorem, $u^{-1}(t) \cap \mathbb{R}^d \setminus S_u$ is a hypersurface of class C^d. Since we assume $\operatorname{supp} \varphi \subset \mathbb{R}^d \setminus S_u$, we may apply the divergence theorem to obtain

$$\int_{U_t} \operatorname{div} \varphi(x) \, dx = \int_{(\partial U_t) \cap \mathbb{R}^d \setminus S_u} \varphi(x) n(x) \, d(\partial U_t)(x)$$

and

$$- \int_{\mathbb{R}^d \setminus U_t} \operatorname{div} \varphi(x) \, dx = \int_{(\partial U_t) \cap \mathbb{R}^d \setminus S_u} \varphi(x) n(x) \, d(\partial U_t)(x),$$

where $n(x)$ is the exterior normal of U_t. We use this in (3) (recall the definition of u_t) to obtain

$$-\int_{\mathbb{R}^d} Du(x)\varphi(x)dx$$
$$= \int_{\mathbb{R}^d} u(x) \operatorname{div} \varphi(x)dx$$
$$= \int_{\mathbb{R}} \int_{\partial U_t \cap \mathbb{R}^d \setminus S_u} \varphi(x)n(x)d(\partial U_t)(x)dt$$
$$\leq \int_{\mathbb{R}} |u^{-1}(t) \cap \mathbb{R}^d \setminus S_u|_{d-1} dt \quad \text{since we assume } |\varphi \leq 1|$$
$$\leq \int_{\mathbb{R}} |u^{-1}(t)|_{d-1} dt.$$

Taking the supremum over all such φ, we obtain

$$\int_{\mathbb{R}^d} |Du(x)| \, dx = \int_{\mathbb{R}^d \setminus S_u} |Du(x)| \, dx \leq \int_{\mathbb{R}} |u^{-1}(t)|_{d-1} dt. \quad \text{(A.4)}$$

(3) We now prove the reverse inequality. We let $l_n : \mathbb{R}^d \to \mathbb{R}$ be piecewise linear maps with

$$\lim_{n \to \infty} \int_{\mathbb{R}^d} |l_n - u| = 0 \quad \text{(A.5)}$$

$$\lim_{n \to \infty} \int_{\mathbb{R}^d} |Dl_n| = \int_{\mathbb{R}^d} |Du|. \quad \text{(A.6)}$$

Let

$$U_t^n := \{x \in \mathbb{R}^d : l_n(x) > t\}.$$

By (A.5), there exists a countable set $T_1 \subset \mathbb{R}$ with the property that for all $t \notin T_1$

$$\lim_{n \to \infty} \int_{\mathbb{R}^d} |\chi_t - \chi_t^n| = 0, \quad \text{(A.7)}$$

where χ_t is the characteristic function of $\{u(x) > t\}$, and χ_t^n the one of $\{l_n(x) > t\}$. As noted above, by Sard's theorem and the implicit function theorem, there exists a null set $T_2 \subset \mathbb{R}$ such that for all $t \notin T_2$, $u^{-1}(t)$ is a smooth hypersurface of class C^d. We put

$$T := T_1 \cup T_2.$$

Let $t \in \mathbb{R}\setminus T$, $\epsilon > 0$. By Lemma 7.1.2, there exists $g \in C_0^\infty(\mathbb{R}^d, \mathbb{R}^d)$ with $|g| \leq 1$ and

$$\left|u^{-1}(t)\right|_{d-1} < \int_{U_t} \operatorname{div} g(x)\,dx + \frac{\epsilon}{2}. \tag{A.8}$$

We let $M := \int_{\mathbb{R}^d} |\operatorname{div} g(x)|\,dx$. We choose n_0 so large that for $n \geq n_0$

$$\int_{\mathbb{R}^d} |\chi_t - \chi_t^n| < \frac{\epsilon}{2M} \quad \text{(cf. (7))}. \tag{A.9}$$

Then for $n \geq n_0$

$$\left| \int_{\{u(x)>t\}} \operatorname{div} g(x)\,dx - \int_{\{l_n(x)>t\}} \operatorname{div} g(x)\,dx \right|$$

$$\leq M \int_{\mathbb{R}^d} |\chi_t - \chi_t^n|\,dx < \frac{\epsilon}{2}. \tag{A.10}$$

(A.8) and (A.10) imply

$$\left|u^{-1}(t)\right|_{d-1} \leq \int_{\{u(x)>t\}} \operatorname{div} g(x)\,dx + \frac{\epsilon}{2}$$

$$\leq \int_{\{l_n(x)>t\}} \operatorname{div} g(x)\,dx + \epsilon$$

$$= \int_{\partial\{l_n(x)>t\}} g(x)n(x)\,d(\partial\{l_n(x) > t\}) + \epsilon,$$

$n(x)$ denoting the exterior normal of $\{l_n(x) > t\}$

$$\leq \left|l_n^{-1}(t)\right|_{d-1} + \epsilon.$$

Thus, for $t \notin T$,

$$\left|u^{-1}(t)\right|_{d-1} \leq \liminf_{n \to \infty} \left|l_n^{-1}(t)\right|_{d-1}.$$

From Fatou's lemma (Theorem 1.2.2), (A.2) and (A.6), we obtain

$$\int_{\mathbb{R}^d} \left|u^{-1}(t)\right|_{d-1} dt \leq \liminf_{n \to \infty} \int_{\mathbb{R}} \left|l_n^{-1}(t)\right|_{d-1} dt$$

$$\leq \liminf_{n \to \infty} \int_{\mathbb{R}^d} |Dl_n(x)|\,dx$$

$$= \int_{\mathbb{R}^d} |Du(x)|\,dx. \tag{A.11}$$

(A.4) and (A.11) easily imply the claim.

q.e.d.

Corollary A.1. Let $u \in C_0^d(\mathbb{R}^d)$, $g : \mathbb{R} \to \bar{\mathbb{R}}$ integrable, $\Omega \subset \mathbb{R}^d$ open. Then

$$\int_\Omega g(u(x)) |Du(x)| \, dx = \int_{-\infty}^\infty g(t) \left| u^{-1}(t) \cap \Omega \right|_{d-1} dt. \tag{A.12}$$

Proof. (A.12) follows from Theorem A.1 if g is the characteristic function of an open set and similarly if g is the characteristic function of a measurable set. By considering

$$g^+(t) := \max(0, g(t))$$
$$g^-(t) := \max(0, -g(t))$$

separately, it suffices to consider the case where $g \geq 0$, since always $g(t) = g^+(t) - g^-(t)$. We thus assume $g \geq 0$. Let now $(\rho_n)_{n \in \mathbb{N}} \subset \mathbb{R}^+$ with

$$\lim_{n \to \infty} \rho_n = 0$$

$$\sum_{n=1}^\infty \rho_n = \infty,$$

and put inductively

$$A_n := \left\{ x \in \mathbb{R} : g(x) \geq \rho_n + \sum_{j<n} \rho_j \chi_{A_j}(x) \right\}.$$

Then for all $x \in \mathbb{R}^d$

$$g(x) = \sum_{n=1}^\infty \rho_n \chi_{A_n}(x). \tag{A.13}$$

Since we observed that (A.12) holds for χ_{A_n} in place of g, the representation in (A.13) in conjunction with Beppo Levi's Theorem 1.2.1 on monotone convergence then implies (A.12) for g.

q.e.d.

Remark A.1. The coarea formula is due to Federer. It holds more generally for Lipschitz functions $u : \mathbb{R}^d \to \mathbb{R}$. See H. Federer, *Geometric Measure Theory*, Springer, New York, 1969, pp. 241–760, 268–71.

Appendix B
The distance function from smooth hypersurfaces

We also need some elementary results about the (signed) distance function from a smooth hypersurface. Let $\Omega \subset \mathbb{R}^d$ be open with nonempty boundary $\partial\Omega$. We put

$$d(x) := \begin{cases} \operatorname{dist}(x, \partial\Omega) & \text{if } x \in \Omega \\ -\operatorname{dist}(x, \partial\Omega) & \text{if } x \in \mathbb{R}^d \setminus \Omega. \end{cases}$$

d is Lipschitz continuous with Lipschitz constant 1. Namely, for $x, y \in \mathbb{R}^d$, we find $\pi_y \in \partial\Omega$ with $d(y) = |y - \pi_y|$, hence

$$d(x) \leq |x - \pi_y| \leq |x - y| + |y - \pi_y| = |x - y| + d(y),$$

and interchanging the rôles of x and y yields

$$|d(x) - d(y)| \leq |x - y|.$$

We now assume that $\partial\Omega$ is of class C^2. Let $x_0 \in \partial\Omega$. Let $n(x_0)$ be the outer normal vector of Ω at x_0, and let T_{x_0} be the tangent plane of $\partial\Omega$ at x_0. We rotate the coordinates of $\mathbb{R}^d$ so that the x^d coordinate axis is pointing in the direction of $-n(x_0)$. In some neighbourhood $U(x_0)$ of x_0, $\partial\Omega$ can then be represented as

$$x^d = f(x') \tag{B.1}$$

with $x' = (x^1, \ldots, x^{d-1})$, where $f \in C^2(T_{x_0} \cap U(x_0))$, $Df(x_0') = 0$. The Hessian $D^2 f(x_0)$ is symmetric, and therefore, after a further rotation of coordinates, it becomes diagonalized,

$$D^2 f(x_0) = \begin{pmatrix} \kappa_1 & & 0 \\ & \ddots & \\ 0 & & \kappa_{d-1} \end{pmatrix}. \tag{B.2}$$

$\kappa_1, \ldots, \kappa_{d-1}$ are the eigenvalues of $D^2 f(x_0)$, and they do not depend on the special position of our coordinates. They are invariants of $\partial\Omega$, and are called the principal curvatures of $\partial\Omega$ at x_0. The mean curvature of $\partial\Omega$ at x_0 is

$$H(x_0) = \frac{1}{d-1} \sum_{i=1}^{d-1} \kappa_i = \frac{1}{d-1} \Delta f(x_0). \tag{B.3}$$

The outer normal vector $n(x)$ at $x \in \partial\Omega \cap U(x_0)$ has components

$$n^i(x) = \frac{\frac{\partial}{\partial x^i} f(x')}{(1 + |Df(x')|)^{\frac{1}{2}}}, \quad i = 1, \ldots, d-1 \tag{B.4}$$

$$n^d(x) = \frac{-1}{(1 + |Df(x')|)^{\frac{1}{2}}} \tag{B.5}$$

$(x' = (x^1, \ldots, x^{d-1}))$. In particular

$$\frac{\partial}{\partial x^j} n^i(x_0) = \kappa_i \delta_{ij} \quad \text{for } i,j = 1, \ldots, d-1. \tag{B.6}$$

Lemma B.1. *Suppose Ω is open in $\mathbb{R}^d$ and that $\partial\Omega$ is bounded and of class C^k with $k \geq 2$. For $\eta \in \mathbb{R}$, put*

$$\Sigma_\eta := \{x \in \mathbb{R}^d : d(x) = \eta\}.$$

There exists $\epsilon_0 > 0$ (depending on $\partial\Omega$) with the property that for

$$|\eta| < \epsilon_0,$$

Σ_η *is a hypersurface of class C^k. Also,*

$$\lim_{\eta \to 0} |\Sigma_\eta| = |\partial\Omega|_{d-1}. \tag{B.7}$$

Proof. Since $\partial\Omega$ is compact and of class C^2, there exists $\epsilon > 0$ with the following property: Whenever $|\eta| < \epsilon$ for each $x_0 \in \partial\Omega$, there exist two unique open balls B_1, B_2 with $B_1 \subset \Omega$, $B_2 \subset \mathbb{R}^d \setminus \bar\Omega$,

$$\bar B_1 \cap \partial\Omega = x_0 = \bar B_2 \cap \partial\Omega$$

of radius $|\eta|$. The eigenvalues of the Hessian $D^2 f(x_0)$ of a normalized representation f of $\partial\Omega$ at x_0 as above then have to lie between $-\frac{1}{\epsilon}$ and $\frac{1}{\epsilon}$, i.e.

$$|\kappa_i| \leq \frac{1}{\epsilon} \tag{B.8}$$

for the principal curvatures $\kappa_1, \ldots, \kappa_{d-1}$. If x is a centre of such a ball, then $x_0 = \pi_x$. Also, by uniqueness, these balls depend continuously on $x_0 \in \partial \Omega$. Thus, if $|\eta| \leq \epsilon$, each $x \in \Sigma_\eta$ is the centre of such a ball, and

$$\pi_x = x + n(x)d(x) \quad \text{with } n(x) := n(\pi_x) \tag{B.9}$$

is the unique point in $\partial \Omega$ with $|x - \pi_x| = |d(x)|$. We once again employ the coordinates used for the definition of f and rewrite (B.9) as

$$x = F(x', d) = (x', f(x')) - n(x', f(x'))d. \tag{B.10}$$

Then $F \in C^{k-1}\left(((T_{x_0} \cap U(x_0)) \times \mathbb{R}), \mathbb{R}^d\right)$ and at the point $(x_0', d(x))$

$$DF = \begin{pmatrix} 1 - \kappa_1 d(x) & & & 0 \\ & \ddots & & \\ & & 1 - \kappa_{d-1} d(x) & \\ 0 & & & 1 \end{pmatrix} \quad \text{by (6)}. \tag{B.11}$$

By (B.8) and since $|\eta| < \epsilon$,

$$\det DF \neq 0.$$

By the inverse function theorem, x' and d therefore locally are C^{k-1}-functions of x (cf. (B.9)). Since

$$d(x) = d(x_0 - \eta n(x_0)) = \eta,$$

we have

$$Dd(x) \cdot n(x_0) = -1.$$

Since d is Lipschitz with Lipschitz constant 1, we conclude

$$|Dd(x)| = 1$$

and

$$Dd(x) = -n(x_0) \in C^{k-1}.$$

Thus $d \in C^k$ locally, and the level hypersurfaces Σ_η are of class C^k. For (B.7), we may w.l.o.g. take $\eta > 0$ as the case $\eta < 0$ succumbs to the same reasoning. We consider the vector field

$$V(x) = Dd(x).$$

The Gauss theorem yields

$$\int_{\{0 < d(x) < \eta\}} \operatorname{div} V(x) = \int_{\Sigma_0} V(x)n(x)d\{\Sigma_0\}(x) + \int_{\Sigma_\eta} V(x)n_\eta(x)d\{\Sigma_\eta\}(x),$$

where n_η is the normal vector of Σ_η pointing in the direction opposite to n. Since the measure of $\{0 < d(x) < \eta\}$ goes to zero with η and

$$V(x) = -n(x) \quad \text{for } x \in \Sigma_0 = \partial\Omega$$

$$V(x) = n_\eta(x) \quad \text{for } x \in \Sigma_\eta,$$

(B.7) easily follows.

q.e.d.

References

D. Gilbarg, N. Trudinger, *Elliptic Partial Differential Equations*, Springer, Berlin, 2nd edition, 1983, pp. 354–6.

8
Bifurcation theory

8.1 Bifurcation problems in the calculus of variations

We wish to consider a variational problem depending on a parameter λ, and to investigate how the space of solutions depends on this parameter. We thus consider

$$I(u, \lambda) := \int_a^b F(t, u(t), \dot{u}(t), \lambda) \, dt.$$

λ is supposed to vary in some open set $\Lambda \subset \mathbb{R}^l$. Often, one has $l = 1$. We assume that

$$F : [a, b] \times \mathbb{R}^d \times \mathbb{R}^d \times \Lambda \to \mathbb{R}$$

is sufficiently often differentiable so that all derivatives taken in the sequel exist. For that purpose, one may simply assume that F is of class C^∞ in all its arguments although that is a little stronger than needed in the sequel.

Remark 8.1.1. One may also impose boundary conditions depending on λ, i.e.

$$u(a) = u_1(\lambda)$$
$$u(b) = u_2(\lambda),$$

and finally, one may vary the boundary points themselves,

$$a = a(\lambda)$$
$$b = b(\lambda).$$

This latter variation, however, can formally be incorporated in the variation of F, by transforming the integral.

8.1 Bifurcation problems in the calculus of variations

Let
$$\tau(\cdot, \lambda) : [a(\lambda_0), b(\lambda_0)] \to [a(\lambda), b(\lambda)]$$
be a bijective linear map, for some fixed λ_0. Then
$$\int_{a(\lambda)}^{b(\lambda)} F(\tau, u(\tau), \dot{u}(\tau))\, d\tau$$
$$= \int_{a(\lambda_0)}^{b(\lambda_0)} F(\tau, u(\tau), \dot{u}(\tau)) \frac{\partial \tau(t, \lambda)}{\partial t}\, dt,$$

and putting
$$v(t) := u(\tau(t, \lambda))$$
$$F(t, v(t), \dot{v}(t), \lambda) := F\left(\tau(t, \lambda), v(t), \dot{v}(t)\left(\frac{\partial \tau(t, \lambda)}{\partial t}\right)^{-1}\right) \frac{\partial \tau(t, \lambda)}{\partial t}$$
$$I(v, \lambda) := \int_{a(\lambda_0)}^{b(\lambda_0)} F(t, v, \dot{v}(t), \lambda)\, dt$$

yields a parameter-dependent variational integral for v with fixed boundary points $a(\lambda_0), b(\lambda_0)$.

As established in Theorem 1.1.1 of Part I, a critical point u of $I(\cdot, \lambda)$ of class C^2 satisfies the Euler–Lagrange equations

$$F_{pp}(t, u(t), \dot{u}(t), \lambda)\ddot{u}(t) + F_{pu}(t, u(t), \dot{u}(t), \lambda)\dot{u}(t) \quad (8.1.1)$$
$$+ F_{pt}(t, u(t), \dot{u}(t), \lambda) - F_u(t, u(t), \dot{u}(t), \lambda) = 0.$$

We abbreviate (8.1.1) as
$$L_\lambda u = 0. \quad (8.1.2)$$

In the light of Theorems 1.2.2 and 1.2.4 and Lemma 1.3.1 of Part I, we shall assume
$$\det F_{pp}(t, u(t), \dot{u}(t), \lambda) \neq 0 \quad (8.1.3)$$
for all functions u occurring in the sequel. Equation (8.1.3) implies that (8.1.1) can be solved for $\ddot{u}$ in terms of u and $\dot{u}$, i.e.

$$\ddot{u} = -F_{pp}(t, u(t), \dot{u}(t), \lambda)^{-1} \{F_{pu}(t, u(t), \dot{u}(t), \lambda)\dot{u}(t)$$
$$+ F_{pt}(t, u(t), \dot{u}(t), \lambda) - F_u(t, u(t), \dot{u}(t), \lambda)\} \quad (8.1.4)$$
$$=: f(t, u(t), \dot{u}(t), \lambda),$$

see (1.2.10) of Part I. (8.1.2) thus is equivalent to

$$\ddot{u}(t) - f(t, u(t), \dot{u}(t), \lambda) = 0. \tag{8.1.5}$$

The topic of bifurcation theory then is to study the space of solutions of (8.1.5) in its dependence on the parameter λ. Before approaching this problem from a general point of view in the next section, we should briefly comment on the relations with the Jacobi theory introduced in Section 1.3 of Part I. For a critical point u of $I(\cdot, \lambda)$ and $\eta \in D_0^1(I, \mathbb{R}^d)$, we had established the expansion

$$I(u + s\eta, \lambda) = I(u, \lambda) + \frac{1}{2}s^2 \delta^2 I(u, \eta, \lambda) + o(s) \text{ for } s \to 0, \tag{8.1.6}$$

with

$$\delta^2 I(u, \eta, \lambda) = Q_\lambda(\eta) := \frac{d^2}{ds^2} \int_a^b F(t, u(t) + s\eta(t), \dot{u}(t) + s\dot{\eta}(t), \lambda) dt_{|s=0}$$

$$= \int_a^b \{F_{p^i p^j}(t, u, \dot{u}, \lambda)\dot{\eta}_i \dot{\eta}_j + 2F_{p^i u^j}(t, u, \dot{u}, \lambda)\dot{\eta}_i \eta_j \tag{8.1.7}$$

$$+ F_{u^i u^j}(t, u, \dot{u}, \lambda)\eta_i \eta_j\} dt,$$

abbreviated as

$$\int_a^b \{F_{\lambda, pp}\dot{\eta}\dot{\eta} + 2F_{\lambda, pu}\dot{\eta}\eta + F_{\lambda, uu}\eta\eta\} dt.$$

Critical points of Q satisfy the Jacobi equations

$$J_\lambda(u)\eta := \frac{d}{dt}(F_{pp}(t, u, \dot{u}, \lambda)\dot{\eta} + F_{pu}(t, u, \dot{u}, \lambda)\eta) \tag{8.1.8}$$

$$- F_{pu}(t, u, \dot{u}, \lambda)\dot{\eta} - F_{uu}(t, u, \dot{u}, \lambda)\eta = 0.$$

$J_\lambda(u)$ is called the Jacobi operator associated with the critical point u of $I(\cdot, \lambda)$. We also observe that

$$J_\lambda(u)\eta = \frac{d}{ds}L_\lambda(u + s\eta)_{|s=0}. \tag{8.1.9}$$

Of course, this is not surprising since L_λ represents the first variation of $I(\cdot, \lambda)$ and J_λ the second one. From the expansion (8.1.6) we see that

$$I(u + s\eta, \lambda) \gtreqless I(u, \lambda) \text{ if } \delta^2 I(u, \eta, \lambda) \gtreqless 0. \tag{8.1.10}$$

No such conclusions can be achieved, however, if

$$\delta^2 I(u, \eta, \lambda) = 0. \tag{8.1.11}$$

8.1 Bifurcation problems in the calculus of variations 269

Now by Lemma 1.3.2 of Part I, for a Jacobi field η that vanishes at the boundary points a and b, (8.1.11) holds. This indicates that Jacobi fields play a decisive rôle for deciding about the minimizing property of a critical point u of $I(\cdot, \lambda)$. Jacobi fields satisfy

$$J_\lambda(u)\eta = 0, \tag{8.1.12}$$

i.e. are solutions of the linearization of the equation $L_\lambda u = 0$ satisfied by u. This also indicates that Jacobi fields will play a decisive rôle in analysing the bifurcation behaviour of $L_\lambda u = 0$ as λ varies. Namely, in finite dimensional problems, the presence of a nontrivial solution of the linearization of a parameter-dependent equation $L_\lambda u = 0$ at some parameter value λ_0 either results from a nontrivial family $u(\tau)$ of solutions of $L_{\lambda_0} u(\tau) = 0$ by differentiating the equation w.r.t the parameter τ, or it indicates a nontrivial bifurcation as λ varies in the vicinity of λ_0. In the next section, we shall see that under appropriate assumptions, the same also holds in the present infinite dimensional context. In fact, the bifurcation problem will be reduced to a finite dimensional one via Lyapunov–Schmid reduction. The reason why this is possible in our variational context is that under our assumption (8.1.3), the space of Jacobi fields is always finite dimensional. Namely, analogously to (8.1.4), (8.1.5), the assumption (8.1.3) implies that (8.1.8) can be solved w.r.t $\ddot{\eta}$, i.e

$$\ddot{\eta} - \varphi(t, u, \dot{u}, \eta, \dot{\eta}, \lambda) = 0. \tag{8.1.13}$$

(Although this is not indicated by the notation, (8.1.13) is a linear equation for η, and so the space of solutions is a linear space.)

Now suppose that we have a sequence $(\eta_n)_{n \in \mathbb{N}}$ of solutions of (8.1.13) (for fixed u, λ) that are bounded in some appropriate function space like $C^2(I)$ or $W^{2,2}(I)$. For concreteness, let us consider $C^2(I)$, i.e. for example

$$\|\eta_n\|_{C^2(I)} \leq 1.$$

By the Arzela–Ascoli theorem, after selection of a subsequence, $(\eta_n)_{n \in \mathbb{N}}$ then converges in $C^1(I)$ to some limit denoted by η_0. (8.1.13) then implies that $(\ddot{\eta})_{n \in \mathbb{N}}$ converges in $C^0(I)$ (as it follows from our assumptions on the differentiability of F that φ is smooth, in particular continuous). Thus (since the uniform limit of derivatives is the derivative of the limit), $(\eta_n)_{n \in \mathbb{N}}$ converges in $C^2(I)$ to η_0, and consequently η_0 also solves (8.1.13). From this compactness result, one easily deduces that the space of solutions of (8.1.13) has finite dimension.

8.2 The functional analytic approach to bifurcation theory

We consider the following general situation. We have Banach spaces V, W, and a parameter space Λ. We assume that Λ is an open subset of some Banach space. We consider a parameter dependent family of equations

$$L_\lambda u = 0, \qquad (8.2.1)$$

with

$$V \times \Lambda \to W$$
$$(u, \lambda) \mapsto L_\lambda u.$$

We assume that $L_\lambda u$ is sufficiently often differentiable w.r.t. to u and λ so that all subsequent expansions are valid. The aim of bifurcation theory is to study the set of solutions u of (8.2.1) as λ varies, to identify the bifurcation values of λ, i.e. those values of λ where the structure of the solution set changes, and to investigate that structure at such bifurcation points. In order to arrive at concrete results, we need an additional assumption. We consider the derivative of $L_\lambda u$ w.r.t. u,

$$J_\lambda(u)v := (D_u L_\lambda(u))v := \frac{d}{dt} L_\lambda(u+tv)_{|t=0} \qquad (8.2.2)$$

for $v \in V$. We assume that J_λ is a Fredholm operator of index 0, i.e. that $\ker J_\lambda$ and $\operatorname{coker} J_\lambda$ are of finite and equal dimension, and furthermore that there exists a canonical isomorphism

$$\ker J_\lambda \cong \operatorname{coker} J_\lambda. \qquad (8.2.3)$$

We first consider the case where

$$L_{\lambda_0} u_0 = 0 \qquad (8.2.4)$$

$$\ker J_{\lambda_0}(u_0) = \{0\} \text{ for some } \lambda_0 \in \Lambda, u_0 \in V. \qquad (8.2.5)$$

We shall see that in this case, no bifurcation can occur at λ_0. Namely, we have:

Theorem 8.2.1. *Let $L_{\lambda_0} u_0 = 0$ for some $\lambda_0 \in \Lambda$, $u_0 \in V$, $\ker J_{\lambda_0}(u_0) = \{0\}$. Then there exist neighbourhoods $U(\lambda_0)$ of λ_0 in Λ and $V(u_0)$ of u_0 in V such that for all $\lambda \in U(\lambda_0)$, there exists a unique $u \in V(u_0)$ with*

$$L_\lambda u = 0.$$

8.2 The functional analytic approach to bifurcation theory

Proof. Since J_{λ_0} is assumed to be a Fredholm operator of index 0, (8.2.5) implies that

$$J_{\lambda_0} : V \to W$$

is an isomorphism. Thus the derivative w.r.t. the variable u of the map

$$V \times \Lambda \to W$$
$$(u, \lambda) \mapsto L_\lambda u$$

is an isomorphism at (u_0, λ_0), and the implicit function Theorem 2.4.1 implies that the equation

$$L_\lambda u = 0$$

can be locally resolved w.r.t. u, i.e. there exist neighbourhoods $U(\lambda_0), V(u_0)$ and a map

$$U(\lambda_0) \to V(u_0)$$
$$\lambda \mapsto u(\lambda)$$

such that

$$L_\lambda u = 0$$

precisely if

$$u = u(\lambda).$$

q.e.d.

We next consider the case where

$$L_{\lambda_0} u_0 = 0 \qquad (8.2.6)$$

$$K := \ker J_{\lambda_0}(u_0) \text{ is one-dimensional.} \qquad (8.2.7)$$

The assumption that this kernel is one-dimensional may look restrictive, but it is typically satisfied in variational problems, and in this situation, we can already see the typical phenomena of bifurcation while avoiding additional technical complications that arise for higher dimensional kernels. In the sequel, we shall assume for simplicity

$$u_0 = 0$$

(which can always be achieved by changing the dependent variables in our equation by a translation). In the sequel, we shall also usually write J_{λ_0} in place of $J_{\lambda_0}(u_0) = J_{\lambda_0}(0)$. We may write

$$V = K \oplus V_1, \tag{8.2.8}$$

and in view of (8.2.3), we may also write

$$W = K \oplus W_1, \tag{8.2.9}$$

with

$$W_1 = J_{\lambda_0}(V) = J_\lambda(V_1). \tag{8.2.10}$$

We denote by

$$\pi : V \to K$$

the projection onto K according to (8.2.8), and we consider $\pi(V)$ as a subspace of W, according to (8.2.9). Thus, if

$$u = \xi + w$$

with $\xi \in K$, $w \in V_1$, then

$$\pi(u) = \xi.$$

In particular,

$$\pi(0) = 0.$$

We consider the map

$$A_{\lambda_0} : V \to W$$
$$u \mapsto L_{\lambda_0} u + \pi(u).$$

Lemma 8.2.1. A_{λ_0} *is a local diffeomorphism, i.e. the derivative* $DA_{\lambda_0} = DA_{\lambda_0}(0) : V \to W$ *is an isomorphism.*

Proof. The derivative is computed as

$$DA_{\lambda_0} v = J_{\lambda_0} v + \pi(v) \text{ for } v \in V. \tag{8.2.11}$$

The Fredholm operator J_{λ_0} yields a bijective continuous linear map between V_1 and W_1 because of the decompositions (8.2.8), (8.2.9), (8.2.10), and its inverse is likewise continuous (by Definition 2.3.1). From the definition of K and π and (8.2.3) we then conclude that DA_{λ_0} is an isomorphism.

q.e.d.

8.2 The functional analytic approach to bifurcation theory

We now consider the map

$$A : V \times \Lambda \to W$$
$$(u, \lambda) \mapsto A_\lambda(u) := L_\lambda(u) + \pi(u).$$

By Lemma 2.3.4, there exists a neighbourhood $V(\lambda_0)$ of λ_0 in Λ such that for all $\lambda \in V(\lambda_0)$, $A_\lambda(0)$ is a local diffeomorphism. We may therefore apply the implicit function Theorem 2.4.1. Consequently, as

$$A(0, \lambda_0) = 0, \qquad (8.2.12)$$

there exist neighbourhoods $U(0)$ of $u_0 = 0$ in V, $U_1(0)$ of 0 in W such that for all $\lambda \in V(\lambda_0)$ and $\xi \in U_1(0)$, there exists a unique $u \in U(0)$ with

$$A(u, \lambda) = \xi, \qquad (8.2.13)$$

i.e.

$$L_\lambda u + \pi(u) = \xi. \qquad (8.2.14)$$

We write

$$u = u(\xi, \lambda)$$

for the solution u of (8.2.13). We have in particular

$$u(0, \lambda_0) = 0, \qquad (8.2.15)$$

since $L_{\lambda_0} 0 = 0$, $\pi(0) = 0$ (remember $u_0 = 0$). In this notation, (8.2.13) is

$$A(u(\xi, \lambda), \lambda)) = \xi.$$

The aim now is to find ξ with

$$\pi(u(\xi, \lambda)) = \xi, \qquad (8.2.16)$$

because (8.2.14) will then give

$$L_\lambda u(\xi, \lambda) = 0, \qquad (8.2.17)$$

which is the equation that we wish to solve. Since the image of π is assumed to be one-dimensional (and in any case finite dimensional as J_λ is supposed to be a Fredholm operator), we have reduced our bifurcation problem to a finite dimensional problem. In the sequel, we shall thus let ξ vary only in K, the image of π. Thus, we may consider ξ as a scalar quantity, $\xi = \alpha \xi_0$, with $\alpha \in \mathbb{R}$, where ξ_0 is a generator of K. We denote the derivative of $u(\alpha\xi_0, \lambda)$ w.r.t. α and λ, respectively, at $\alpha = 0, \lambda = \lambda_0$ by

$$\partial_\alpha u \text{ and } \partial_\lambda u, \text{ respectively.}$$

(Note that λ in general is not a scalar quantity, as we do not assume that Λ is one-dimensional.) Differentiating (8.2.14) w.r.t. α yields

$$J_{\lambda_0}\partial_\alpha u + \pi(\partial_\alpha u) = \xi_0. \tag{8.2.18}$$

Since $\xi_0 \in K$, also

$$J_{\lambda_0}\xi_0 + \pi(\xi_0) = \xi_0. \tag{8.2.19}$$

Lemma 8.2.1 then implies

$$\partial_\alpha u = \xi_0. \tag{8.2.20}$$

We are now ready for the essential point, namely the asymptotic expansion of the equation (8.2.16), i.e.

$$\pi(u(\xi, \lambda)) = \xi \tag{8.2.21}$$

near $\xi = 0, \lambda = \lambda_0$.

We let $\partial_\alpha^2 u, \partial_\lambda^2 u$ be the second derivatives of $u(\alpha\xi_0, \lambda)$ w.r.t. α and λ, respectively, at $\alpha = 0, \lambda = \lambda_0$, and likewise $\partial_{\alpha,\lambda}^2 u$ be the mixed second derivative w.r.t. α and λ. Higher derivatives will be denoted similarly by corresponding symbols. The Taylor expansion of (8.2.16) then is

$$\xi = \pi(u(\xi, \lambda)) = \pi(0) + \pi(\partial_\alpha u)\alpha + \pi(\partial_\lambda u)\mu$$
$$+ \frac{1}{2}\pi(\partial_\alpha^2 u)\alpha^2 + \pi(\partial_{\alpha,\lambda}^2 u)\alpha\mu + \frac{1}{2}\pi(\partial_\lambda^2 u)(\mu, \mu)$$
$$+ \text{ terms of higher order in } \alpha \text{ and } \mu. \tag{8.2.22}$$

Since $\pi(0) = 0$ and since, by (8.2.20), $\partial_\alpha u = \xi_0$, hence $\pi(\partial_\alpha u)\alpha = \alpha\xi_0 = \xi$, we may write (8.2.22) as

$$0 = \pi(\partial_\lambda u)\mu + \frac{1}{2}\pi(\partial_\alpha^2 u)\alpha^2 + \text{ higher order terms in } \alpha \text{ only} \tag{8.2.23}$$
$$+ \pi(\partial_{\alpha,\lambda}^2 u)\alpha\mu + \text{ higher order terms that also involve } \mu.$$

Remark 8.2.1. In order to interpret the terms in this expansion, we differentiate (8.2.14), i.e

$$L_\lambda u(\xi, \lambda) + \pi(u(\xi, \lambda))\xi = \alpha\xi_0 \tag{8.2.24}$$

twice w.r.t. α. One differentiation yields

$$J_\lambda(u)\partial_\alpha u + \pi(\partial_\alpha u) = \xi_0, \tag{8.2.25}$$

8.2 The functional analytic approach to bifurcation theory

and differentiating once more gives

$$DJ_\lambda(\partial_\alpha u)^2 + J_\lambda \partial_\alpha^2 u + \pi(\partial_\alpha^2 u) = 0. \tag{8.2.26}$$

We put $\lambda = \lambda_0$ and project onto K in the decomposition (8.2.9). We may also denote that projection by π, and we then have $\pi \circ J_{\lambda_0} = 0$. Also, from (8.2.20), $\partial_\alpha u = \xi_0$, and so we get

$$\pi(DJ_\lambda \xi_0^2) = -\pi(\partial_\alpha^2 u). \tag{8.2.27}$$

Thus, the first term in the expansion of Q in (8.2.24) can be expressed via DJ_λ. In a variational context, J_λ represents the second variation, and so DJ_λ represents the third variation of the variational integral. Likewise, if $\partial_\alpha^2 u$ vanishes, i.e. if the third variation vanishes on the Jacobi field ξ_0, then $\pi(\partial_\alpha^3 u)$ can be expressed by the fourth variation, and so on.

We now discuss the simplest case of a bifurcation, namely where

$$\pi(\partial_\alpha^2 u) \neq 0. \tag{8.2.28}$$

We put $\alpha := t\tau$, $\mu =: t^2 \bar{\mu}$, $a_0 := \pi(\partial_\lambda u)\bar{\mu}$, $a_1 := \frac{1}{2}\pi(\partial_\alpha^2 u)$, and (8.2.23) becomes

$$0 = a_0 t^2 + a_1 t^2 \tau^2 + t^2 \Sigma(t, \tau, \bar{\mu}) \tag{8.2.29}$$

with

$$\Sigma(t, \tau, \bar{\mu}) = O(t) \text{ for fixed } \tau, \bar{\mu} \text{ for } t \to 0. \tag{8.2.30}$$

For $t \neq 0$, (8.2.29) is equivalent to

$$0 = a_0 + a_1 \tau^2 + \Sigma(t, \tau, \bar{\mu}). \tag{8.2.31}$$

We shall now see by a simple application of the implicit function theorem that the bifurcation behaviour of equation (8.2.31) is equivalent to the one of

$$0 = a_0 + a_1 \tau^2. \tag{8.2.32}$$

We assume $a_0 \neq 0$; as will be discussed below (see Lemma 8.2.2), this can be derived from a suitable assumption about the variation of L_λ as a function of λ. (8.2.28) of course means that $a_1 \neq 0$. If $\frac{a_0}{a_1} > 0$, then there is no solution τ of (8.2.32), whereas for $\frac{a_0}{a_1} < 0$, we have two solutions τ_1, τ_2. We keep $\bar{\mu}$ fixed for the moment and write (8.2.31) as

$$0 = a_0 + a_1 \tau^2 + \Sigma(t, \tau, \bar{\mu}) =: \Phi(t, \tau). \tag{8.2.33}$$

We consider the case $\frac{a_0}{a_1} < 0$ with the solutions τ_1, τ_2 of (8.2.32). As $\Sigma(0, t, \bar{\mu}) = 0$, we have

$$\Phi(0, \tau_i) = 0 \text{ for } i = 1, 2, \tag{8.2.34}$$

whereas

$$\frac{\partial}{\partial \tau} \Phi(0, \tau_i) \neq 0, \text{ because of } a_0, a_1 \neq 0 \text{ and } (8.2.30). \tag{8.2.35}$$

The implicit function theorem then implies the existence of (locally unique) functions

$$\tau_i(t) : (-\varepsilon, \varepsilon) \to \mathbb{R}$$

for $i = 1, 2$, $0 < |t| < \varepsilon$, for some $\varepsilon > 0$ that satisfy

$$\Phi(t, \tau_i(t)) = 0. \tag{8.2.36}$$

We have thus found two solutions $\tau_1(t), \tau_2(t)$ of (8.2.33), hence (8.2.22), hence (8.2.16), hence (8.2.17), i.e. (8.2.1) for $t \neq 0$, for the parameters

$$\lambda_t = \lambda_0 + t^2 \bar{\mu}. \tag{8.2.37}$$

In the other case, $\frac{a_0}{a_1} > 0$, (8.2.30) implies that for sufficiently small $|t| \neq 0$, there is no solution of (8.2.33), i.e. of (8.2.1). Thus, as promised, the bifurcation behaviour in case $\pi(\partial_\alpha^2 u) \neq 0$ ((8.2.28)) is completely described by the simple quadratic equation (8.2.32). Of course, replacing $\bar{\mu}$ by $-\bar{\mu}$ changes the sign of a_0 and thus interchanges the cases $\frac{a_0}{a_1} > 0$ and $\frac{a_0}{a_1} < 0$.

We summarize our result in:

Theorem 8.2.2. *We consider a parameter dependent family of equations*

$$L_\lambda u = 0 \tag{8.2.38}$$

as above,

$$V \times \Lambda \to W$$
$$(u, \lambda) \mapsto L_\lambda u,$$

where V, W are Banach spaces and Λ is an open subset of some Banach space, and $L_\lambda u$ is smooth in u and λ. We suppose that

$$L_{\lambda_0} 0 = 0,$$

and we wish to find the solutions of $L_\lambda u = 0$ in the vicinity of 0 as λ

8.2 The functional analytic approach to bifurcation theory

varies in the vicinity of λ_0.
With
$$J_\lambda(u)v := (D_u L_\lambda(u))v = \frac{d}{dt} L_\lambda(u + tv)_{|t=0},$$
we assume that there is a canonical isomorphism
$$\ker J_\lambda \cong \operatorname{coker} J_\lambda \quad (see\ (8.2.3)), \tag{8.2.39}$$
and we let
$$\pi : V \to \ker J_{\lambda_0} \quad (J_{\lambda_0} = J_{\lambda_0}(0))$$
be a projection as defined above (see (8.2.8)–(8.2.10)). We assume furthermore that
$$\dim \ker J_{\lambda_0} = 1 \quad (see\ (8.2.7)). \tag{8.2.40}$$
We assume that there exists some $\bar{\mu}$ with
$$a_0 := \pi(\partial_\lambda u)\bar{\mu} \quad (= \frac{d}{dt}\pi(u(0, \lambda_0 + t\bar{\mu})))_{|t=0}) \neq 0 \tag{8.2.41}$$
(see Lemma 8.2.2 below), and also
$$2\,a_1 := \pi(\partial_\alpha^2 u) \neq 0 \tag{8.2.42}$$
(nonvanishing of the third variation, see Remark 8.2.1). Then there exist $\varepsilon > 0$ and a variation $\lambda_t = \lambda_0 + t^2 \bar{\mu}$ of λ_0 with the property that for $0 < t < \varepsilon$, there exists a neighbourhood U_t of 0 in V such that the number of solutions $u \in U_t$ of
$$L_{\lambda_t} u = 0 \tag{8.2.43}$$
equals the number of solutions of the quadratic equation
$$a_0 + a_1 \tau^2 = 0. \tag{8.2.44}$$
q.e.d.

Remark 8.2.2. Since $\ker J_{\lambda_0}$, the image of π, is assumed to be one-dimensional, we have simply considered $\pi(\partial_\lambda u)$, $\pi(\partial_\alpha^2 u)$ as scalar quantities.

We now consider the case where
$$\pi(\partial_\alpha^2 u) = 0, \tag{8.2.45}$$
but
$$\pi(\partial_\alpha^3 u) \neq 0. \tag{8.2.46}$$

(8.2.23) then becomes

$$0 = \pi(\partial_\lambda u)\mu + \frac{1}{6}\pi(\partial_\alpha^3 u)\alpha^3 + \pi(\partial_{\alpha,\lambda}^2 u)\alpha\mu + \text{higher order terms.} \quad (8.2.47)$$

For a complete description of the bifurcation behaviour, this time we need to consider a two parameter variation. We assume that there exist μ_1, μ_2 with

$$\pi(\partial_\lambda u)\mu_1 \neq 0 \quad \text{(see Lemma 8.2.2 below)} \quad (8.2.48)$$

and

$$\pi(\partial_{\alpha,\lambda}^2 u)\mu_2 \neq 0, \text{ but } \pi(\partial_\lambda u)\mu_2 = 0. \quad (8.2.49)$$

We put $\alpha := t\tau$, $\mu = t^3 b_1 \mu_1 + t^2 b_2 \mu_2$, with parameters b_1, b_2, and rewrite (8.2.47) as

$$0 = t^3(\pi(\partial_\lambda u)\mu_1 b_1 + \pi(\partial_{\alpha,\lambda}^2 u)\mu_2 b_2 \tau \quad (8.2.50)$$

$$+ \frac{1}{6}\pi(\partial_\alpha^3 u)\tau^3 + \Sigma(t, \tau, \mu_1, \mu_2))$$

$$=: c_0 t^3(a_0 + a_1\tau + \tau^3 + \Sigma(t, \tau, \mu_1, \mu_2)),$$

with $c_0 = \frac{1}{6}\pi(\partial_\alpha^3 u) \neq 0$ (see (8.2.46)). For $t \neq 0$, this is equivalent to

$$0 = a_0 + a_1\tau + \tau^3 + \Sigma(t, \tau, \mu_1, \mu_2). \quad (8.2.51)$$

Again

$$\Sigma(t, \tau, \mu_1, \mu_2) = O(t) \text{ as } t \to 0, \text{ for fixed } \tau, \mu_1, \mu_2. \quad (8.2.52)$$

As before, we may thus invoke the implicit function theorem to conclude that the qualitative description of the bifurcation behaviour is furnished by the solution structure of the cubic equation

$$0 = a_0 + a_1\tau + \tau^3. \quad (8.2.53)$$

In particular, locally there exist at most three solutions. We summarize our result in:

Theorem 8.2.3. *As in Theorem 8.2.2, assume the general conditions (8.2.38)–(8.2.40). Furthermore, we assume that there exist parameter variations μ_1, μ_2 with*

$$\pi(\partial_\lambda u)\mu_1 \neq 0, \quad (8.2.54)$$

$$\pi(\partial_{\alpha,\lambda}^2 u)\mu_2 \neq 0, \text{ but } \pi(\partial_\lambda u)\mu_2 = 0 \quad \text{(see (8.2.48), (8.2.49)).} \quad (8.2.55)$$

8.2 The functional analytic approach to bifurcation theory

Then there exist $\varepsilon > 0$ and a two-parameter variation of λ_0,

$$\lambda_t = \lambda_0 + t^3 b_1 \mu_1 + t^2 b_2 \mu_2, \tag{8.2.56}$$

such that for $0 < t < \varepsilon$, there exists a neighbourhood U_t of 0 in V for which the number of solutions $u \in U_t$ of

$$L_{\lambda_t} u = 0 \tag{8.2.57}$$

equals the number of solutions of the cubic equation

$$a_0 + a_1 \tau + \tau^3 = 0. \tag{8.2.58}$$

$(a_0 = 6\pi(\partial_\lambda u) b_1 \mu_1 / \pi(\partial_\alpha^3 u), a_1 = 6\pi(\partial_{\alpha,\lambda}^2 u) b_2 \mu_2 / \pi(\partial_\alpha^3 u)$, noting Remark 8.2.2 again.) q.e.d.

What we are seeing in Theorem 8.2.3 is the so-called cusp catastrophe (in the language of R. Thom's theory of catastrophes), the bifurcation of the zero set of a cubic polynomial depending on the parameters a_0, a_1. In the same manner, one may also identify conditions where the bifurcation behaviour is described by other so-called elementary catastrophes, as classified by R. Thom (see e.g. Th. Bröcker, *Differentiable Germs and Catastrophes*, LMS Lect. Notes 17, Cambridge Univ. Press, Cambridge, 1975). The higher the order of the polynomial involved, however, the more independent parameters one needs. The general idea is that the singular behaviour at a bifurcation point, in particular the nonsmooth structure of the solution set at such a point, is simply the result of the projection of a smooth hypersurface in the product of the solution space and the parameter space onto the solution space. The singularity arises because that hypersurface happens to have a vertical tangent plane over the solution space at the bifurcation point.

In order to discuss the assumption (8.2.41), (8.2.54), we provide

Lemma 8.2.2. *Assume that for every $\beta \in \mathbb{R}$, there exists some μ with*

$$\beta = \pi(D_\lambda L_{\lambda_0} u(0, \lambda_0))(\mu) \; (:= \pi(\frac{1}{dt} L_{\lambda_0 + t\mu} u(0, \lambda_0))_{|t=0}). \tag{8.2.59}$$

(Again, we write β in place of $\beta \xi_0$ and consider it as a scalar quantity, as the image of π is assumed to be one-dimensional.) Then for every $\beta \in \mathbb{R}$, there exists some μ with

$$\pi((\partial_\lambda u)\mu) = \beta. \tag{8.2.60}$$

Proof. We abbreviate

$$\lambda_t = \lambda_0 + t\mu \; (\text{ as } \Lambda \text{ is open, } \lambda_0 + t\mu \in \Lambda \text{ for sufficiently small } |t|).$$

By (8.2.14)
$$L_{\lambda_t} u(\xi, \lambda_t) + \pi(u(\xi, \lambda_t)) = \xi.$$
Taking the derivative w.r.t. t at $t = 0$ and $\xi = 0$ gives
$$\pi((\partial_\lambda u)\mu) = -\frac{d}{dt}(L_{\lambda_t} u(0, \lambda_t))_{|t=0}$$
$$= -(\frac{d}{dt} L_{\lambda_t}) u(0, \lambda_0)_{|t=0}$$
$$- (D_u L_{\lambda_0}) \frac{d}{dt} u(0, \lambda_t)_{|t=0}.$$
Since $D_u L_{\lambda_0} = J_{\lambda_0}$, and $\pi \circ J_{\lambda_0} = 0$ by definition of π, applying π to both sides of the preceding equation gives
$$\pi((\partial_\lambda u)\mu) = -\pi(D_\lambda L_{\lambda_0}) u(0, \lambda_0),$$
and by assumption (8.2.59), we may find μ for which the right-hand side becomes $\beta \xi_0$. (We take $-\beta$ in place of β in (8.2.59).)

q.e.d.

The approach to bifurcation theory presented here originated with L. Lichtenstein, Untersuchung über zweidimensionale reguläre Variationsprobleme, *Monatsh. Math. Phys.* **28** (1917), and was developed in X. Li-Jost, Eindeutigkeit und Verzweigung von Minimalflächen, Thesis, Bonn, 1991, see also X. Li-Jost, Bifurcation near solutions of variational problems with degenerate second variation, *Manuscr. Math.* **86** (1995), 1–14, J. Jost, X. Li-Jost, X. W. Peng, Bifurcation of minimal surfaces in Riemannian manifolds, *Trans. AMS* **347** (1995), 51–62, Correction ibid. **349** (1997), 4689–90.

The reduction of a bifurcation problem in an infinite dimensional setting to a finite dimensional one is an example of the Lyapunov–Schmid reduction which we now wish to discuss.

As before, we consider a parameter dependent family of equations
$$L_\lambda u = 0 \tag{8.2.61}$$
with
$$V \times \Lambda \to W$$
$$(u, \lambda) \mapsto L_\lambda u.$$
(V, W Banach spaces, Λ an open subset of some Banach space) near (u_0, λ_0) with
$$L_{\lambda_0} u_0 = 0. \tag{8.2.62}$$

8.2 The functional analytic approach to bifurcation theory

Again, we assume that $J_\lambda(u) = D_u L_\lambda(u)$ is a Fredholm operator. Thus
$$V_0 := \ker J_{\lambda_0}(u_0)$$
is finite dimensional, and we have decompositions
$$V = V_0 \oplus V_1 \tag{8.2.63}$$
$$W = W_0 \oplus W_1, \text{ with } W_1 = R(J_{\lambda_0}(u_0)), W_0 \text{ finite dimensional.} \tag{8.2.64}$$

(R denotes the range of an operator as in Definition 2.3.1.) We let
$$\pi : W \to W_0$$
be the projection defined by the decompostion (8.2.64). Then our equation $L_\lambda u = 0$ is equivalent to
$$\pi L_\lambda u = 0 \text{ and } (Id - \pi) L_\lambda u = 0. \tag{8.2.65}$$

We first consider
$$(Id - \pi) : V_1 \times V_0 \times \Lambda \to W_1,$$
and with $X := V_0 \times \Lambda$, we write
$$L_\lambda(v'' + v') = g(v'', v', \lambda) \quad \text{with } v' \in V_0, v'' \in V_1, \lambda \in \Lambda.$$
Then, at $v'' + v' = u_0$,
$$D_{v''} g(v'', v', \lambda_0) = D_{v''} L_{\lambda_0}(v'' + v') : V_1 \to W_1$$
is an isomorphism by definition of V_1, W_1; namely it is simply $J_{\lambda_0}(u_0)$, considered as a map from V_1 to W_1. Therefore, by the implicit function Theorem 2.4.1, near (n_0, λ_0), we may find a unique
$$v'' = \varphi(v', \lambda)$$
with
$$(Id - \pi) L_\lambda(v' + \varphi(v', \lambda)) = 0. \tag{8.2.66}$$
Thus $u = v' + \varphi(v', \lambda)$ solves $L_\lambda u = 0$ if and only if
$$\pi L_\lambda(v' + \varphi(v', \lambda)) = 0. \tag{8.2.67}$$

Equation (8.2.67) is a finite dimensional system of equations, because the image of π, W_0, is finite dimensional. This is a Lyapunov–Schmid reduction, and we have seen an instance of this in detail in the preceding for the case where V_0 and W_0 are one-dimensional. A general reference for this and other topics and methods in bifurcation theory is S. N. Chow, J. Hale, *Methods of Bifurcation Theory*, Springer, New York, 1982.

8.3 The existence of catenoids as an example of a bifurcation process

We consider the variational problem

$$I(u) = \int_a^b F(t, u(t), \dot{u}(t))\, dt \tag{8.3.1}$$

with

$$F(t, u, \dot{u}) = u\sqrt{1 + \dot{u}^2}. \tag{8.3.2}$$

This variational problem is of the type considered in Section 1.1 of Part I. $I(u)$ with F given by (8.3.2) is the area of the surface of revolution obtained by rotating the curve $u(t)$, $a \leq t \leq b$, about the t-axis. Critical points are so-called minimal surfaces of revolution. According to Theorem 1.1.1 of Part I, the corresponding Euler–Lagrange equation is computed as

$$\frac{d}{dt} F_p\Big(t, u(t), \dot{u}(t)\Big) - F_u\Big(t, u(t), \dot{u}(t)\Big)$$
$$= F_{pp}\Big(t, u(t), \dot{u}(t)\Big)\ddot{u}(t) + F_{pu}\Big(t, u(t), \dot{u}(t)\Big)\dot{u}(t)$$
$$+ F_{pt}\Big(t, u(t), \dot{u}(t)\Big) - F_u\Big(t, u(t), \ddot{u}(t)\Big)$$
$$= 0$$

which in the present case becomes

$$\frac{d}{dt}\left(\frac{u\dot{u}}{\sqrt{1+\dot{u}^2}}\right) - \sqrt{1+\dot{u}^2} =$$
$$\frac{u\ddot{u}}{(\sqrt{1+\dot{u}^2})^3} + \frac{\dot{u}^2}{\sqrt{1+\dot{u}^2}} - \sqrt{1+\dot{u}^2} = 0, \tag{8.3.3}$$

or equivalently

$$u\ddot{u} - \dot{u}^2 - 1 = 0. \tag{8.3.4}$$

By (1.1.7) of Part I, we have $F - \dot{u}F_p = \text{constant}$, since $F_t \equiv 0$, hence in our case $F = u\sqrt{1 + \dot{u}^2}$,

$$\frac{u}{\sqrt{1+\dot{u}^2}} \equiv \text{constant} =: \lambda.$$

Therefore, for $\lambda \neq 0$, the general solution of (8.3.4) is

$$u(t) = \lambda \cosh\left(\frac{t - t_0}{\lambda}\right)$$

8.3 Example: bifurcation of catenoids

with parameters λ, t_0. Here $\lambda \neq 0$, and we may assume $\lambda > 0$ as the case $\lambda < 0$ is symmetric to the case $\lambda > 0$. Also, since t_0 just represents a translation of the independent variables, we may assume $t_0 = 0$, i.e.

$$u(t) = \lambda \cosh\left(\frac{t}{\lambda}\right). \tag{8.3.5}$$

The curve $u(t)$ is called a catenary, and the minimal surface of revolution obtained by revolving $u(t)$ about the t-axis is called a catenoid. For the sake of normalization, we consider the interval $I = [-1, 1]$. In order to use the general theory of Section 8.2, we need to choose appropriate Banach spaces V, W and $\Lambda = \mathbb{R}$ and consider the operator

$$L_\lambda : V \times \Lambda \to W \tag{8.3.6}$$

$$(u, \lambda) \mapsto \left(\frac{d}{dt}\left(\frac{u\dot{u}}{\sqrt{1+\dot{u}^2}}\right) - \sqrt{1+\dot{u}^2}, u(1) - \lambda \cosh \tfrac{1}{\lambda}, u(-1)\right.$$
$$\left. - \lambda \cosh \tfrac{1}{\lambda}\right).$$

On the right hand side, we have a differential operator of second order and a Dirichlet boundary condition. The boundary values are real numbers, and so W should contain $\mathbb{R}^2$ as a factor as we have two boundary points. Otherwise, V and W should differ by two orders of differentiability. Thus, possible choices are Sobolev spaces

$$V = W^{k+2,p}(I), \quad W = W^{k,p}(I) \times \mathbb{R}^2 \text{ for some } k, p$$

or spaces of differentiable functions

$$V = C^{k+2}(I), \quad W = C^k(I) \times \mathbb{R}^2.$$

Here, we shall take

$$V = W^{2,2}(I), \quad W = L^2(I) \times \mathbb{R}^2, \tag{8.3.7}$$

but the reader should also convince herself or himself that the other choices work as well, although the space L^2 will always play some auxiliary rôle.

In the sequel, we shall denote the scalar product in $L^2(I)$ by $(\cdot, \cdot)_{L^2}$, i.e.

$$(w_1, w_2)_{L^2} = \int_I w_1(t) w_2(t)\, dt$$

for $w_1, w_2 \in L^2(I)$. The scalar product on $W = L^2(I) \times \mathbb{R}^2$ for $w_1 = (w_1, s_1)$, $w_2 = (w_2, s_2)$ ($w_1, w_2 \in L^2(I)$, $s_1, s_2 \in \mathbb{R}^2$),

$$(w_1, w_2)_W = (w_1, w_2)_{L^2} + s_1 \cdot s_2$$

is obtained from the scalar products on $L^2(I)$ and on $\mathbb{R}^2$.

The Jacobi operator is given by

$$J_\lambda(u)v = D_u L_\lambda(u)v$$
$$= \left(\left(\frac{1}{\cosh^2 \frac{t}{\lambda}} \dot{v}\right)^{\bullet} + \frac{1}{\lambda^2} \frac{1}{\cosh^2 \frac{t}{\lambda}} v, v(1), v(-1)\right) \quad (8.3.8)$$

by (8.3.5). In order to determine the kernel of $J_\lambda(u)$, we need to solve the equation

$$J_\lambda(u)v = 0. \quad (8.3.9)$$

This is equivalent to

$$\ddot{v}(t) - \tfrac{2}{\lambda} \tanh \tfrac{t}{\lambda} \dot{v}(t) + \tfrac{1}{\lambda^2} v(t) = 0 \quad (8.3.10)$$
$$v(-1) = v(1) = 0. \quad (8.3.11)$$

The space of solutions of (8.3.10) is generated by

$$v_1(t) = -\sinh \tfrac{t}{\lambda}$$
$$v_2(t) = \cosh \tfrac{t}{\lambda} - \tfrac{t}{\lambda} \sinh \tfrac{t}{\lambda}.$$

(These solutions are simply obtained by differentiating the general solution $\lambda \cosh\left(\frac{t-t_0}{\lambda}\right)$ of (8.3.4) w.r.t. the parameters λ and t_0 (at $t_0 = 0$), cf. Theorem 1.3.3 of Part I.) The boundary condition (8.3.11) cannot be satisfied by v_1, and so we have to find out for which values of λ

$$v(t) := v_2(t) = \cosh \tfrac{t}{\lambda} - \tfrac{t}{\lambda} \sinh \tfrac{t}{\lambda} \quad (8.3.12)$$

satisfies $v(1) = v(-1) = 0$. This is the case precisely if

$$\lambda = \tanh \lambda. \quad (8.3.13)$$

We agreed above to consider only positive values of λ, and this equation has precisely one positive solution which we denoted by λ_0, and likewise, we put

$$u_0(t) = \lambda_0 \cosh\left(\frac{t}{\lambda_0}\right),$$

cf. (8.3.5).

8.3 Example: bifurcation of catenoids

The only solutions of (8.3.10), (8.3.11) are $\alpha v(t)$ with $\alpha \in \mathbb{R}$ and $v(t)$ given in (8.3.12), and so we have

$$\dim \ker J_{\lambda_0}(u_0) = 1. \tag{8.3.14}$$

We call v a weak solution of the Jacobi equation if

$$\int_I v(t) \left(\frac{1}{\cosh^2 \frac{t}{\lambda}} \dot{\eta}(t) \right)^{\cdot} dt + \int_I \frac{1}{\lambda^2} \frac{1}{\cosh^2 \frac{t}{\lambda}} v(t) \eta(t) \, dt = 0 \tag{8.3.15}$$

for all $\eta \in C_0^\infty(I)$.

In the sequel, we shall need a little regularity result, namely that any solution v of (8.3.15) of class $L^2(I)$ is automatically smooth, in fact of class $C^\infty(I)$. As we are dealing with a one-dimensional problem here, this result is not too hard to demonstrate, but since that would lead us too far astray, we omit the proof. It can be found in most good books on differential equations or functional analysis, e.g. K. Yosida, *Functional Analysis*, Springer-Verlag, Berlin, 5th edition, 1978, pp. 177–82.

Of course, if v is of class C^2, (8.3.15) is equivalent to

$$\int_I \left\{ \left(\frac{1}{\cosh^2 \frac{t}{\lambda}} \dot{v}(t) \right)^{\cdot} + \frac{1}{\lambda^2} \frac{1}{\cosh^2 \frac{t}{\lambda}} v(t) \right\} \eta(t) \, dt = 0 \tag{8.3.16}$$

for all $\eta \in C_0^\infty(I)$,

and by Lemma 1.1.1 of Part I, this is equivalent to v being a solution of the Jacobi equation.

We shall now identify $\ker J_{\lambda_0}(u_0)$ and $\operatorname{coker} J_{\lambda_0}(u_0)$ as required in (8.2.3). We shall simply write J_{λ_0} in place of $J_{\lambda_0}(u_0)$. According to our choice (8.3.7), we consider J_{λ_0} as an operator

$$J_{\lambda_0} : W^{2,2}(I) \to L^2(I) \times \mathbb{R}^2.$$

If $w \in R_0(J_{\lambda_0}) := R(J_{\lambda_0}|_{H_0^{2,2}(I)})$, i.e. if there exists $v \in H_0^{2,2}(I)$ with $J_{\lambda_0} v = w$, then for all $\varphi \in \ker J_{\lambda_0}$

$$(w, \varphi)_W = (J_{\lambda_0} v, \varphi)_{L^2} = (v, J_{\lambda_0} \varphi)_{L^2}$$
$$= 0$$

(in the same manner as the equivalence of (8.3.15) and (8.3.16) and noting that φ is smooth and v and φ both vanish on ∂I.) Thus if $w \in R_0(J_{\lambda_0})$, then also $w \in (\ker J_{\lambda_0})^\perp$, where $\perp$ denotes the orthogonal complement in the Hilbert space $L^2(I)$, as in Corollary 2.2.4. Consequently, if we denote the closure of a linear subspace M of $L^2(I) \times \mathbb{R}^2$

by $\overline{M}$, then also
$$\overline{R_0(J_{\lambda_0})} \subset (\ker J_{\lambda_0})^\perp.$$
Conversely, if $w \in R_0(J_{\lambda_0})^\perp (= (\overline{R_0(J_{\lambda_0})})^\perp)$, then
$$(w, J_{\lambda_0} v)_W = 0 \quad \text{for all } v \in H_0^{2,2}(I).$$
By the regularity result mentioned above, this implies that w is smooth, and so we may integrate by parts to get
$$(w, J_{\lambda_0} v)_W = (J_{\lambda_0} w, v)_W \quad \text{for all } v \in H_0^{2,2}(I)$$
and hence $w \in \ker J_{\lambda_0}$. Altogether, we have shown that
$$\ker J_{\lambda_0} = R_0(J_{\lambda_0})^\perp.$$
Since, according to Corollary 2.2.4, we have the decomposition
$$L^2(I) = \overline{R_0(J_{\lambda_0})}^{L^2} \oplus R_0(J_{\lambda_0})^\perp,$$
we may also consider $R_0(J_{\lambda_0})^\perp$ as coker J_{λ_0}, and so we get the required identification
$$\ker J_{\lambda_0} \cong \operatorname{coker} J_{\lambda_0}. \tag{8.3.17}$$
We note that this depends on the fact that J_{λ_0} is formally self-adjoint in the sense that
$$(v, J_{\lambda_0} w) = (J_{\lambda_0} v, w) \tag{8.3.18}$$
if e.g. $v, w \in H_0^{2,2}(I)$.

Remark 8.3.1. The situation here is slightly different from the one in Section 8.2 inasmuch as we identify coker J_{λ_0} here with $R_0(J_{\lambda_0})^\perp$ and not with $R(J_{\lambda_0})^\perp$. Therefore, in the present situation, if π denotes the orthogonal projection onto $\ker J_{\lambda_0} \cong \operatorname{coker} J_{\lambda_0}$, we have
$$\pi(J_{\lambda_0} v) = 0$$
only for $v \in H_0^{2,2}(I)$, but not for all $v \in H_0^{2,2}(I)$. This is for example relevant for the argument of the proof of Lemma 8.2.2.

Regularity theory also implies that $R(J_{\lambda_0})$ is closed. Namely, if for $(v_n)_{n \in \mathbb{N}} \subset (\ker J_{\lambda_0})^\perp \subset W^{2,2}(I)$, we have
$$J_{\lambda_0} v_n = f_n,$$
and f_n converges to f_0 in $L^2(I) \times \mathbb{R}^2$, then $\|v_n\|_{W^{2,2}(I)}$ is bounded.

8.3 Example: bifurcation of catenoids

By Rellich's Theorem 3.4.1, after selection of a subsequence, v_n then converges in $W^{1,2}(I)$. From the equation, i.e.

$$\frac{1}{\cosh^2 \frac{t}{\lambda}} \ddot{v}_n + \frac{d}{dt}\left(\frac{1}{\cosh^2 \frac{t}{\lambda}}\right) \dot{v}_n + \frac{1}{\lambda^2} \frac{1}{\cosh^2 \frac{t}{\lambda}} v_n = f_n,$$

we then see that $\ddot{v}_n$ also converges in $L^2(I)$. Thus, v_n converges in $W^{2,2}(I)$, and the limit v_0 then satisfies

$$J_{\lambda_0} v_0 = f_0.$$

Thus, the image of J_{λ_0} is closed. Thus, J_{λ_0} is a Fredholm operator of index 0.

Our aim is now to check that the assumptions of Theorem 8.2.2 hold. In order to verify (8.2.42), i.e. $\pi(\partial_\alpha^2 u) \neq 0$, according to Remark 8.2.1, we need to compute dJ_{λ_0}, i.e. the second derivative of L_{λ_0}. Starting from (8.3.3) and inserting (8.3.6), i.e. $u_0 = \lambda_0 \cosh t/\lambda_0$, we obtain

$$dJ_{\lambda_0}(u_0)(v,v) = \frac{d}{dt}\left(\frac{2}{\cosh^3 \frac{t}{\lambda}} v\dot{v} - \frac{3\lambda \tanh \frac{t}{\lambda}}{\cosh^3 \frac{t}{\lambda}} \dot{v}^2\right) - \frac{1}{\cosh^3 \frac{t}{\lambda}} \dot{v}^2.$$

By (8.2.27), we have to project this onto the kernel of $J_{\lambda_0}(u_0)$ and check that the result is nonzero, for our Jacobi field v given in (8.3.12), i.e. $v = \cosh t/\lambda - t/\lambda \sinh t/\lambda$. Since here the projection π is given by the orthogonal projection in the Hilbert space $L^2(I) \times \mathbb{R}^2$ onto $\ker J_{\lambda_0}(u_0)$, which is generated by the Jacobi field v, we simply have to verify that the L^2-product of $dJ_{\lambda_0}(u_0)(v,v)$ with v is nonzero. Thus, we compute

$$(dJ_{\lambda_0}(u_0)(v,v), v)_{L^2} = \int_I \left\{\frac{d}{dt}\left(\frac{2}{\cosh^3 \frac{t}{\lambda}} v(t)\dot{v}(t) - \frac{3\lambda \tanh \frac{t}{\lambda}}{\cosh^3 \frac{t}{\lambda}} \dot{v}(t)^2\right)\right.$$
$$\left. - \frac{1}{\cosh^3 \frac{t}{\lambda}} \dot{v}(t)^2\right\} v(t)\, dt$$

$$= \int_I \left\{\frac{3\lambda \tanh \frac{t}{\lambda}}{\cosh^3 \frac{t}{\lambda}} \dot{v}(t)^3 - \frac{3}{\cosh^3 \frac{t}{\lambda}} \dot{v}(t)^2 v(t)\right\} dt$$

by an integration by parts

$$= \int_I \frac{3\dot{v}(t)^2}{\cosh^3 \frac{t}{\lambda}} \left(\lambda \tanh \frac{t}{\lambda} \dot{v}(t) - v(t)\right) dt.$$

Now with $v = \cosh \frac{t}{\lambda} - \frac{t}{\lambda} \sinh \frac{t}{\lambda}$, we have

$$\lambda \tanh \frac{t}{\lambda} \dot{v}(t) - v(t) = -\cosh \frac{t}{\lambda},$$

and so
$$(dJ_{\lambda_0}(u_0)(v,v),v)_{L^2} = -3\int_I \frac{\dot{v}(t)^2}{\cosh^2 \frac{t}{\lambda}} < 0.$$

Thus, indeed
$$\pi(\partial_\alpha^3 u) > 0. \tag{8.3.19}$$

We finally consider (8.2.41). Thus, we have to verify that

$$\pi(\partial_\lambda u) \neq 0, \text{ with } \partial_\lambda u = \frac{d}{dt}u(0,\lambda_t)_{|t=0} \text{ for a suitable family } \lambda_t \text{ of parameters.} \tag{8.3.20}$$

We start with (8.2.14), i.e. in the notations of Section 8.2
$$L_{\lambda_t} u(\xi, \lambda_t) + \pi(u(\xi, \lambda_t)) = \xi. \tag{8.3.21}$$

In the present case L_λ is given by (8.3.6), and π is the orthogonal projection in $L^2(I)$ onto $\ker J_{\lambda_0}$, the one-dimensional space generated by $v(t) = \cosh t/\lambda_0 - t/\lambda_0 \sinh t/\lambda_0$ (see (8.3.12)), where λ_0 is so chosen that $v(1) = v(-1) = 0$. Thus, this v can be taken as the ξ_0 of Section 8.2. However, since

$$\frac{d}{d\lambda}\left(\lambda \cosh \tfrac{1}{\lambda}\right)_{|\lambda=\lambda_0} = 0$$

by choice of λ_0 (see (8.3.13)), we shall need to employ a variation of the parameter somewhat different from the family $\lambda_t = \lambda + t\mu$ employed in Section 8.2. Here, we put

$$\nu_0 := \lambda_0 \cosh \tfrac{1}{\lambda_0}$$

and choose the family λ_t such that
$$\lambda_t \cosh \tfrac{1}{\lambda_t} = \nu_0 + t\mu. \tag{8.3.22}$$

We then differentiate (8.3.21) w.r.t. t at $t = 0$, $\xi = 0$ to obtain
$$J_{\lambda_0}(u_0)\partial_\lambda u + \frac{1}{\|v\|^2_{L^2(I)}}(\partial_\lambda u, v)_{L^2} v = 0 \tag{8.3.23}$$
$$\partial_\lambda u(1) = \partial_\lambda u(-1) = \mu.$$

Then
$$(J_{\lambda_0}(u_0)\partial_\lambda u, v)_{L^2} = \int \left(\frac{1}{\cosh^2 \frac{t}{\lambda_0}}(\partial_\lambda u(t))^\bullet\right)^\bullet$$
$$+ \frac{1}{\lambda_0^2}\frac{1}{\cosh^2 \frac{t}{\lambda_0}}\partial_\lambda u(t))v(t)\,dt$$

$$= -\frac{1}{\cosh^2 \frac{t}{\lambda_0}} \partial_\lambda u(t) \dot{v}(t)\Big|_{-1}^{1},$$

integrating by parts and using $J_{\lambda_0}(u_0)v = 0$ $v(1) = 0 = v(-1)$.

$$= \frac{2}{\lambda_0 \cosh \frac{1}{\lambda_0}} \mu$$

$\gtreqless 0$ for $\mu \gtreqless 0$.

Equation (8.3.23) then implies

$$\pi(\partial_\lambda u) = \frac{1}{\|v\|^2_{L^2(I)}} (\partial_\lambda u, v) v \neq 0,$$

i.e. (8.3.20).

We thus have verified all the assumptions of Theorem 8.2.2 (for the family λ_t defined by (8.3.22) in place of the family $\lambda_t = \lambda_0 + t\mu$). Theorem 8.2.2 thus describes the bifurcation behaviour of the solutions of (8.3.3) or (8.3.4), i.e. the critical points of (8.3.1), (8.3.2) near $u_0(t) = \lambda_0 \cosh \frac{t}{\lambda_0}$: For boundary values $u(1) = u(-1) < \lambda_0 \cosh \frac{1}{\lambda_0}$, there is no solution (at least in the vicinity of u_0), whereas for $u(1) = u(-1) > \lambda_0 \cosh \frac{1}{\lambda_0}$, we may find two solutions. Of course, this may also be verified directly without going through all the abstract machinery of Section 8.2, but hopefully this example can serve to illustrate the general scheme. The catenoids are frequently discussed in books on the calculus of variations, e.g. O. Bolza, *Vorlesungen über Variationsrechnung*, Teubner, Leipzig, Berlin, 1909, or M. Giaquinta, St. Hildebrandt, *Calculus of Variations*, Springer, Berlin, 1996, I, p. 366 and II, pp. 263–70. A discussion in terms of bifurcation theory also in the case of not necessarily symmetric boundary conditions (i.e. not requiring $u(1) = u(-1)$) is given by H. Wenk, Extremverhalten der Stabilität von Catenoiden als Rotationsminimalfläche, Diplom thesis, Bochum, 1994.

Exercises

8.1 How many parameters are needed for a complete description of the bifurcation behaviour of the roots of a fourth-order polynomial?

8.2 Consider the problem of finding critical points

$$I(u) = \int u(t)\sqrt{1+\dot{u}(t)^2}\,dt$$

$$u(\kappa) = u(-\kappa) = 1$$

for a parameter $\kappa > 0$. Determine the value κ_0 for which a bifurcation occurs.

(Hint: This problem can be reduced to the one considered in Section 8.3.)

8.3 Consider geodesics on S^2 as in Chapter 2 of Part I. More precisely, we take two points $p, q \in S^2$ with distance $d(p,q) = \lambda$, and consider geodesic arcs between p and q of length λ, i.e. length minimizing arcs. What happens at $\lambda = \pi$? Does this fit into the framework described in Section 8.2?

9
The Palais–Smale condition and unstable critical points of variational problems

9.1 The Palais–Smale condition

In this chapter, we take up a direction that has already been presented in Chapter 3 of Part I, namely the search for nonminimizing critical points of variational problems. This chapter will consequently be independent of Chapters 4–8 of the present Part II. In Section 3.1 of Part I, we presented existence results for unstable critical points of functionals F of class C^1 on some finite dimensional Euclidean space $\mathbb{R}^d$. We only needed a coercivity condition on the functional guaranteeing that a critical sequence $(x_n)_{n \in \mathbb{N}}$ (i.e. satisfying $DF(x_n) \to 0$, $|F(x_n)|$ bounded) stayed in a bounded set. The local compactness of $\mathbb{R}^d$ then allowed the extraction of a convergent subsequence whose limit x_0 satisfied $DF(x_0) = 0$, because of the continuity of DF. In Sections 2.3 and 3.2 of Part I, we also presented examples where variational problems could be reduced to such finite dimensional problems. The domain was a little more complicated than $\mathbb{R}^d$, but being finite dimensional, it was still locally compact so that we had no difficulties finding limits of subsequences for critical sequences. In the remainder of this book, however, we have had ample opportunity to realize that variational problems are typically and naturally posed on some infinite-dimensional Hilbert or Banach space. Such a space is not locally compact anymore w.r.t. its Hilbert or Banach space topology, so that the previous strategy encounters a serious problem. Also weak topologies do not help much as the functionals under consideration typically are not continuous w.r.t. the weak topology. If one searches for minimizers, this problem can be overcome by introducing convexity assumptions as we have seen in Chapters 4 and 8, but any convexity assumption excludes the existence of critical points other than minima.

Nevertheless, the lack of compactness of the underlying space must be compensated by an assumption on the functional that guarantees the appropriate compactness of critical sequences. In other words we do not require the compactness of arbitrary bounded sequences on our space — which is impossible as argued — but only of critical sequences. This is the idea of the **Palais–Smale condition** which we now formulate:

Definition 9.1.1. *Let* $(V, \|\cdot\|)$ *be a Banach space*, $F : V \to \mathbb{R}$ *a functional of class* C^1. *We say that* F *satisfies the* Palais–Smale condition, *abbreviated as* (PS), *if any sequence* $(x_n)_{n \in \mathbb{N}} \subset V$ *satisfying*

(i) $|F(x_n)| \leq c$ *for some constant* c

(ii) $\|DF(x_n)\| \to 0$ *for* $n \to \infty$

contains a convergent subsequence.

Note that a limit x_0 of such a subsequence satisfies $DF(x_0) = 0$ (i.e. is a critical point of F) because DF is continuous.

A direct consequence of the definition is:

Lemma 9.1.1. *Suppose* $F : V \to \mathbb{R}$ *satisfies* (PS). *Then for any* $\alpha \in \mathbb{R}$

$$K_\alpha := \{x \in V : F(x) = \alpha, DF(x) = 0\}$$

(*the set of critical points of* F *with value* α) *is compact.*

q.e.d.

We also have:

Lemma 9.1.2. *Suppose* $F : V \to \mathbb{R}$ *satisfies* (PS). *For* $\alpha \in \mathbb{R}$, *we put*

$$U_{\alpha,\rho} := \bigcup_{x \in K_\alpha} \{z \in V \mid \|x - z\| < \rho\} \quad (\rho > 0),$$

$$N_{\alpha,\delta} := \{y \in V \mid |F(y) - \alpha| < \delta,\ \|DF(y)\| < \delta\} \quad (\delta > 0).$$

Then the families $(U_{\alpha,\rho})_{\rho>0}$ *and* $(N_{\alpha,\delta})_{\delta>0}$ *are fundamental systems of neighbourhoods of* K_α (*i.e. each neighbourhood of* K_α *contains some* $U_{\alpha,\rho}$ *and some* $N_{\alpha,\delta}$).

Proof. If is clear that $U_{\alpha,\rho}$ and $N_{\alpha,\delta}$ are neighbourhoods of K_α for $\rho > 0$ respectively $\delta > 0$. It follows from the compactness of K_α that each neighbourhood of K_α contains some $U_{\alpha,\rho}$. Concerning the same property of the $N_{\alpha,\delta}$, let us assume on the contrary that there exist a neighbourhood U of K_α and a sequence $(y_n)_{n \in \mathbb{N}}$ with $y_n \in N_{\alpha, \frac{1}{n}} \setminus (U \cap N_{\alpha, \frac{1}{n}})$

9.1 The Palais-Smale condition

for all n. (PS) implies that a subsequence of $(y_n)_{n \in \mathbb{N}}$ converges to some $y_0 \in K_\alpha \subset U$, contradicting the openness of U.

q.e.d.

In our applications below, we shall also encounter the situation where we want to find critical points of the restriction of some functional F to the level hypersurface $G(x) = \beta$ of some other functional G. For that purpose, we shall need a relative version of the Palais–Smale condition which we shall formulate only for the case of a Hilbert space:

Definition 9.1.2. *Let* $(H, <\cdot,\cdot>)$ *be a Hilbert space,* $F, G : H \to \mathbb{R}$ *functionals of class* C^1, $\beta \in \mathbb{R}$. *Suppose*

$$DG(x) \neq 0 \quad \text{for all } x \text{ with } G(x) = \beta.$$

We say that F satisfies (PS) relative to $G = \beta$ if every sequence $(x_n)_{n \in \mathbb{N}} \subset H$ with $G(x_n) = \beta$ and satisfying

(i) $|F(x_n)| \leq c$ *for some constant* c

(ii) $\left\| DF(x_n) - \dfrac{\langle DF(x_n), DG(x_n) \rangle}{\|DG(x_n)\|^2} DG(x_n) \right\| \to 0 \quad \text{for } n \to \infty$

contains a convergent subsequence.

A limit x_0 of such a subsequence then satisfies

$$G(x_0) = \beta \tag{9.1.1}$$

and

$$DF(x_0) - \frac{\langle DF(x_0), DG(x_0) \rangle}{\|DG(x_0)\|^2} DG(x_n) = 0, \tag{9.1.2}$$

i.e. is a critical point of the restriction of F to $G(x) = \beta$. Of course, results analogous to Lemmas 9.1.1 and 9.1.2 hold in the relative case. One simply intersects the corresponding sets with $\{G(x) = \beta\}$ and replaces DF by its projection to that level set.

As in Sections 2.3, 3.1, 3.2 of Part I, in order to find critical points of a functional, one needs to construct (local) deformations that decrease the value of the functional except at or at least away from critical points. We shall now do so in stages of increasing generality.

We start with a functional

$$F : H \to \mathbb{R}$$

of class C^2 on some Hilbert space $(H, \langle \cdot, \cdot \rangle)$ that satisfies (PS). For each

$u \in H$, $DF(u)$ is a linear functional on H, and by Corollary 2.2.3, it can therefore be identified with an element $\nabla F(u)$ of H, called the gradient of F at u. Thus, $\nabla F(u)$ satisfies

$$DF(u)(\nabla F(u)) = \|DF(u)\|^2 \tag{9.1.3}$$

$$\|\nabla F(u)\| = \|DF(u)\|. \tag{9.1.4}$$

Since F is assumed to be of class C^2, DF and hence ∇F are of class C^1 in their dependence on u. In particular, ∇F is locally Lipschitz.
We now consider the (negative) gradient flow induced by F:

$$\frac{\partial}{\partial t}\psi(u,t) = -\nabla F(\psi(u,t)) \quad \text{for } t \geq 0 \tag{9.1.5}$$

$$\psi(u,0) = u. \tag{9.1.6}$$

Because of the Lipschitz property, by Theorem 2.4.2 and Corollary 2.4.2, for small $t \geq 0$, there exists a unique solution $\psi(u,t)$ satisfying the semigroup property

$$\psi(u, t+s) = \psi(\psi(u,t), s) \tag{9.1.7}$$

for sufficiently small $s, t \geq 0$.
Moreover,

$$\psi(u,t) = u \quad \text{for all } u \text{ with } \nabla F(u) = 0, \text{ i.e. for all critical points of } F. \tag{9.1.8}$$

Finally

$$\begin{aligned}
F(\psi(u,t)) &= F(u) + \int_0^t \frac{d}{d\tau} F(\psi(u,\tau)) d\tau \\
&= F(u) + \int_0^t DF(\psi(u,t)) \frac{\partial}{\partial \tau}\psi(u,\tau) d\tau \\
&= F(u) - \int_0^t \|DF(\psi(u,\tau))\|^2 d\tau \quad \text{by (9.1.5), (9.1.3)} \\
&< F(u) \quad \text{for } t > 0, \text{ if } DF(u) = DF(\psi(u,0)) \neq 0, \quad (9.1.9)
\end{aligned}$$

i.e. if u is not a critical point of F.

Thus, we have found the prototype of a deformation that decreases the value of F except at its critical points. For technical reasons, however, the above flow will need some modifications and generalizations.
First of all, a solution of (9.1.5) need not exist for all $t \geq 0$ because it

9.1 The Palais–Smale condition

may become unbounded in finite 'time' t. This can be easily remedied by using the Lipschitz function

$$\eta : \mathbb{R}^+ \to \mathbb{R}^+$$

$$\eta(s) = \begin{cases} 1 & \text{for } 0 \leq s \leq 1 \\ \dfrac{1}{s} & \text{for } s \geq 1 \end{cases}$$

and putting

$$\tilde{\nabla} F(u) := \eta(\|\nabla F(u)\|) \nabla F(u)$$

(i.e. $\tilde{\nabla} F(u) = \nabla F(u)$ for $\|\nabla F(u)\| \leq 1$ and $\left\|\tilde{\nabla} F(u)\right\| \leq 1$ for all u) and replacing (9.1.5) by

$$\frac{\partial}{\partial t} \psi(u,t) = -\tilde{\nabla} F(\psi(u,t)). \tag{9.1.10}$$

Of course, we still use (9.1.6).

Since $\left\|\tilde{\nabla} F(u)\right\| \leq 1$ for all u, the solution of (9.1.10), (9.1.6) now exists for all $t \geq 0$, and satisfies (9.1.7) for all $s, t \geq 0$. Equation (9.1.8) also still holds, and as in the derivation of (9.1.9), we get

$$F(\psi(u,t)) = F(u) - \int_0^t \eta(\|\nabla F(\psi(u,\tau))\|) \, \|DF(\psi(u,\tau))\|^2 \, d\tau$$
$$< F(u) \quad \text{for } t > 0,$$

if u is not a critical point of F. More generally, we have

$$F(\psi(u,t)) \leq F(\psi(u,s)) \quad \text{whenever } 0 \leq s \leq t, \text{ for all } u.$$

Next, we wish to localize the construction near a level α. Thus, for given $\epsilon_0 > 0$ and a neighbourhood U of K_α we want to have a flow $\psi(u,t)$ with (9.1.7), (9.1.8) and also

$$\psi(u,t) = u \quad \text{if } |F(u) - \alpha| \geq \epsilon_0, \tag{9.1.11}$$

and the following more explicit local decrease of the value of F: For $\alpha \in \mathbb{R}$, we put

$$F_\alpha := \{v \in H \mid F(v) < \alpha\}.$$

We want to find $0 < \epsilon < \epsilon_0$ with

$$\psi(F_{\alpha+\epsilon} \setminus U, 1) \subset F_{\alpha-\epsilon} \tag{9.1.12}$$
$$\psi(U, 1) \subset F_{\alpha-\epsilon} \cup U, \tag{9.1.13}$$

and of course also
$$F(\psi(u,t)) \leq F(\psi(u,s)) \quad \text{if } 0 \leq s \leq t \quad \text{for all } u. \tag{9.1.14}$$

We let $\varphi : \mathbb{R} \to \mathbb{R}$ be Lipschitz continuous with
$$\varphi(s) = 0 \quad \text{for } |\alpha - s| \geq \epsilon_0$$
$$\varphi(s) = 1 \quad \text{for } |\alpha - s| \leq \frac{\epsilon_0}{2}$$
$$0 \leq \varphi(s) \leq 1 \quad \text{for all } s$$

and replace (9.1.10) by
$$\frac{\partial}{\partial t}\psi(u,t) = -\varphi(F(\psi(u,t)))\tilde{\nabla} F(\psi(u,t)). \tag{9.1.15}$$

Again, a solution $\psi(u,t)$ exists for all $t \geq 0$ and satisfies (9.1.7) for all $s,t \geq 0$, as well as (9.1.8) and (9.1.14) (for the latter it was necessary to require $\varphi \geq 0$). (9.1.11) also is clear from the choice of φ. We now verify (9.1.12), (9.1.13). If $0 < \epsilon \leq \frac{\epsilon_0}{2}$ and $u \in F_{\alpha+\epsilon}$ and if $F(\psi(u,1)) \geq \alpha - \epsilon$, from (9.1.14)
$$|F(\psi(u,t)) - \alpha| \leq \epsilon \quad \text{for all } 0 \leq t \leq 1,$$

and therefore
$$\varphi(F(\psi(u,t))) = 1 \quad \text{for all } 0 \leq t \leq 1. \tag{9.1.16}$$

As before, we may now compute
$$F(\psi(u,1))$$
$$= F(u) + \int_0^1 \frac{d}{d\tau} F(\psi(u,\tau))d\tau$$
$$= F(u) - \int_0^1 \varphi(F(\psi(u,\tau)))\eta(\|\nabla F(\psi(u,\tau))\|)\|DF(\psi(u,\tau))\|^2 d\tau$$
$$< \alpha + \epsilon - \int_0^1 \min(1, \|DF(\psi(u,\tau))\|^2)d\tau \tag{9.1.17}$$

since we assume $u \in F_{\alpha+\epsilon}$, using (9.1.16) and the properties of η.

By Lemma 9.1.2, we may find $\delta, \rho > 0$ with
$$N_{\alpha,\delta} \subset U_{\alpha,\rho} \subset U_{\alpha,2\rho} \subset U \tag{9.1.18}$$

(here, we are using (PS)!). From the definition of $N_{\alpha,\delta}$, thus $\|DF(\psi(u,\tau))\|^2 \geq \delta^2$ whenever $\psi(u,\tau) \notin N_{\alpha,\delta}$. Without loss of generality $\delta \leq 1$. (9.1.17) then yields
$$F(\psi(u,1)) < \alpha + \epsilon - (\text{meas } \{0 \leq \tau \leq 1 \mid \psi(u,\tau) \notin N_{\alpha,\delta}\})\delta^2. \tag{9.1.19}$$

9.1 The Palais–Smale condition

From (9.1.18), we have for $v \in H \setminus U$

$$\text{dist}(v, N_{\alpha,\delta}) \left(:= \inf_{w \in N_{\alpha,\delta}} \|v - w\| \right) > \rho.$$

Since

$$\left\| \frac{\partial}{\partial t} \psi(u, t) \right\| \leq 1, \qquad (9.1.20)$$

therefore, if $u \notin U$, then also $\psi(u, \tau) \notin N_{\alpha,\delta}$ for $0 \leq \tau \leq \rho$, and similarly, if $\psi(u, 1) \notin U$, then also $\psi(u, \tau) \notin N_{\alpha,\delta}$ for $1 - \rho \leq \tau \leq 1$. Therefore, if either $u \notin U$ or $\psi(u, 1) \notin U$, then

$$\text{meas} \{0 \leq \tau \leq 1 \mid \psi(u, \tau) \notin N_{\alpha,\delta}\} \geq \rho.$$

Thus, from (9.1.19), if $u \notin U$ or if $\psi(u, 1) \notin U$,

$$F(\psi(u, 1)) < \alpha + \epsilon - \rho \delta^2$$

$$\leq \alpha - \epsilon \quad \text{if we choose } \epsilon \leq \frac{1}{2} \rho \delta^2. \qquad (9.1.21)$$

Thus, for $0 < \epsilon \leq \min(\frac{1}{2}\epsilon_0^2, \frac{1}{2}\rho\delta^2)$, we get (9.1.12), (9.1.13).

In conclusion, we have shown the following deformation result:

Theorem 9.1.1. *Let $F : H \to \mathbb{R}$ be a C^2 functional on a Hilbert space H, satisfying (PS). Let $\alpha \in \mathbb{R}$, and put*

$$F_\alpha := \{v \in H : F(v) < \alpha\},$$

$$K_\alpha := \{v \in H : F(v) = \alpha, DF(v) = 0\}.$$

Let $\epsilon_0 > 0$ and a neighbourhood U of K_α be given. Then there exist $0 < \epsilon < \epsilon_0$ and a continuous family

$$\psi : H \times [0, \infty) \to H$$

with the semigroup property $\psi(\psi(u, s), t) = \psi(u, s + t)$ for all $s, t \geq 0$, $u \in H$ and with

(i) $\psi(u, 0) = u$ for all $u \in H$
(ii) $F(\psi(u, t))$ is nonincreasing in t for all $u \in H$
(iii) $\psi(u, t) = u$ for all t whenever $DF(u) = 0$, in particular for $u \in K_\alpha$
(iv) $\psi(u, t) = u$ whenever $|F(u) - \alpha| \geq \epsilon_0$, for all t
(v) $\psi(F_{\alpha+\epsilon} \setminus U, 1) \subset F_{\alpha-\epsilon}, \; \psi(F_{\alpha+\epsilon}, 1) \subset F_{\alpha-\epsilon} \cup U$
(vi) *If $F(u)$ is even (i.e. $F(u) = F(-u)$ for all u), then also $F(\psi(u, t))$ is even in u for all t (i.e. $F(\psi(u, t)) = F(\psi(-u, t)))$.*

(Property (vi) follows from the construction: All the auxiliary functions are invariant under replacing u by $-u$ if F is even, and $\nabla F(-u) = -\nabla F(u)$ in the even case.)   q.e.d.

Corollary 9.1.1. *If under the preceding assumptions, F has no critical point with value α, i.e. $K_\alpha = \emptyset$, then there exist a deformation ψ with the preceding properties and*

$$\psi(F_{\alpha+\epsilon}, 1) \subset F_{\alpha-\epsilon} \quad \text{for some } \epsilon > 0. \tag{9.1.22}$$

Proof. If $K_\alpha = \emptyset$, we may choose $U = \emptyset$ in Theorem 9.1.1.

q.e.d.

We shall now extend Theorem 9.1.1 in two directions. First, we consider the relative case, where in addition to F, we have another C^2 functional $F : H \to \mathbb{R}$ with

$$DF(x) \neq 0 \quad \text{for all } x \text{ with } G(x) = \beta,$$

for some given value $\beta \in \mathbb{R}$. We wish to find critical points of the restriction of F to $G = \beta$. We assume that F satisfies the relative (PS) condition of Definition 9.1.2 on $G = \beta$. We then perform the preceding construction with

$$\nabla^G F(u) := \nabla F(u) - \frac{\langle \nabla F(u), \nabla G(u) \rangle}{\|\nabla G(u)\|^2} \nabla G(u) \tag{9.1.23}$$

in place of $\nabla F(u)$. We then have

$$\frac{d}{dt} G(\psi(u,t))$$
$$= -\varphi(F(\psi(u,t))) \eta(\|\nabla^G F(u)\|) \langle \nabla^G F(\psi(u,t)), \nabla G(\psi(u,t)) \rangle$$

from the chain rule and the analogue of (9.1.15)

$$= 0,$$

since $\langle \nabla^G F(v), \nabla G(v) \rangle = 0$ for all $v \in H$. Therefore, the flow $\psi(u,t)$ now leaves $G = \beta$ invariant. We obtain:

Theorem 9.1.2. *Let $F, G : H \to \mathbb{R}$ be C^2 functionals on a Hilbert space $(H, \langle \cdot, \cdot \rangle)$ with F satisfying (PS) relative to $G = \beta$. Let $\alpha \in \mathbb{R}$,*

$$F_\alpha^{G,\beta} := \{ v \in H \mid F(v) < \alpha, G(v) = \beta \},$$

$$K_\alpha^{G,\beta} := \{ v \in H \mid F(v) = \alpha, G(v) = \beta, \nabla^G F(v) = 0 \}.$$

9.1 The Palais-Smale condition

Let $\epsilon_0 > 0$, and let U be a neighbourhood of $K_\alpha^{G,\beta}$ in $\{G(v) = \beta\}$. Then there exist $\epsilon > 0$ and a continuous semigroup family

$$\psi : \{G(v) = \beta\} \times [0, \infty) \to \{G(v) = \beta\}$$

satisfying

(i) $\psi(u, 0) = u$ for all $u \in \{G(v) = \beta\}$
(ii) $F(\psi(u, t))$ is nonincreasing in t
(iii) $\psi(u, t) = u$ for all $u \in K_\alpha^{G,\beta}$
(iv) $\psi(u, t) = u$ for all t if $|F(u) - \alpha| \geq \epsilon_0$
(v) $\psi(F_{\alpha+\epsilon}^{G,\beta} \setminus U, 1) \subset F_{\alpha-\epsilon}^{G,\beta}$, $\psi(F_{\alpha+\epsilon}^{G,\beta}, 1) \subset F_{\alpha-\epsilon}^{G,\beta} \cup U$
(vi) If F and G are even, so is $F(\psi(\cdot, t))$ for all t.

Secondly, we wish to extend the preceding construction to functionals on Banach spaces. For a functional on a Banach space, in general one does not have a good notion of a gradient. We therefore need to introduce Palais' concept of a pseudo-gradient:

Definition 9.1.3. *Let $(V, \|\cdot\|)$ be a Banach space, $U \subset V$, $F : U \to \mathbb{R}$ a functional of class C^1. A pseudo-gradient vector field for F is a locally Lipschitz continuous vector field $v : U \to V$ satisfying*

(i) $\|v(u)\| \leq \min(1, \|DF(u)\|)$
(ii) $DF(u)(v) \geq \frac{1}{2} \min(\|DF(u)\|, \|DF(u)\|^2)$

for all $u \in U$.

Lemma 9.1.3. *Let $F : V \to \mathbb{R}$ be a functional of class C^1 on the Banach space V. Then F admits a pseudo-gradient vector field on*

$$V' := \{u \in V \mid DF(u) \neq 0\}.$$

Proof. For each $u \in V'$, we can find $w = w(u)$ with

$$\|w\| < \min(1, \|DF(u)\|) \tag{9.1.24}$$

$$DF(u)(w) > \frac{1}{2} \min(\|DF(u)\|, \|DF(u)\|^2). \tag{9.1.25}$$

Since DF is continuous (as we assume $F \in C^1$), w satisfies (9.1.24), (9.1.25) also for all v in some neighbourhood N_u of u. Since $\{N_u : u \in V'\}$ is an open covering of V', it possesses a locally finite refinement $\{M_\alpha\}_{\alpha \in I}$†. Let

$$\rho_\alpha(v) := \text{dist}(v, V' \setminus M_\alpha).$$

---

† This holds for any open covering of a paracompact set, see e.g. J. Dieudonné, *Grundzüge der Modernen Analysis*, 2, Vieweg, Braunschweig, second edition, 1987, pp. 26–9; V' is paracompact for example because it is metrizable.

ρ_α is Lipschitz continuous, and $\rho_\alpha(v) = 0$ for $v \notin M_\alpha$. We put
$$\varphi_\alpha(v) := \frac{\rho_\alpha(v)}{\sum_\beta \rho_\beta(v)}.$$
Since each v is only contained in finitely many M_β, because of the local finiteness of the covering, the denominator of φ_α is a finite sum. $(\varphi_\alpha)_{\alpha \in I}$ is a partition of unity subordinate to $\{M_\alpha\}$, i.e. $0 \leq \varphi_\alpha \leq 1$, $\varphi_\alpha = 0$ outside M_α, $\sum_{\alpha \in I} \varphi_\alpha \equiv 1$. Also, the φ_α are Lipschitz continuous. Then
$$v(u) := \sum_{\alpha \in I} \varphi_\alpha(u) w(u_\alpha) \quad \text{for some } u_\alpha \in M_\alpha$$
is a convex combination of vectors satisfying (9.1.24), (9.1.25) and hence satisfies these relations, too.
$v(u)$ thus is a pseudo-gradient vector field for F.

q.e.d.

Note that we only need to require $F \in C^1$, and not $F \in C^2$, in order to construct a locally Lipschitz pseudo-gradient field. We then have the following deformation for C^1-functionals on Banach spaces.

Theorem 9.1.3. *Let $F : V \to \mathbb{R}$ be a C^1-functional on a Banach space V satisfying (PS). Let $\alpha \in \mathbb{R}$, $\epsilon_0 > 0$, U a neighbourhood of K_α as in Theorem 9.1.1. Then there exist $0 < \epsilon < 1$ and a continuous family $\psi : V \times [0, \infty) \to V$ satisfying the semigroup property w.r.t. $t \geq 0$, and*

(i) $\psi(u, 0) = u$ for all $u \in V$
(ii) $F(\psi(u, s)) \leq F(\psi(u, t))$ whenever $0 \leq t \leq s$, $u \in H$
(iii) $\psi(u, t) = u$ for all t whenever $DF(u) = 0$
(iv) $\psi(u, t) = u$ whenever $|F(u) - \alpha| \geq \epsilon_0$, for all t
(v) $\psi(F_{\alpha+\epsilon} \setminus U, 1) \subset F_{\alpha-\epsilon}$, $\psi(U, 1) \subset F_{\alpha-\epsilon} \cup U$
(vi) If $F(\cdot)$ is even, so is $F(\psi(\cdot, t))$ for all t.

Proof. The proof is the same as the one of Theorem 9.1.1, replacing $\nabla F(u)$ by a pseudo-gradient vector field $v(u)$ — except for the following technical point: Lemma 9.1.3 asserts the existence of a pseudo-gradient field only on $\{x \in V \mid DF(x) \neq 0\}$. We therefore have to choose another Lipschitz continuous cut-off function $\gamma : V \to \mathbb{R}$ with $0 \leq \gamma \leq 1$, $\gamma(u) = 0$ if $u \in N_{\alpha, \frac{\delta}{2}}$, $\gamma(u) = 1$ for $u \in V \setminus N_{\alpha, \delta}$. We may then consider
$$\frac{\partial \psi(u, t)}{\partial t} = -\gamma(\psi(u, t)) \varphi(F(\psi(u, t))) \eta(\|v(\psi(u, t))\|) v(\psi(u, t)) \quad (9.1.26)$$
with φ, η as before. This has the additional effect that
$$\frac{\partial \psi(u, t)}{\partial t} = 0,$$

whenever $\psi(u,t) \in N_{\alpha,\frac{\delta}{2}}$, which is a neighbourhood of K_α, while the evolution is the same as before (with $v(u)$ in place of $\nabla F(u)$) outside $N_{\alpha,\delta}$. This cut-off near K_α does not affect the rest of the construction. If F is even, we may also choose γ even.

However, there might still exist critical points of F in $F_{\alpha+\epsilon} \setminus N_{\alpha,\delta}$. In order to take account of those, we strengthen the requirements on the above cut-off function φ to

$$\varphi(s) = 0 \quad \text{for } |\alpha - s| \geq \min(\epsilon_0, \frac{\delta}{4})$$
$$\varphi(s) = 1 \quad \text{for } |\alpha - s| \leq \min(\frac{\epsilon_0}{2}, \frac{\delta}{8}).$$

With such a φ, the right-hand side of (9.1.26) vanishes near any critical point of F, and it is therefore defined on all of V. If we then also impose the additional restriction

$$\epsilon \leq \frac{\delta}{8},$$

everything works out as before.

q.e.d.

It is possible, and not overly difficult, to extend Theorem 9.1.3 to the relative case and to obtain a result analogous to Theorem 9.1.2. Here, however, we refrain from doing so.

9.2 The mountain pass theorem

With the help of the deformation theorems of the previous section, one may easily derive existence results for critical points of a functional satisfying (PS). To illustrate this point, we start with the trivial

Lemma 9.2.1. *Let $F : V \to \mathbb{R}$ be a C^1 functional on a Banach space satisfying (PS). If*

$$\alpha := \inf_{u \in V} F(u) > -\infty,$$

then F possesses a critical point u_0 with value α (i.e. $F(u_0) = \alpha$, $DF(u_0) = 0$).

Proof. Suppose that $K_\alpha = \emptyset$. Then $U = \emptyset$ is a neighbourhood of K_α. Let $\epsilon_0 > 0$ be arbitrary. Choose ϵ as in Theorem 9.1.3. From the definition of α,

$$F_{\alpha+\epsilon} \neq \emptyset, \ F_{\alpha-\epsilon} = \emptyset.$$

Therefore, it is impossible that as Theorem 9.1.3 (v) asserts, the deformation $\psi(\cdot, 1)$ maps $F_{\alpha+\epsilon}$ into $F_{\alpha-\epsilon}$. This contradiction implies $K_\alpha \neq \emptyset$, which means the existence of the desired critical point.

q.e.d.

Of course, the methods presented in Chapter 4 yield more general existence results for minimizers of variational problems. The strength of the Palais–Smale approach rather lies in its capability of producing nonminimizing critical points. To demonstrate this, we now present the mountain pass theorem of Ambrosetti–Rabinowitz.

Theorem 9.2.1. *Let $F : V \to \mathbb{R}$ be a C^1 functional on a Banach space $(V, ||\cdot||)$ satisfying (PS). Suppose $F(0) = 0$ and*

(i) $\exists \rho > 0, \beta > 0 : F(u) \geq \beta$ *for all u with* $||u|| = \rho$
(ii) $\exists u_1$ *with* $||u_1|| > \rho$ *and* $F(u) < \beta$.

We let
$$\Gamma := \{\gamma \in C^0([0,1], V) \mid \gamma(0) = 0, \gamma(1) = u_1\}.$$

Then
$$\alpha := \inf_{\gamma \in \Gamma} \sup_{\tau \in [0,1]} F(\gamma(\tau)) \quad (\geq \beta)$$

is a critical value of F (i.e. there exists u_0 with $F(u_0) = \alpha$, $DF(u_0) = 0$).

Proof. Suppose again that $K_\alpha = \emptyset$, and take the neighbourhood $U = \emptyset$ of K_α. We let $\epsilon_0 = \min(\beta, \beta - E(u_1))$. Choose ϵ as in Theorem 9.1.3. From the definition of α, there exists $\gamma_0 \in \Gamma$ with
$$\sup_{\tau \in [0,1]} F(\gamma_0(\tau)) < \alpha + \epsilon,$$
while no such γ can satisfy
$$\sup_{t \in [0,1]} F(\gamma(t)) < \alpha - \epsilon,$$
i.e. satisfy $\gamma([0,1]) \subset F_{\alpha-\epsilon}$. However, if we apply the deformation $\psi(\cdot, 1)$ of Theorem 9.1.3, we obtain a path
$$\gamma(\tau) := \psi(\gamma_0(\tau), 1) \subset F_{\alpha-\epsilon}$$
with $\gamma(0) = \gamma_0(0) = 0$ and $\gamma(1) = \gamma_0(1) = u_1$ by choice of ϵ_0. This contradiction implies $K_\alpha \neq \emptyset$, i.e. the existence of the desired critical point.

q.e.d.

9.2 The mountain pass theorem

Let us summarize the essential features of the preceding reasoning:

(1) One chooses a family of sets, here Γ, that exploits some properties of F and is invariant under the deformation $\psi(\cdot, 1)$.
(2) This family yields a minimax value α.
(3) α can be estimated from above with the help of any member of our family Γ ($\alpha \leq \sup_{\tau \in [0,1]} F(\gamma(t))$ for any $\gamma \in \Gamma$), and from below through the constraints that the members of Γ have to satisfy (in Theorem 9.2.1, every $\gamma \in \Gamma$ intersects $\partial B(0, \rho)$, and therefore $\alpha \geq \beta > 0$, and therefore in particular, the critical point produced is different from 0).
(4) A reasoning by contradiction, based on the deformation theorem, shows that α is a critical value.

As an application of the mountain pass theorem, we consider the following example:

Theorem 9.2.2. *Let $\Omega \subset \mathbb{R}^d$ be a bounded domain, $2 < p < \frac{2d}{d-2}$ (respectively $< \infty$ for $d = 1, 2$). Then the Dirichlet problem*

$$\Delta u + |u|^{p-2} u = 0 \quad \text{in} \quad \Omega \tag{9.2.1}$$

$$u = 0 \quad \text{on} \quad \partial\Omega \tag{9.2.2}$$

admits at least two nontrivial (weak) solutions ('nontrivial' means not identically 0).

Proof. If u is a solution, so is $-u$. Therefore, it suffices to verify the existence of one nontrivial solution. (9.2.1), (9.2.2) are the Euler–Lagrange equations in $H_0^{1,2}(\Omega)$ for the functional

$$F(u) = \frac{1}{2} \int_\Omega |Du|^2 - \frac{1}{p} \int_\Omega |u|^p. \tag{9.2.3}$$

This functional is a continuous functional on $H_0^{1,2}(\Omega)$, because $\int_\Omega |Du|^2$ clearly is continuous there, and $\int |u|^p$ too, because of the Sobolev Embedding Theorem 3.4.3 as we assume $p < \frac{2d}{d-2}$. F is also differentiable, with

$$DF(u)(\varphi) = \int_\Omega Du \cdot D\varphi - \int_\Omega |u|^{p-2} u\varphi. \tag{9.2.4}$$

Again

$$\varphi \mapsto \int_\Omega Du \cdot D\varphi$$

clearly is continuous on $H_0^{1,2}(\Omega)$, whereas
$$\varphi \mapsto \int_\Omega |u|^{p-2} u\varphi$$
is continuous, because we have by Hölder's inequality
$$\left|\int_\Omega |u|^{p-2} u\varphi\right| \leq \left(\int_\Omega |u|^p\right)^{\frac{p-1}{p}} \left(\int_\Omega |\varphi|^p\right)^{\frac{1}{p}} \qquad (9.2.5)$$
$$\leq c_0 \|u\|_{H^{1,2}(\Omega)}^{p-1} \|\varphi\|_{H^{1,2}(\Omega)} \qquad (9.2.6)$$
by the Sobolev Embedding Theorem 3.4.3

for some constant c_0. Thus $F: H_0^{1,2}(\Omega) \to \mathbb{R}$ is of class C^1.

We shall verify the Palais–Smale condition for F: Suppose $(u_n)_{n\in\mathbb{N}} \subset H_0^{1,2}(\Omega)$ satisfies
$$|F(u_n)| \leq c_1 \quad \text{for some constant } c_1 \qquad (9.2.7)$$
$$DF(u_n) \to 0 \quad \text{for } u \to \infty. \qquad (9.2.8)$$

Thus
$$\left|\frac{1}{2}\int_\Omega |Du_n|^2 - \frac{1}{p}\int |u_n|^p\right| \leq c_1 \qquad (9.2.9)$$
and
$$\sup_{\varphi \in H_0^{1,2}, \|\varphi\|_{H_0^{1,2}} \leq 1} \left|\int_\Omega Du_n \cdot D\varphi - \int |u_n|^{p-2} u_n \varphi\right| \to 0 \quad \text{for } n \to \infty. \qquad (9.2.10)$$

In (9.2.10), we use $\varphi = \frac{-u_n}{\|u_n\|_{H^{1,2}}}$ and obtain
$$-\int |Du_n|^2 + \int |u_n|^p \leq c_2 \|u_n\|_{H^{1,2}}. \qquad (9.2.11)$$

Since $p > 2$, (9.2.9) and (9.2.11) imply
$$\int |Du_n|^2 \leq c_3 \|u_n\|_{H^{1,2}} + c_4. \qquad (9.2.12)$$

Since by the Poincaré inequality (Theorem 3.4.2)
$$\|u_n\|_{H^{1,2}(\Omega)}^2 = \int_\Omega |u_n|^2 + \int_\Omega |Du_n|^2 \leq c_5 \int |Du_n|^2, \qquad (9.2.13)$$
we conclude from (9.2.12)
$$\|u_n\|_{H^{1,2}(\Omega)} \leq c_6. \qquad (9.2.14)$$

Thus, any 'critical sequence' $(u_n)_{n\in\mathbb{N}}$ is bounded. We now claim that such a sequence $(u_n)_{n \in \mathbb{N}}$ contains a convergent subsequence, thereby completing the verification of (PS). We need to show that, after selection of a subsequence,

$$\int |Du_n - Du_m|^2 \to 0 \quad \text{for } n, m \to \infty \tag{9.2.15}$$

(using again the Poincaré inequality as in (9.2.13)). Now

$$\int Du_n D(u_n - u_m) - \int |u_n|^{p-2} u_n(u_n - u_m) \to 0 \quad \text{for } n, m \to \infty \tag{9.2.16}$$

by (9.2.10), (9.2.14).
By the Rellich–Kondrachev theorem (Corollary 3.4.1), we may also assume (by selecting a subsequence) that $(u_n)_{n \in \mathbb{N}}$ is a Cauchy sequence in $L^p(\Omega)$. Then, using Hölder's inequality as in (9.2.5),

$$\left| \int |u_n|^{p-2} u_n(u_n - u_m) \right| \le \left(\int |u_n|^p \right)^{\frac{p-1}{p}} \left(\int |u_n - u_m|^p \right)^{\frac{1}{p}} \to 0$$
$$\text{for } n, m \to \infty. \tag{9.2.17}$$

Equation (9.2.16) then implies

$$\int Du_n \cdot D(u_n - u_m) \to 0 \quad \text{for } m, n \to \infty,$$

which implies (9.2.15). We have thus verified (PS) for F. We shall now check the remaining assumptions of Theorem 9.2.1. First of all,

$$F(0) = 0.$$

Recalling that by the Sobolev Embedding Theorem 3.4.3 (and the Poincaré inequality, see (9.2.13))

$$\left(\int_\Omega |u|^p \right)^{\frac{1}{p}} \le c_7 \left(\int_\Omega |Du|^2 \right)^{\frac{1}{2}},$$

we have

$$F(u) \ge \left(\frac{1}{2} - c_8 \|u\|_{H_0^{1,2}(\Omega)}^{p-2} \right) \|u\|_{H_0^{1,2}(\Omega)}^2 \ge \beta > 0$$

if $\|u\|_{H_0^{1,2}(\Omega)} = \rho$ is sufficiently small.
Finally, take any $u_2 \in H_0^{1,2}(\Omega)$ with $\int_\Omega |u_2|^p \ne 0$. Then for sufficiently large $\lambda > 0$, $u_1 = \lambda u_2$ satisfies

$$F(u_1) = \frac{\lambda^2}{2} \int |Du_2|^2 - \frac{\lambda^p}{p} \int_\Omega |u_2|^p < 0.$$

We have now verified all the assumptions of the mountain pass Theorem 9.2.1, and we consequently get a critical point u of F with

$$F(u) \geq \beta > 0.$$

This is the desired nontrivial solution. (In fact, regularity theory implies that any weak solution of (9.2.1) is smooth in Ω, see e.g. Gilbarg–Trudinger, loc. cit.)

q.e.d.

Remark 9.2.1. By the same method, we can also treat the equation

$$\Delta u - \lambda u + |u|^{p-2} u = 0 \quad \text{for any } \lambda \geq 0. \tag{9.2.18}$$

9.3 Topological indices and critical points

In Section 3.2 of Part I, we have seen an example where a topological construction permitted to deduce the existence of more than one (unstable) critical point of a functional. In the present section, we first give an axiomatic approach to such constructions and then apply this in conjunction with the Palais–Smale condition to a concrete variational problem to show the existence of infinitely many solutions.

Such global topological constructions originated with the work of Lyusternik. Contributors also include Schnirelman, and, more recently, Rabinowitz, and many others. The reader will find detailed references in the monographs quoted at the end of this chapter.

Definition 9.3.1. *Let X be a topological space, $F : X \to \mathbb{R}$ continuous. $x \in X$ is called a special point for F, with value α,*

$$x \in \text{spec}_\alpha F \quad (\alpha \text{ then is called a special value})$$

if x is contained in all $A \subset X$ with the following property: For each open $U \supset A$ there exist $\epsilon = \epsilon(U) > 0$ and a continuous

$$\psi : X \times [0, 1] \to X$$

satisfying

(i) $\psi(y, 0) = y$ for $y \in X$
(ii) $F(\psi(y, t)) \leq F(\psi(y, s))$ for all $y \in X$, $0 \leq s \leq t \leq 1$

9.3 Topological indices and critical points

(iii) *For every $y \in X \setminus U$ with*

$$F(y) \leq \alpha + \epsilon,$$

we have

$$F(\psi(y, 1)) \leq \alpha - \epsilon.$$

Of course, the ψ of the preceding definition is an abstract version of the deformations constructed in Section 9.1, and the notion of special point is a topological version of the notion of critical point.

Remark 9.3.1. Since the composition of any two deformations ψ_1, ψ_2 satisfying the properties of Definition 9.3.1 continues to satisfy these properties, the intersection of any two sets A_1, A_2 still satisfies the property expressed in Definition 9.3.1 if A_1, A_2 do. Therefore, if $\text{spec}_\alpha F = \emptyset$, we may take $U = A = \emptyset$ in Definition 9.3.1 and find a deformation ψ that satisfies (i)–(iii) for all $y \in X$.

In order to illustrate the notion of special point as well as the topological constructions to follow, we now present the simple:

Lemma 9.3.1. *Let $F : X \to \mathbb{R}$ be a continuous function on the topological space X. Let $\mathcal{M}$ be a (nonempty) class of nonempty subsets of X. If $\text{spec}_\alpha F = \emptyset$, we require that $\mathcal{M}$ is invariant under the deformations considered in Definition 9.3.1:*

$$\text{If } A \in \mathcal{M}, \text{ then also } \psi(A, 1) \in \mathcal{M}. \tag{9.3.1}$$

Suppose

$$-\infty < \alpha = \inf_{A \in \mathcal{M}} \sup_{y \in A} F(y) < \infty. \tag{9.3.2}$$

Then α is a special value for F, i.e. there exists

$$x_0 \in \text{spec}_\alpha F.$$

Proof. Suppose $\text{spec}_\alpha F = \emptyset$. According to the preceding remark, we may then take $U = \emptyset$ and find $\psi : X \times [0, 1] \to X$ and $\epsilon > 0$ with

$$F(\psi(y, 1)) \leq \alpha - \epsilon \quad \text{whenever } F(y) \leq \alpha + \epsilon. \tag{9.3.3}$$

We may find $A_0 \in \mathcal{M}$ with

$$\sup_{y \in A_0} F(y) \leq \alpha + \epsilon, \tag{9.3.4}$$

but no $A \in \mathcal{M}$ can satisfy

$$\sup_{y \in A} F(y) \leq \alpha - \epsilon. \qquad (9.3.5)$$

However, if we take $A_1 := \psi(A_0, 1)$ then $A_1 \in \mathcal{M}$ by assumption, and by (9.3.3)

$$\sup_{y \in A_1} F(y) \leq \alpha - \epsilon,$$

contradicting (9.3.5). Thus, $\operatorname{spec}_\alpha F \neq \emptyset$.

q.e.d.

In order to obtain the existence of further special points, we now shall introduce the notion of a (topological) index. Such an index is based on symmetry or invariance properties of the functional under consideration. Here, we only consider the case of the simplest nontrivial symmetry group, namely $\mathbb{Z}_2$, although the subsequent constructions easily generalize to any compact group G. We thus make the following *symmetry assumptions*:

- X is a topological space with a nontrivial involution, i.e. there exists a continuous map $j : X \to X$, $j \neq \operatorname{id}$, with

$$j^2 = \operatorname{id}.$$

- $F : X \to \mathbb{R}$ is continuous and even, i.e.

$$F(j(x)) = F(x) \quad \text{for all } x \in X.$$

- $\mathcal{M} := \{A \subset X \mid j(A) = A \text{ and for all } x \in A, j(x) \neq x$ (i.e. A contains no fixed points of j)$\}$.

We now also require $\psi(j(x), t) = j(\psi(x, t))$ for all the deformations of Definition 9.3.1.

Definition 9.3.2. *An index for (X, F) is a map*

$$i : \mathcal{M} \to \{0, 1, 2, \ldots, \infty\}$$

satisfying for all $A, A_1, A_2 \in \mathcal{M}$:

(i) $i(A) = 0 \Leftrightarrow A = \emptyset$
(ii) A *finite* $(A \neq \emptyset) \Rightarrow i(A) = 1$
(iii) $i(A_1 \cup A_2) \leq i(A_1) + i(A_2)$
(iv) $A_1 \subset A_2 \Rightarrow i(A_1) \leq i(A_2)$
(v) $i(A) \leq i(\overline{j(A)})$

(vi) A compact $\Rightarrow \exists$ neighbourhood U of A in X with $\overline{U} \in \mathcal{M}$,
$$i(A) = i(\overline{U}) < \infty.$$
For $n \in \{0,1,2,\ldots,\infty\}$, we put
$$\mathcal{M}_n := \{A \in \mathcal{M} \mid i(A) \geq n\}.$$
Remark 9.3.2. More precisely, one should call an i as in Definition 9.3.2 an index for $(X, F, \mathbb{Z}_2)$, in order to specify the symmetry group involved.

For $n \in \{0, 1, 2, \ldots, \infty\}$, we define
$$\alpha_n := \inf_{A \in \mathcal{M}_n} \sup_{y \in A} F(y).$$

Theorem 9.3.1. *Suppose the above symmetry assumptions hold, an index i for (X, F) exists, and*
$$-\infty < \alpha_n < \infty\dagger$$

(i) *Then*
$$\operatorname{spec}_{\alpha_n} F \neq \emptyset \qquad (9.3.6)$$

(ii) *If furthermore for some $k \geq 1$, $\alpha_n = \alpha_{n+1} = \ldots = \alpha_{n+k}$, then $\operatorname{spec}_{\alpha_n} F$ is infinite.*

Proof. We note that property (v) of Definition 9.3.2 implies that $\mathcal{M}_n$ is invariant under (symmetric) deformations ψ. Therefore, Lemma 9.3.1 implies $\operatorname{spec}_{\alpha_n} F \neq \emptyset$. For the second statement, we claim that for $A_0 = \operatorname{spec}_{\alpha_n} F$,
$$i(A_0) \geq k+1. \qquad (9.3.7)$$

If $k \geq 1$, property (ii) of Definition 9.3.2 then implies the existence of infinitely many special points with value α_n.

Suppose on the contrary that
$$i(A_0) \leq k. \qquad (9.3.8)$$

By Definition 9.3.2 (vi), we may find a neighbourhood U of A_0 with $\overline{U} \in \mathcal{M}$ and
$$i(A_0) = i(\overline{U}). \qquad (9.3.9)$$

Since A_0 consists of special points, we may find a (symmetric) deformation ψ with
$$F(\psi(y,1)) \leq \alpha_n - \epsilon \quad \text{for all } y \in X \setminus U \quad \text{with} \quad F(y) \leq \alpha_n + \epsilon$$

† Since the infimum over an empty set is ∞, this contains the assumption $\mathcal{M}_n \neq \emptyset$.

for some $\epsilon > 0$. Since $\alpha_n = \alpha_{n+k}$, we may find $A \in \mathcal{M}_{n+k}$ with

$$\sup_{y \in A} F(y) \leq \alpha_n + \epsilon,$$

hence

$$\sup_{z \in \psi(A \setminus U, 1)} F(z) \leq \alpha_n - \epsilon. \qquad (9.3.10)$$

We have

$$i(A \setminus U) \geq i(A) - i(\overline{U}) \quad \text{by (iii)}$$
$$\geq n + k - k, \quad \text{using (9.3.8), (9.3.9), } A \in \mathcal{M}_{n+k}$$
$$= n.$$

Thus

$$A \setminus U \in \mathcal{M}_n,$$

hence $A \setminus U \neq \emptyset$ by (i). Since, as noted in the beginning, $\mathcal{M}_n$ is invariant under ψ, we get

$$\psi(A \setminus U, 1) \in \mathcal{M}_n,$$

hence

$$\sup_{y \in \psi(A \setminus U, 1)} F(y) \geq \alpha_n,$$

contradicting (9.3.10).

q.e.d.

In order to apply the preceding considerations, we need to construct an index with the properties listed in Definition 9.3.2. We shall present here Coffman's version of the genus of Krasnoselskij.

Definition 9.3.3. *Suppose the symmetry assumptions stated before Definition 9.3.2 hold. The genus of $A \neq \emptyset$, $A \in \mathcal{M}$ is defined as follows:*

$$\text{gen}(A) := \inf \{n \in \{1, 2, 3, \ldots, \infty\} \mid \exists \text{ continuous } f : A \to \mathbb{R}^n \setminus \{0\}$$

$$\text{with} \quad f(j(x)) = -f(x) \quad \text{for all } x \in A\}$$

while $\text{gen}(\emptyset) := 0$.

As an example, we state:

Lemma 9.3.2. *The genus of the unit sphere $S^{n-1} = \{||x|| = 1\}$ in $\mathbb{R}^n$ (with involution $j(x) = -x$) is equal to n.*

9.3 Topological indices and critical points 311

Proof. The inclusion map $S^{n-1} \hookrightarrow \mathbb{R}^n$ satisfies the properties of Definition 9.3.3, and so $\text{gen}(S^{n-1}) \leq n$. If $n \geq 2$, S^{n-1} is connected, and therefore, by the mean value theorem, there is no continuous map $f : S^{n-1} \to \mathbb{R}^1 \setminus \{0\}$ with $f(-x) = -f(x)$ for all x. Hence $\text{gen}(S^{n-1}) \geq 2$. In fact, by the Borsuk–Ulam theorem†, there is no such continuous map to $\mathbb{R}^m \setminus \{0\}$ with $m < n$. Therefore, $\text{gen}(S^{n-1}) \geq n$.

q.e.d.

Corollary 9.3.1. *The genus of the unit sphere* $S := \{x \in V : \|x\| = 1\}$ *in an infinite dimensional Banach space* $(V, \|\cdot\|)$ *is* ∞.

Proof. For any n-dimensional subspace V^n of V,

$$\text{gen}(S) \geq \text{gen}(S \cap V^n) \geq n \quad \text{by Lemma 9.3.2}.$$

q.e.d.

Theorem 9.3.2. *The genus as defined in Definition 9.3.3 is an index in the sense of Definition 9.3.2.*

Proof. We need to check the properties (i)–(vi) of Definition 9.3.2.

(i) is obvious.

(ii) If $A \in \mathcal{M}$ is finite, then A is of the form $\{x_\nu, j(x_\nu) \mid \nu = 1, \ldots, k\}$ for some k. We define $f : A \to \mathbb{R}^1 \setminus \{0\}$ by $f(x_\nu) = 1$, $f(j(x_\nu)) = -1$ for all ν (of course, we may assume $x_\mu \neq j(x_\nu)$ for all μ, ν).

(iii) Let $\text{gen}(A_\nu) = n_\nu < \infty$, $\nu = 1, 2$, and let the continuous $f_\nu : A_\nu \to \mathbb{R}^{n_\nu} \setminus \{0\}$ satisfy $f_\nu(j(x)) = -f_\nu(x)$ for all x. By the Tietze extension theorem‡, f_ν can be continuously extended to

$$\tilde{f}_\nu : X \to \mathbb{R}^{n_\nu}.$$

By considering $\frac{1}{2}(\tilde{f}_\nu(x) - \tilde{f}_\nu(j(x)))$ in place of $\tilde{f}_\nu$, we may assume that the extension still satisfies

$$\tilde{f}_\nu(j(x)) = -\tilde{f}_\nu(x) \quad \text{for all } x.$$

The map $(\tilde{f}_1, \tilde{f}_2) : A_1 \cup A_2 \to \mathbb{R}^{n_1+n_2} \setminus \{0\}$ then shows that

$$\text{gen}(A_1 \cup A_2) \leq n_1 + n_2 = \text{gen}(A_1) + \text{gen}(A_2).$$

(iv) is obvious.

(v) follows, since $f \circ j$ shares the necessary properties with f.

---

† See e.g. E. Zeidler, *Nonlinear Functional Analysis and its Applications*, I, Springer, New York, 1984, p. 708, for a proof.
‡ See E. Zeidler, loc. cit., p. 49.

(vi) Let $A \in \mathcal{M}$ be compact. Since $j(x) \neq x$ for all $x \in A$ (by the properties of $\mathcal{M}$), for each $x \in A$, we may find a neighbourhood $U(x)$ with $U(x) \cap j(U(x)) = \emptyset$. Since A is compact, it can be covered by finitely many such neighbourhoods U_ν, $\nu = 1, \ldots, n$. For each U_ν, we choose a continuous function $\varphi_\nu : X \to \mathbb{R}$ with $\varphi_\nu(x) > 0$ for $x \in U_\nu$, $\varphi_\nu(x) = 0$ for $x \in X \setminus U_\nu$. We then define $h = (h^1, \ldots, h^n) : A \to \mathbb{R}^n \setminus \{0\}$ by

$$h^\nu(x) := \begin{cases} \varphi_\nu(x) & \text{for } x \in U_\nu \\ -\varphi_\nu(x) & \text{for } x \in A \setminus U_\nu, \text{ in particular for } x \in j(U_\nu). \end{cases}$$

(Since every $x \in A$ is contained in some U_ν, we have $h(x) \neq 0$ for all $x \in A$.)
Thus $\text{gen}(A) \leq n < \infty$.
If $A \in \mathcal{M}$ is compact with $\text{gen}(A) = n$, and

$$f : A \to \mathbb{R}^n \setminus \{0\} \quad \text{is continuous with } f(j(x)) = -f(x),$$

we may extend f as before to $\bar{f} : X \to \mathbb{R}^n$ (with the same symmetry property). Since A is compact, so is $f(A)$, and therefore, we may find an open neighbourhood V of $f(A)$ with $\bar{V} \subset \mathbb{R}^n \setminus \{0\}$. Then $U := \bar{f}^{-1}(A)$ satisfies

$$n = \text{gen}(A)$$
$$\leq \text{gen}(\bar{U}) \quad \text{by (iv)}$$
$$\leq n \quad \text{since } \bar{f}(\bar{U}) \text{ is contained in } \bar{V} \subset \mathbb{R}^n \setminus \{0\}.$$

Thus $\text{gen}(\bar{U}) = \text{gen}(A)$ as required.

q.e.d.

We may now obtain a general existence theorem for critical points of functionals satisfying (PS):

Theorem 9.3.3. *Let $F, G : H \to \mathbb{R}$ be C^2 functionals on a Hilbert space $(H, \langle \cdot, \cdot \rangle)$ that are even, i.e. $F(x) = F(-x)$, $G(x) = G(-x)$ for all $x \in H$. Suppose F satisfies (PS) relative to $G = \beta$, and is bounded from below. Let*

$$\mathcal{M} := \{A \subset \{G(x) = \beta\} \mid 0 \notin A \quad \text{and} \quad (x \in A \Leftrightarrow -x \in A)\}.$$

Let $\gamma_0 := \sup\{\text{gen}(K) \mid K \in \mathcal{M} \text{ compact}\}$ ($\leq \infty$). Then F possesses at least γ_0 critical points relative to $G = \beta$.

Proof. Since (PS) holds, by Theorem 9.1.2, all special points (in the

9.3 Topological indices and critical points

sense of Definition 9.3.1) for the restriction of F to $X := \{x \in H \mid G(x) = \beta\}$ are critical points for F relative to $G = \beta$. Hence, it suffices to produce γ_0 special points of F on X. Let

$$\alpha_n := \inf_{A \in \mathcal{M}, \text{gen}(A) \geq n} \sup_{x \in A} F(x).$$

Since F is bounded below, and since in the definition of γ_0, we only consider compact sets, we have

$$-\infty < \alpha_n < \infty \quad \text{whenever } n \leq \gamma_0.$$

By Theorem 9.3.2, we may apply Theorem 9.3.1 to the genus as an index. We have in fact

$$-\infty < \alpha_1 \leq \alpha_2 \leq \cdots \leq \alpha_n \leq \cdots < \infty \quad \text{whenever } n \leq \gamma_0.$$

If we always have strict equality, then the

$$x_n \in \text{spec}_{\alpha_n} F$$

produced by Theorem 9.3.2 (i) are all different, because their values $F(x_n)$ are all different. If however any two such numbers α_{n-1} and α_n are equal, then by Theorem 9.3.2 (ii) we even obtain infinitely many special points. Thus, in any case, we have at least γ_0 special, hence critical points.

q.e.d.

As an application of Theorem 9.3.3, we consider the example of the previous section:

Corollary 9.3.2. *Let $\Omega \subset \mathbb{R}^d$ be a bounded domain, $2 < p < \frac{2d}{d-2}$ (respectively $< \infty$ for $d = 1, 2$). Then for any $\lambda \geq 0$, the Dirichlet problem*

$$\Delta u - \lambda u + |u|^{p-2} u = 0 \quad \text{in } \Omega \tag{9.3.11}$$

$$u = 0 \quad \text{on } \partial \Omega \tag{9.3.12}$$

admits infinitely many (weak) solutions.

Proof. We consider the even functionals

$$F(u) = \frac{1}{2} \int_\Omega \left(|Du|^2 + \lambda u^2 \right)$$

$$G(u) = \frac{1}{p} \int_\Omega |u|^p.$$

We claim that F satisfies (PS) relative to $G = 1$. The proof is similar to the argument employed for the demonstration of Theorem 9.2.2: let $(u_n)_{n \in \mathbb{N}}$ be a critical sequence, i.e.

$$F(u_n) \leq c_1 \qquad (9.3.13)$$

$$\left\| DF(u_n) - \frac{\langle DG(u_n), DF(u_n) \rangle}{\|DG(u_n)\|^2} DG(u_n) \right\| \to 0 \quad \text{for } n \to \infty \qquad (9.3.14)$$

where all norms and scalar products are from $H_0^{1,2}(\Omega)$. From (9.3.13) (and the Poincaré inequality in case $\lambda = 0$), we obtain

$$\|u_n\|_{H_0^{1,2}(\Omega)} \leq c_2. \qquad (9.3.15)$$

We obtain as in the proof of Theorem 9.2.2 (cf. (9.2.5)), by using Hölder's inequality, that

$$|DG(u_n)(u_n - u_m)| = \left| \int |u_n|^{p-2} u_n (u_n - u_m) \right|$$

$$\leq \left(\int |u_n|^p \right)^{\frac{p-1}{p}} \left(\int |u_n - u_m|^p \right)^{\frac{1}{p}}. \qquad (9.3.16)$$

Since $p < \frac{2d}{d-2}$, from (9.3.15) and Sobolev's Embedding Theorem 3.4.3, we conclude that $\int |u_n|^p$ is bounded, whereas (9.3.15) and the Rellich–Kondrachev theorem (Corollary 3.4.1) imply that $(u_n)_{n \in \mathbb{N}}$ is a Cauchy sequence in $L^p(\Omega)$. Thus, from (9.3.16)

$$DG(u_n)(u_n - u_m) \to 0 \quad \text{for } n, m \to \infty. \qquad (9.3.17)$$

Also

$$\|DG(u_n)\| = \sup_{w \in H_0^{1,2}(\Omega)} \frac{|DG(u_n)(w)|}{\|w\|_{H_0^{1,2}}}$$

$$\geq \frac{|DG(u_n)(u_n)|}{\|u_n\|_{H_0^{1,2}}}$$

$$= \frac{\int |u_n|^p}{\|u_n\|_{H_0^{1,2}}} \qquad (9.3.18)$$

$$\geq \mu$$

$$> 0 \quad \text{from (9.3.15) and}$$

$$\frac{1}{p} \int |u_n|^p = G(u_n) = 1. \qquad (9.3.19)$$

9.3 Topological indices and critical points

From (9.3.17), (9.3.18) we conclude that there exist $h_{nm} \in H_0^{1,2}(\Omega)$ with

$$DG(u_n)(u_n - u_m + h_{nm}) = 0 \quad \text{for all } n, m \tag{9.3.20}$$

$$\|h_{nm}\|_{H_0^{1,2}} \to 0 \quad \text{for } n, m \to \infty. \tag{9.3.21}$$

Therefore, from (9.3.14)

$$DF(u_n)(u_n - u_m + h_{nm}) \to 0,$$

i.e.

$$\int \left(Du_n \left(D(u_n - u_m) + Dh_{nm} \right) + \lambda u_n (u_n - u_m + h_{nm}) \right) \to 0$$

$$\text{for } n, m \to \infty$$

and because of (9.3.21) then also

$$\int \left(Du_n (D(u_n - u_m)) + \lambda u_n (u_n - u_m) \right) \to 0.$$

This implies

$$\int \left(|(D(u_n - u_m)|^2 + \lambda |(u_n - u_m)|^2 \right) \to 0 \quad \text{for } n, m \to \infty,$$

and consequently, $(u_n)_{n \in \mathbb{N}}$ is a Cauchy sequence in $H_0^{1,2}(\Omega)$. This verifies (PS) relative to $G = 1$.

In order to apply Theorem 9.3.2, we thus only need to check that in the present case, $\gamma_0 = \infty$. However,

$$\left\{ u \in H_0^{1,2}(\Omega) : \frac{1}{p} \|u\|_{L^p}^p = 1 \right\}$$

is the intersection of a sphere centered at the origin in $L^p(\Omega)$ with the subspace $H_0^{1,2}(\Omega)$. Therefore, the argument of Lemma 9.3.2 easily implies $\gamma_0 = \infty$. Theorem 9.3.2 thus produces infinitely many solutions of

$$DF(u_n) - \frac{\langle DG(u_n), DF(u_n) \rangle}{\|DG(u_n)\|^2} DG(u_n) = 0,$$

i.e. with

$$\mu_n := \frac{\langle DG(u_n), DF(u_n) \rangle}{\|DG(u_n)\|^2},$$

weak solutions of

$$\Delta u_n - \lambda u_n + \mu_n |u_n|^{p-2} u_n = 0 \quad \text{in } \Omega$$

$$u_n = 0 \quad \text{on } \partial\Omega.$$

If we choose v_n with $v_n^{p-2}\mu_n = 1$, then $v_n := v_n u_n$ solves (9.3.11), (9.3.12) weakly. Again, we remark that elliptic regularity theory implies that all u_n and v_n are smooth in Ω, so that in fact we obtain classical solutions of (9.3.11), (9.3.12).

q.e.d.

In Theorem 9.2.2 and in Corollary 9.3.1, we had imposed the restriction
$$p < \frac{2d}{d-2} \quad (\text{in case } d \geq 3),$$
and the reader may wonder whether this is necessary. To pursue this question, we shall now discuss the theorem of Pohozaev:

Theorem 9.3.4. *Let $\Omega \subset \mathbb{R}^d$ be a smooth domain which is strictly star shaped w.r.t. $0 \in \mathbb{R}^d$ (this means that the outer normal ν of Ω satisfies $\langle x, \nu(x)\rangle > 0$ for all $x \in \partial\Omega$). Then for $\lambda \geq 0$, any solution of*

$$\Delta u - \lambda u + |u|^{\frac{d+2}{d-2}} u = 0 \quad \text{in } \Omega \tag{9.3.22}$$

$$u = 0 \quad \text{on } \partial\Omega \tag{9.3.23}$$

vanishes identically.

We shall present a complete *proof* only for $\lambda > 0$ and for smooth solutions u (elliptic regularity implies that any weak solution of (9.3.21), (9.3.22) is automatically smooth† on $\bar{\Omega}$, but the present book does not treat this topic):
We multiply (9.3.22) by $\sum_{i=1}^d x^i \frac{\partial u}{\partial x^i}$ and obtain

$$0 = \left(\Delta u - \lambda u + |u|^{\frac{d+2}{d-2}} u\right) \sum x^i \frac{\partial u}{\partial x^i} \tag{9.3.24}$$

$$= \operatorname{div}\left(Du \sum x^i \frac{\partial u}{\partial x^i} - x\frac{|Du|^2}{2} - \frac{\lambda}{2}x|u|^2 + \frac{d-2}{2d} x|u|^{\frac{2d}{d-2}}\right)$$

$$+ \frac{d-2}{2}|Du|^2 + \frac{\lambda d}{2}|u|^2 - \frac{d-2}{2}|u|^{\frac{2d}{d-2}}. \tag{9.3.25}$$

By (9.3.23), we have $u = 0$ on $\partial\Omega$, hence also $\sum x^i \frac{\partial u}{\partial x^i} = \sum x^i \nu^i \frac{\partial u}{\partial \nu}$ ($\nu = (\nu^1, \ldots, \nu^d)$ is the exterior normal of Ω). Integrating (9.3.25) therefore yields

$$\frac{d-2}{2}\int_\Omega |Du|^2 + \frac{\lambda d}{2}\int_\Omega |u|^2 - \frac{d-2}{2}\int_\Omega |u|^{\frac{2d}{d-2}} + \frac{1}{2}\int_{\partial\Omega} \left|\frac{\partial u}{\partial \nu}\right|^2 \sum x^i \nu^i = 0. \tag{9.3.26}$$

---

† See for example Appendix B in M. Struwe, *Variational Methods*, Springer, Berlin, 2nd edition, 1996.

9.3 Topological indices and critical points

On the other hand, multiplying (9.3.22) by u leads to

$$\int_\Omega |Du|^2 + \lambda \int_\Omega |u|^2 - \int_\Omega |u|^{\frac{2d}{d-2}} = 0. \tag{9.3.27}$$

Equations (9.3.26) and (9.3.27) imply

$$2\lambda \int_\Omega |u|^2 + \int_{\partial\Omega} \left|\frac{\partial u}{\partial \nu}\right|^2 \sum x^i \nu^i = 0. \tag{9.3.28}$$

If $\lambda > 0$, this implies $u \equiv 0$, hence the result. (If $\lambda = 0$, one still concludes that $\frac{\partial u}{\partial \nu} \equiv 0$ on $\partial\Omega$. Since also $u = 0$ on $\partial\Omega$ by (9.3.23) one may invoke a unique continuation theorem for solutions of elliptic equations to obtain $u \equiv 0$ in Ω. We omit the details.)

q.e.d.

Theorem 9.3.4 implies that for $p = \frac{2d}{d-2}$ in Theorem 9.2.2 and Corollary 9.3.2, the Palais–Smale condition no longer holds. Namely, if it did, the proofs of those results would yield the existence of nontrivial solutions. It also shows that if the Palais–Smale condition fails the whole scheme developed in the present chapter for producing critical points breaks down.

Since for $p < \frac{2d}{d-2}$, (PS) does hold, the case $p = \frac{2d}{d-2}$ can be considered as as limit case for (PS). In fact, such limit cases of the Palais–Smale condition occur in many variational problems that are of importance in Riemannian geometry, e.g. the Yang–Mills functional on a four-dimensional Riemannian manifold, two-dimensional harmonic maps, surfaces of constant mean curvature, the Yamabe functional etc. The interested reader is for example referred to

K. C. Chang, *Infinite Dimensional Morse Theory and Multiple Solution Problems*, Birkhäuser, Boston, 1993,

J. Jost, *Riemannian Geometry and Geometric Analysis*, Springer, Berlin, 2nd edition, 1998,

M. Struwe, *Variational Methods*, Springer, Berlin, 2nd edition, 1996,

and the references contained therein.

The basic references that have been used in writing the present chapter are the monograph of M.Struwe just quoted, as well as

P. Rabinowitz, *Minimax Methods in Critical Point Theory with Applications to Differential Equations*, CBMS Reg. Conf. Ser. 65, AMS, Providence, 1986

and

E. Zeidler, *Nonlinear Functional Analysis and its Applications*, III, Springer, Berlin, 1984.

These three monographs contain not only detailed bibliographical references — which the reader is urged to consult in order to find the original sources of the results of the present chapter — but also many further results and examples concerning the Palais–Smale condition and index theories.

Exercises

9.1 Why is Theorem 9.2.1 called 'mountain pass theorem'? Hint: Try to find an analogy between the statement of that result and the geometry of mountain passes.

9.2 Try to find conditions for a function

$$f : \Omega \times \mathbb{R} \to \mathbb{R}$$

so that the reasoning of Theorem 9.2.2 can be extended to the Dirichlet problem

$$\Delta u(x) = f(x, u(x)) \quad \text{for } x \in \Omega$$
$$u(x) = 0 \quad \text{for } x \in \partial\Omega$$

in a smooth bounded domain Ω. (An answer can be found in Theorem 6.2 of the quoted monograph of M.Struwe.)

9.3 Develop an index theory for a general compact group G in place of $\mathbb{Z}_2$.

9.4 Extend Theorem 9.1.3 to the relative case as indicated at the end of Section 9.1.

Index

$x \cdot y = \sum_{i=1}^{d} x^i y^i = x^i y^i$, xv
$|x|^2 = x \cdot x$, xv
$\dot{u}(t) = \frac{d}{dt} u(t)$, xv
$C^k(\Omega)$, xv
$C^k(\Omega, \mathbb{R}^d)$, xvi
$C^\infty(\Omega)$, xvi
$C_0^\infty(\Omega)$, xvi
$C_0^k(\Omega)$, xvi
$I(u) = \int_a^b F(t, u(t), \dot{u}(t)) \, dt$, 3
Df, 4
$\eta \in C_0^1([a,b], \mathbb{R}^d)$, 5
F_u, 5
F_p, 5
$\delta I(u, \eta) := \frac{d}{ds} I(u+s\eta)_{,-s=0}$ 10
$AC([a,b])$, 11
$D^1(I, \mathbb{R}^d)$, 11
D_0^1, 13
$\delta^2 I(u, \eta) := \frac{d^2}{ds^2} I(u+s\eta)_{,-s=0}$ 19
$F_{p^i p^j} \dot{\eta}_i \dot{\eta}_j = \sum_{i,j=1}^{d} F_{p^i p^j} \dot{\eta}_i \dot{\eta}_j$, 19
$\dot{c}(t) := \frac{dc}{dt}(t)$, 32
$L(c) := \int_0^T |\dot{c}(t)| \, dt =$
$\int_0^T \left(\sum_{\alpha=1}^{d} (\dot{c}^\alpha)^2 \right)^{\frac{1}{2}} dt$, 32
$E(c) := \frac{1}{2} \int_0^T |\dot{c}(t)|^2 \, dt =$
$\frac{1}{2} \int_0^T \sum_{\alpha=1}^{d} (\dot{c}^\alpha)^2 \, dt$, 32
$(g^{ij})_{i,j=1,\ldots,n}$, 39
$g_{ij,k} := \frac{\partial}{\partial z^k} g_{ij}$, 39
$\Gamma^i_{jk} := \frac{1}{2} g^{il} (g_{jl,k} + g_{kl,j} - g_{jk,l})$, 39
$S^n :=$
$\left\{ (x^1, \ldots, x^{n+1}) \in \mathbb{R}^{n+1}, \sum_{i=1}^{n+1} (x^i)^2 = 1 \right\}$, 39
$d(p, q) := \inf\{L(c) | \, c : [a,b] \to M$ rectifiable curve with $c(a) = p, c(b) = q\}$, 51

$\dot{x}^i := \frac{dx^i}{dt}$, 79
meas, 118
$\int_A f(x) dx$ of f, 120
$\mathcal{L}^1(A)$, 120
$\varphi \in C_0^0(\mathbb{R}^d)$, 122
$\|\cdot\|$, 125
$(\cdot, \cdot)$, 126
$\mathbb{R}^+ := \{t \in \mathbb{R} \mid t \geq 0\}$, 130
$\|f\|_* := \sup_{x \neq 0} \frac{|f(x)|}{\|x\|}$, 133
$V^* := \{f : V \to \mathbb{R} \text{ linear with } \|f\|_* < \infty\}$, 133
$(V^*)^* =: V^{**}$, 133
$x_n \rightharpoonup x$, 135
$M^\perp :=$
$\{x \in H : (x,y) = 0 \quad \text{for all } y \in M\}$, 141
$\|T\| := \sup_{x \neq 0} \frac{\|Tx\|}{\|x\|} \in \mathbb{R}^+ \cup \{\infty\}$, 144
$L(V, W)$, 145
$\ker T := \{x \in V : Tx = 0\}$, 145
$V = V_1 \oplus V_2$, 146
$coker$, 147
$R(T)$, 147
$\text{ind } T$, 147
$F(V, W)$, 147
$DF(u)$, 150
C^1, 150
C^2, 150
$D^2 F(u)$, 150
ODE, 155
$\|y\|_{C^0} := \sup_{t \in I} \|y(t)\|$, 156
$\|f\|_p := \|f\|_{L^p(A)} := \left(\int_A |f(x)|^p \, dx \right)^{\frac{1}{p}}$, 159
$\text{ess sup}_{x \in A} f(x) := \inf \{ \lambda \in \mathbb{R} \mid f(x) \leq \lambda \text{ for almost all } x \in A \}$, 162
$f_h(x) := \frac{1}{h^d} \int_{\mathbb{R}^d} \varrho \left(\frac{x-y}{h} \right) f(y) dy$, 167
$C_0^\infty(\Omega)$, 166
$\text{supp } \varphi$, 166

319

320 Index

$\Omega' \subset\subset \Omega$, 167
$f_h \rightrightarrows f$, 167
$\boldsymbol{\alpha} := (\alpha_1, \ldots, \alpha_d)$, 171
$|\boldsymbol{\alpha}| := \sum_{i=1}^{d} \alpha_i$, 171
$D_{\boldsymbol{\alpha}} := \left(\frac{\partial}{\partial x^1}\right)^{\alpha_1} \cdots \left(\frac{\partial}{\partial x^d}\right)^{\alpha_d}$, 171
$v := D_{\boldsymbol{\alpha}} u$, 171
$W^{k,p}(\Omega)$, 171
$\|u\|_{W^{k,p}(\Omega)} := \left(\sum_{|\boldsymbol{\alpha}| \le k} \int_\Omega |D_{\boldsymbol{\alpha}} u|^p\right)^{\frac{1}{p}}$, 171
$H^{k,p}(\Omega)$, 171
$H_0^{k,p}(\Omega)$, 171
$D_i u$, 172
Du, 172
lsc, 185
$F^\lambda(x) := \inf_{y \in X}(\lambda F(y) + d^2(x, y))$, 190
$J^\lambda(x)$, 191
$\left(\Delta := \sum_{i=1}^{d} \frac{\partial^2}{(\partial x^i)^2}\right)$, 199
$sc^- F$, 208
$\imath_A$, 210
$q^- f$, 216
$\Gamma\text{-}\lim_{n \to \infty} F_n$, 225
$BV(\Omega)$, 242
$\|Du\|$, 242
$\|u\|_{BV(\Omega)} := \|u\|_{L^1(\Omega)} + \|Du\|(\Omega)$, 242
$|S|_{d-1}$, 243
$P(E, \Omega) := \|D\chi_E\|(\Omega)$, 244
$\rho_h * u(x)$, 246
$Gl(d, \mathbb{R})$, 257
$O(d, \mathbb{R})$, 257
$J_\lambda(u)v$, 270
$(\cdot, \cdot)_{L^2}$, 283
(PS), 292
K_α, 292
$\nabla F(u)$, 294
spec_α, 306
$\text{gen}(A)$, 310
accessory variational problem, 19
accumulation point, 185, 208
Ambrosetti, 302
angular momentum, 26, 28, 30
arc-length, 3
Arzela–Ascoli theorem, 176

Banach fixed point theorem, 150, 152
Banach space, 126, 129, 132–134, 138, 145, 161, 162, 270, 291, 292, 299–301
Banach spaces, 150
Bellman equation, 105, 108
Bellman function, 105, 107
Bellman's method, 106
bifurcation theory, 268, 270
Borel measure, 118
Borel set, 117

Borel σ-algebra, 117
brachystochrone, 4

canonical equation, 85, 89, 95, 97, 99–101, 111
canonical equations, 80, 93
canonical system, 80
canonical transformation, 95–100, 103
Cantor diagonalization, 135
catenary, 283
catenoid, 283
Cauchy sequence, 126
characteristic function, 119, 211, 243
Christoffel symbols, 39
classical calculus of variations, 3
closed geodesic, 67
coarea formula, 250, 257
coercive, 186
coercivity condition, 291
Coffman, 310
cokernel, 147
compactness condition, 183
compactness of critical sequences, 292
complementary subspace, 146
complete, 126, 134
complete integral, 84, 93
completely integrable, 100
conjugate, 22, 24
conjugate point, 43
conservation law, 26
conserved quantities, 26
constant of motion, 80, 99
continuous linear functional, 133
continuous linear operator, 144
control condition, 109
control equation, 106, 108, 109, 111, 207
control parameter, 104
control problem, 109
control restriction, 105
control variable, 111, 207
converge, 125
convex, 68, 127, 130, 143, 186, 191, 193, 214, 219, 222
convex combination, 142
convex curve, 68
convex function, 122
convex functional, 188
convexity, 291
coordinate transformation, 36
cost, 105
cost function, 207
countable base, 184
countably additive, 118
critical family, 75
critical point, 5, 62, 66, 293, 294, 298, 301, 303, 306, 307, 312, 317
critical sequence, 291, 292, 304, 314

Index

critical value, 302, 303
cusp catastrophe, 279

de Giorgi, 225
deformation, 293, 294, 297, 298, 302, 307–309
dense, 169
diffeomorphism, 34, 95
differentiable, 150
differentiable map, 150
differentiation under the integral, 124
Dirac delta distribution, 173
Dirac distribution, 166
direct method, 183
Dirichlet boundary condition, 3, 26, 183, 190
Dirichlet principle, 199
Dirichlet's integral, 199, 203
distance, 51
distance function from a smooth hypersurface, 262
distributional derivative, 173
dual space, 133, 163

eiconal, 82
eiconal equation, 83, 86, 90
elementary catastrophes, 279
ellipticity assumption, 198
energy, 26, 30, 32, 34
ϵ-minimizer, 229
equivalence classes of functions, 159
essential supremum, 162
Euler–Lagrange equation, 6, 8–10, 16, 17, 19, 21–23, 29, 38, 60, 79, 80, 83, 88, 89, 111, 197, 267, 282, 303
example of Bolza, 206
extension, 130

Federer, 261
feedback control, 109
Fermat's principle, 4
field of geodesics, 46
field of solutions, 90, 93
finite perimeter, 244
first axiom of countability, 137, 184, 185, 209, 225, 227, 228
first conjugate point, 23
first integral of motion, 30
flow, 298
foliated by tori, 100
Fréchet differentiable, 150
Fredholm alternative, 149
Fredholm operator, 147–149, 270, 281, 287
free boundary condition, 26
Friedrichs mollifier, 166

fundamental lemma of the calculus of variations, 5

Γ-convergence, 225, 227, 229, 231
generating function, 100
genus, 310, 311, 313
genus of Krasnoselskij, 310
geodesic, 39, 43, 45, 50, 51, 55, 57, 58, 60, 88, 102
geodesic distance, 82, 90, 93
geodesic parallel coordinates, 45, 49
geometric optics, 86
gradient, 294, 299
gradient flow, 294
great circle, 42

Hölder continuous, 179
Hölder's inequality, 160, 163
Hahn–Banach theorem, 129, 134, 137, 143, 166
Hamilton–Jacobi equation, 83–86, 89, 92, 93, 101
Hamilton–Jacobi theory, 111
Hamiltonian, 80, 89
Hamiltonian flow, 95, 98
harmonic, 199, 201
harmonic oscillator, 87
Hessian, 4
Hilbert space, 126, 128, 141, 162, 293, 297
Hilbert's invariant integral, 92
homogenization, 232

implicit function theorem, 151, 152
index, 147, 308, 311, 313, 318
indicator function, 210
inner radius, 70
insulating layer, 235
integrable, 120
integral, 155
integral of motion, 27
integral of the Hamiltonian flow, 99
invariant integral, 93
inverse function theorem, 154
inverse operator theorem, 145
involution, 308
isometry, 34

Jacobi, 22
Jacobi equation, 20, 24, 268
Jacobi field, 20–22, 24, 269
Jacobi identity, 103
Jacobi operator, 268, 284
Jacobi's method, 99
Jensen's inequality, 122
Jordan curve, 35
Jordan curve Theorem, 68

Kakutani, 139
Kepler problem, 102
Kolmogorov–Arnold–Moser theory, 100
Kondrachev, 175

Lagrange multiplier, 9
Laplace operator, 199, 200
Lebesgue integral, 117, 120
Lebesgue measure, 117, 118
Legendre condition, 20, 112
Legendre transformation, 79, 88
length, 32, 34
length minimizing curve, 8
light ray, 4
limit cases of the Palais–Smale condition, 317
linear functional, 132, 133, 241
linear functionals, 129
linear operator, 144
Lipschitz continuous, 155, 203
local chart, 25, 32
local minimum, 22
lower semicontinuity, 184
lower semicontinuous, 185, 186, 188, 193, 208, 230
lower semicontinuous w.r.t. weak convergence, 187
lower semicontinuous envelope, 208
Lyapunov–Schmid, 280
Lyapunov–Schmid reduction, 269
Lyusternik, 306
Lyusternik–Schnirelman, 67

mean curvature, 263
mean value property, 201
measurable, 118–120
measure, 117
metric tensor, 33, 47
minimal hypersurface, 255
minimal hypersurfaces, 203
minimal surface of revolution, 282
minimax value, 303
minimizer, 4–6, 12, 183, 186, 229, 291, 302
minimizer of a convex variational problem, 189
minimizing, 3
minimizing sequence, 183
Minkowski functional, 143
Minkowski's inequality, 160
Modica, 254
Möbius strip, 75, 76
mollification, 167, 174, 175, 200, 245
momenta, 80
momentum, 26, 28, 30
monotonically increasing sequence, 122
Moreau–Yosida approximation, 190

Moreau–Yosida transform, 212
Morrey, 222
mountain pass theorem, 302, 303, 306, 318

neighbourhood system, 184
Newtonian motion, 81
Noether, 26
nonminimizing critical point, 291, 302
norm, 125
norm convergence, 125, 132
norm of a linear functional, 133
normed space, 125
null class, 159
null function, 159

optimal control theory, 111, 207
ordinary differential equation, 155
ordinary differential equations in Banach spaces, 155
orthogonal, 90
orthogonal complement, 141

Palais, 299
Palais–Smale condition, 77, 292, 293, 304, 306, 312, 317
parallel surfaces, 92
parallelogram identity, 128
parameterization invariant, 34
parameterized by arc-length, 8, 35, 36, 43, 88
parameterized proportionally to arc-length, 35, 38, 55, 89
perimeter, 244
phase space, 98, 100
phase transition, 254
Picard–Lindelöf theorem, 155
Poincaré inequality, 177, 304
Poisson bracket, 102
polar coordinate, 49
polar coordinates, 48
Pontryagin function, 110, 111
Pontryagin maximum principle, 110–112
principal curvature, 263
projection theorem, 142
proper, 62
pseudo-gradient, 299, 300

quasiconvex, 219, 222
quasilinear partial differential equation, 198

Rabinowitz, 302, 306
Radon measure, 118, 241
range, 147
rectifiable, 35

reflexive, 134, 135, 137–139, 163, 174, 186
regularity, 11
regularity theory, 198, 286, 306, 316
regularizing term, 255
relative minimum, 62, 66
relatively compact, 167
relaxation, 208
relaxed function, 208, 214
relaxed functional, 209
Rellich, 175
Rellich–Kondrachev theorem, 305
reparameterization, 8
Riccati equation, 86, 108
Riemannian manifold, 43, 52, 53
Riemannian normal coordinates, 48
Riemannian polar coordinate, 49, 51, 60
Riesz representation theorem, 241
rotational invariance, 200

Sard's theorem, 250, 257
scalar product, 126
Schnirelman, 306
Schwarz inequality, 35, 127
second axiom of countability, 184
second variation, 18, 23
semigroup family, 299
semigroup property, 157, 294, 297, 300
separable, 135, 169, 173, 184, 186
shortest geodesic, 52, 53, 55
shortest length, 50
σ-algebra, 117
signed measure, 242
simple function, 119
smoothing kernel, 166
Sobolev Embedding Theorem, 175, 179, 303, 305
Sobolev inequalities, 179
Sobolev space, 171, 173
special point, 306–309, 312
special value, 306, 307
sphere, 39
star shaped, 316
state variable, 207
step function, 110
strictly normed, 157
strong convergence, 125, 174
submanifold, 24, 32, 43, 52, 53
summation convention, xv, 19
support, 166
surface of revolution, 60, 282
symmetry assumption, 308–310
symplectic geometry, 96
symplectomorphism, 97

Taylor expansion, 274
test functions, 166
theorem of B. Levi, 122

theorem of Clarkson, 164
theorem of de Giorgi and Nash, 198
theorem of E. Noether, 28
theorem of Fatou, 123
theorem of Fubini, 122
theorem of Helly, 132
theorem of Jacobi, 84, 93, 101
theorem of Kondrachev, 180
theorem of Lebesgue, 123
theorem of Liouville, 98
theorem of Lyusternik–Schnirelman, 67
theorem of Mazur, 142
theorem of Milman, 139
theorem of Modica–Mortola, 248
theorem of Morrey, 179
theorem of Picard–Lindelöf, 39, 155
theorem of Pohozaev, 316
theorem of Rellich, 175
theorem of Riesz, 141
theorem of Riesz-Fischer, 161
theorem of Sobolev, 179
theorem on dominated convergence, 123
theory of catastrophes, 279
Thom, 279
topological space, 185
translation invariance, 118
transversality condition, 110
triangle inequality, 125, 126, 159

uniform convergence, 126, 168
uniformly continuous, 168
uniformly convex, 127, 129, 139, 157, 164
unstable critical point, 291

variational problem, 9
volume preserving, 98

weak convergence, 135–137, 142, 174, 186, 214
weak* convergence, 135
weak* convergent, 135
weak derivative, 171, 172
weak limit, 138
weak solution, 306, 315
weak solution of the Jacobi equation, 285
weak topology, 291
weak* topology, 137
weakly convergent, 135, 136
weakly lower semicontinuous, 222
weakly proper, 185
Weierstraß, 46
Weierstrass approximation theorem, 170
Weierstraß condition, 112
Weyl's lemma, 199

Young's inequality, 160

Zorn's lemma, 131